RADIO AND LINE TRANSMISSION

VOLUME 2

TELECOMMUNICATION TECHNICIAN SERIES

RADIO AND LINE TRANSMISSION

VOLUME 2

G. L. DANIELSON

M.SC.(TECH.), B.SC., A.M.I.E.E.

Head of Telecommunications and Electronics Department
Norwood Technical College

R. S. WALKER

GRAD.I.E.E., GRAD.BRIT.I.R.E.

Lecturer, Norwood Technical College

LONDON

ILIFFE BOOKS LTD

First published in 1963 by Iliffe Books Ltd
Dorset House, Stamford Street, London, S.E.1

Printed and bound in England by
The Chapel River Press Ltd., Andover, Hants

BKS 4249

Contents

Preface

This volume continues, at a rather more advanced level, the aim of RADIO AND LINE TRANSMISSION, VOLUME 1, that is, to provide a background knowledge of the world of telecommunications together with some detailed information about the components and principles used in communications equipment. Students who intend to specialise in either radio communications or line communications should, it is believed, in the early stages of their studies give consideration to all aspects of communication in order to appreciate how their particular service integrates with the wider whole.

RADIO AND LINE TRANSMISSION, VOLUME 2 covers the syllabus of the City and Guilds of London Technician's Certificate examination in 'Radio and Line Transmission B' and should be of interest and help to all communications students who have completed a one-year full-time course or a two-year part-time course in their subject. In some aspects the book will still be of use in the third year of a full-time course and the fourth year of a part-time course. It is assumed that readers have a knowledge of mathematics to G.C.E.(O), or equivalent, level and are familiar with basic technical electricity including the theory of alternating current to the standard reached in Volume 1. Symbols and nomenclature are in accordance with the recommendations of the British Standards Institution. The M.K.S. system of units is used throughout.

There are many worked examples in the text and questions are provided at the end of each chapter. It is suggested that the student use these to test the efficiency of his study. A day or two should elapse between the end of the study of a chapter and the attempt on the relevant questions. Many questions are taken by kind permission from past examination papers of the City and Guilds of London Institute.

The serious student is recommended to maintain an adequate notebook. Such notes are essential for revision and the effort demanded by their making maintains concentration and increases learning efficiency.

Acknowledgement is made to those firms who have helped by supplying information about their products, and to Mr. W. A. L. Smith who gave constructive criticism of the chapters dealing with telephony practice. Thanks are also due to the General Editor, Mr. S. W. Amos, for most helpful advice and discussion during the preparation of the initial manuscript.

London, 1963

G.L.D.
R.S.W.

1

Propagation of Radio Waves

1.1. Introductory

In *Radio and Line Transmission, Volume* 1, Section 4.2, the applications of different orders of radio frequency carrier waves are discussed. The ways in which these waves travel from their transmitters to reception points are now explained in more detail. The frequency bands considered are:

v.l.f. (up to 30 kc/s),
l.f. (30 kc/s to 300 kc/s),
m.f. (300 kc/s to 3 Mc/s), and
h.f. (3 Mc/s to 30 Mc/s).

1.2. The V.L.F. and L.F. Bands

1.2.1. GROUND WAVES

Radiation from a transmitting aerial may occur not only in all directions in the horizontal plane but also at any angle to the vertical. That part of the radiated signal which reaches the receiver after travelling along the earth's surface is called the *ground wave.* Just as a ray of light tends to bend round the edge of a solid object (a process called *diffraction*), a radio wave keeps to the earth's surface instead of taking a straight-line path off into space. Signals in the v.l.f. and l.f. bands are usually transmitted as ground waves. High-power l.f. transmitters produce a strong ground wave signal at a distance of one or two thousand miles. The range of a ground wave in these bands depends directly on the conductivity of the earth, the power of the transmitter and inversely on the frequency used. The sea, which has a conductivity about five thousand times that of dry soil, is the best type of terrain for ground-wave propagation. Dry rocky land attenuates the ground wave energy quickly. The signal travels not merely over the earth but also in it and may penetrate the ground to a depth of ten or more metres.

1.2.2. SKY WAVES

Although at the transmitter it is not easy to differentiate between ground wave and sky wave, at the receiver the sky wave is definable as that part of the signal which reaches the receiving point after being reflected back to earth from some part of the ionised atmosphere. The rarefied gases of the earth's atmosphere are partially ionised from an approximate height of 30 miles to more than 250 miles. Ionisation density is particularly strong at some heights where layers of ionisation can be detected. The layers are referred to by letters; viz.:

D layer from approximately 30 to 60 miles in height,
E layer from approximately 60 to 90 miles in height,
F_1 layer from approximately 90 to 150 miles in height,
F_2 layer at approximately 200 miles height,
G layer at approximately 250 miles height.

L.F. signals and v.l.f. signals are returned to earth by the D and E layers which behave as reflecting layers for them. Where the density of ionisation in the ionosphere changes markedly within a distance of less than a wavelength of the propagated signal an abrupt change of direction can occur. Long waves therefore are easily reflected at acute angles of reflection as though they were light waves striking the surface of a mirror. V.L.F. waves can travel by a process of multiple reflection between the ionosphere and earth all the way round the world provided the transmitted power is sufficient.

1.3. The M.F. Band

1.3.1. MODES OF PROPAGATION

Communication on m.f. is usually over ranges of the order of 100 to 200 miles and relies chiefly on the ground wave signal. Earth losses at these frequencies are higher than at l.f. or v.l.f. and the difference between the range of a given transmitter over the land and the same transmitter sending over the sea is more pronounced. For example a ship's radio officer on watch in a ship to the south of the Isle of Man in the Irish Sea may hear quite strong signals from a ship to the south in the St. George's Channel but signals from a transmitter of similar power in a ship the same distance to the east and off the north-east coast of England may be inaudible to him.

The ground range of an m.f. wave is also affected by the relative permittivity of the terrain over which it travels. The greater the permittivity, the greater is the range for a given frequency and power transmitted. Sea water has a relative permittivity of 80. The permittivity of dry soil may be about 4. This is an additional reason for increased ranges when waves are propagated over the sea.

At night especially, m.f. signals are received as sky waves reflected back to earth by the E layer. After reflection the sky wave may return to earth several hundred miles away from the transmitter so that the ranges of stations are much increased at night. This may be an advantage to ship stations for the transmission of messages to distant coast stations and certainly to ships in distress trying to gain acknowledgement of a distress signal, but for broadcast stations the strong sky wave can be a disadvantage. The allocation of sufficient carrier frequencies to the many m.f. broadcasting stations in Europe is difficult if mutual interference is to be avoided.

Transmitters to which adjacent or similar carrier frequencies are given, are spaced geographically so that their programmes do not become confused in the listeners' receivers. This geographic separation becomes of little effect at night when the strong skywave increases the ranges of transmitters on the same frequency, causing interference. The sky wave may also cause large errors in m.f. direction finding so that radio direction finder bearings tend to be less accurate at night than during the day unless the receiver is so close to the transmitter that the ground wave strength swamps the sky wave.

1.3.2. FADING

Reception on m.f. at night at certain distances from the transmitter is liable to deep fading of signal strength. The fading is worst in those areas where the sky wave and the ground wave have nearly equal amplitudes. The sky wave and the ground wave travel by different paths of different distances to reach the same place. If the path difference is by chance equal to any whole number of wavelengths then the two signals arrive at the receiver aerial in phase and so reinforce one another. If the path difference is an odd number of half wavelengths the two waves reach the receiver in opposite phase and tend to cancel. The effective reflection height of the *E* layer from which the sky wave is returned to earth fluctuates so that the relative phase relationship between ground and sky waves is not constant. The two waves drift in and out of phase and the resultant signal strength varies from a value given by the sum of the two to a value equal to their difference. The deep fading belt is at a distance of possibly one or two hundred miles from the transmitter. Inside this belt the ground wave strength masks the effect of the sky wave and outside this belt the sky wave is strong compared with the ground wave. Many broadcast transmitters use aerials which beam their radiation at low angles to the horizontal to minimise sky-wave radiation, thus localising the programme to the intended service area.

1.4. Effects of the Ionosphere

1.4.1. ATTENUATION

A radio wave is a combination of an alternating magnetic and an alternating electric field of the same frequency. When this combination passes into the region of ionised atmosphere called the ionosphere, it applies an alternating force to the free charges of the ionised gas. (Any electric charge in an electric field has a force exerted on it by the field tending to move it in the direction of the field.) The positive gas ions present have a large mass compared with that of an electron so that whereas the electrons are set oscillating at the frequency of the wave, the movement of the ions is negligible. The oscillating electrons re-radiate electromagnetic waves in the same way as they do when they oscillate within the bounds of a conductor. However, if the oscillating electrons collide with positive gas ions and combine with them, the energy which they could re-radiate is lost as heat and the total energy re-radiated from the ionosphere is less than that reaching it.

The total loss increases if the number of such electron-ion collisions increases and is larger if the probability of electron-ion collision is greater. This probability is higher if the number of gas ions per unit volume is larger. *Thus an increase in the density of ionisation results in an increase of attenuation of the radio wave.* The probability of electron-ion collision is also increased if the path length of displacement of the oscillating electrons is increased. A lower frequency of wave with a longer period allows time for the electric field to cause a longer path of electron oscillation so that the number of electron-ion collisions increases. *Thus attenuation in the ionosphere is greater at low frequencies than at higher ones.*

1.4.2. REFRACTION

An electromagnetic wave passing from one medium into another in which its velocity is different, suffers a change in direction of travel. Most people have observed that a pool of water often appears less deep than it really is. This is due to the fact that the rays of light undergo a change of direction when they pass from water into air. The bending of light or radio waves in this way is called *refraction.*

A radio wave passing from a non-ionised gas into an ionised gas or from an ionised gas into a region of greater density of ionisation experiences an increase of *phase velocity.* (The phase velocity is the product of the frequency of the wave and the wavelength in a given medium.) As a result a radio wave passing into the ionosphere at an angle to the vertical bends farther away from the vertical as it moves into a region of greater density of ionisation. If the rate of change of ionisation density with height is sufficient, the wave may be refracted back towards the earth. *The sharpness of refraction becomes greater if the density of ionisation increases or if the frequency of the radio wave becomes lower.*

1.4.3. CRITICAL FREQUENCY

At a given time and place there is a maximum frequency for which a wave radiated vertically upwards can be returned to earth by the ionosphere. This frequency is called the *critical frequency.* It may be of the order of 4 or 5 Mc/s but depends upon the density of the ionosphere at the time of measurement. If a wave is capable of being refracted through 180 degrees it follows that it can also be returned to earth at angles of incidence for which the required change of direction is smaller.

1.4.4. SKIP AREA

Let us assume an aerial which is radiating equally well in all directions and at all angles to the horizontal. If the frequency radiated is below the critical frequency then radio waves leaving the aerial at all angles to the horizontal may be turned back to earth by the ionosphere and the whole area surrounding the transmitter will receive some returned sky wave. If the frequency radiated is higher than the critical frequency the directions of radiation which must suffer the greatest change of direction in order to return to earth will not be sufficiently refracted and will penetrate the ionosphere. Thus the radiation at and near to the vertical is not reflected back to earth. This leaves a circular area round the transmitter to which no sky wave can return. This is called the *skip area.* The distance from the transmitter to the edge of the skip area is the *skip distance* (*see* Fig. 1.1). The skip area depends upon the frequency of the radiated wave and on the prevalent density of the ionosphere. The higher the frequency the less the refraction effect and the wider the skip

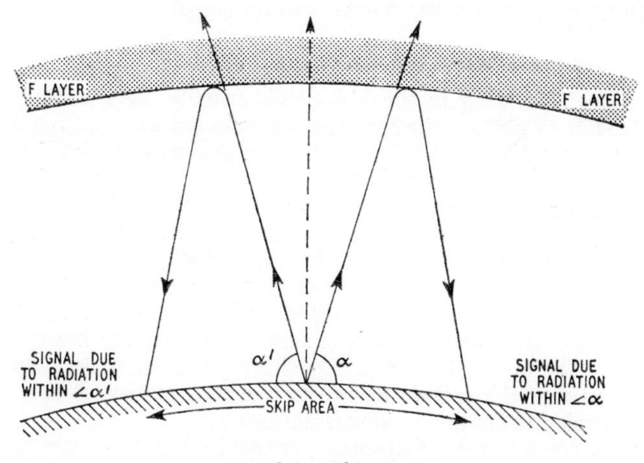

Fig. 1.1. Skip area

area. Because the ionosphere density depends to a large extent on the intensity of the sun's rays, skip areas vary considerably with the time of day and with the season. Skip distances are greater at night than during daylight hours and have a minimum soon after midday. Winter skip distances are on average longer than summer skip distances. A variation of sunspot activity which has a minimum and a maximum value approximately every 11 years, also affects communication. A peak in sunspot activity occurred during the International Geophysical Year which was the period from July 1957 to December 1958. Solar flares radiate energy which alters the ionosphere density and therefore skip distances. Skip distances are maximum when sunspot activity is minimum and vice-versa.

1.5. H.F. Propagation

1.5.1. PROPAGATION PATHS

Transmitters in the h.f. bands produce little ground-wave signal because at such frequencies the attenuation of the ground wave is rapid and its range short. H.F. signals rely on refraction in the ionosphere in the F_1 and F_2 layers to return them to earth to reach their intended reception areas 1,000 or more miles away. Signals may reach receivers after travelling all round the earth by a series of multiple reflections between earth and the F layer. Transmitters providing point to point communication in the h.f. bands may beam their radiation not only on the correct azimuth bearing but also at the best angle to the horizontal for optimum ionosphere refraction to the destination of the signal. Perhaps the chief disadvantage of h.f. communication is its liability to fade out at times of peak sunspot activity which increases the ionosphere density to a degree which absorbs all the useful energy of the signal.

1.5.2. CHOICE OF FREQUENCY FOR H.F. TRANSMISSION

Stations operating in the h.f. bands usually have a number of frequencies available of which the one selected at any time will be that most suitable for the time of day and season and the distance to be covered by the transmission.

In selecting a frequency the first relevant factor to keep in mind is that in the ionosphere the highest frequencies suffer the least attenuation (*see* Section 1.4.1). The highest frequency which may be used is, however, dependent on values of skip distances and the distance between transmitting and receiving stations. The frequency chosen must be low enough for the skip area not to include the receiving station. The higher the frequency chosen, the wider the skip area. It follows that the farther away the receiving station is the higher the transmission frequency which can be used.

The *maximum usable frequency* (for brevity m.u.f.) over a particular distance depends on the time of day and season since this affects the state of the ionosophere. In general frequencies in the 12 to 25 Mc/s bands are used during day time conditions while those in the 4 to 12 Mc/s bands are used at night.

In spite of the lowering of frequencies at night in order to counter the expansion of skip areas caused by the absence of the sun, signal strengths are generally stronger than during daylight hours. This is partly due to the reduction in ionic density of the D and E layers through which the h.f. signals must pass on their journey to and from the F layers. The h.f. signals are attenuated in passing through these layers far less at night.

Forecasts of maximum usable frequencies for different times of day are published monthly in technical journals (*see* Fig. 1.2).

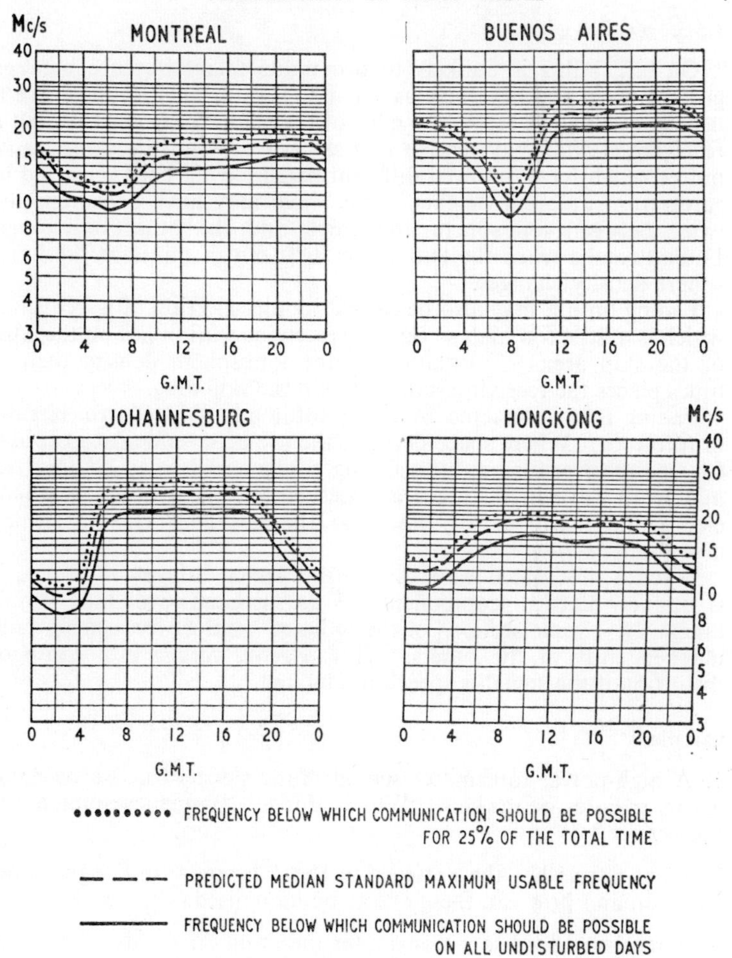

Fig. 1.2. *Frequency prediction charts for communication between London and four world centres for a summer month* (*July*)

In order to ensure continuous working despite unforeseen fluctuations in ionosphere conditions, a frequency is often selected which is about 85 per cent of the m.u.f. and which is termed the *optimum traffic frequency* (o.t.f.).

In the absence of other guidance it is useful to remember that propagation paths are reciprocal so that a guide to the best frequency of transmission from point *A* to point *B* is to listen on a radio receiver to transmissions originating at point *B* and reaching the point *A*. The frequencies which are received the best are probably those which are best for transmission.

1.5.3. FADING ON H.F.

On h.f. fading is unlikely to occur due to interference between ground wave and sky wave as for m.f. because the ground wave is usually attenuated before the edge of the skip area is reached, but a receiving station may receive two sky waves from the same transmitter which have followed different paths having been refracted to earth from different heights. One wave may have been reflected twice between ionosphere and earth and the other only once. Fading results when the two waves drift out of phase as the ionosphere density changes.

Fading on h.f. may also be caused by the selection of a frequency which is a little too high so that the receiving station lies on the edge of the skip area. Fluctuation of the ionosphere density then at times places the receiving station within the skip area.

Fading is counteracted by using automatic gain control circuits in receivers. Where space permits, the use of several spaced aerials for one receiver also reduces fading. The aerials may be approximately 10 wavelengths apart and designed to receive signals whose electric fields are at any angle to the normal. This is termed *diversity reception.*

When a modulated wave is received under fading conditions, it is unlikely that all components of its sidebands will fade simultaneously. Some sections of the radiated band of frequencies fade independently of the others. This *selective fading* is a cause of distortion if the signal is speech modulated.

Questions

1. A high-power transmitter sending time signals for use by navigators radiates on 16 kc/s. Why is this low frequency suitable for this service?

2. What effects has the presence of the E layer on m.f. signal propagation and how can these effects be minimised?

3. A receiver tuned to a transmitter radiating on 16 Mc/s and situated only 50 miles away from the transmitter hears only very weak signals though another receiver 2,000 miles away receives a usable signal strength from this same transmitter. Discuss this.

4. Why are the traffic frequencies used at night time on h.f. generally lower than those used during the day and why are signals received at night often much stronger than those received from the same stations during daylight?

5. What factors influence the choice of transmission frequency to be used between two stations using the h.f. band? What do you understand by the abbreviation o.t.f.?

6. Why are broadcast signals which are received on h.f. liable to fading and distortion?

2

Lines and Cables

2.1. Purposes for which Cables are Required

Excepting those parts of a communication circuit or channel which employ radiated electromagnetic waves to convey information, all sections of a telephone or telegraph route require suitable electrical conductors to carry signal currents of the correct frequencies between the sender and the receiver. Telephone or teleprinter instruments are connected by conductors to the nearest telephone or telegraph exchange. Suitable linking conductors and switching devices at the exchange interconnect or join instruments to lines connecting with other exchanges. Minor exchanges are connected with group exchanges and these in turn with zone exchanges. Trunk lines connect the large zone exchanges. For international circuits, conductors link zone exchanges with an international exchange. Submarine telegraph and telephone cables provide conductive links between the international exchanges of different countries. Alternatively, lines may link an international exchange with a radio terminal from which transmission to and reception from other countries by radio waves occurs. Conductors are then required in the field or area where the transmitting or receiving aerials are erected, and these last conductors are called aerial feeders. Pairs of conductors, assembled together and insulated from one another within a common binding or sheathing are referred to as a *cable*. Cable constructions depend on the order of frequency to be employed, the physical stresses they must endure, and the number of channels of communication to be carried.

2.2. Desirable Attributes of a Cable

The following electrical properties are desirable:

1. Currents should travel along the cable with a minimum loss of energy.
2. All frequencies to be transmitted should suffer the same degree of attenuation which should be small.
3. Currents of all frequencies transmitted should as far as possible travel along the cable with the same velocity otherwise a distortion of their resultant waveform occurs.

4. The cable should not radiate power as electro-magnetic waves because this would not only waste power but also cause interference with other channels of communication.
5. Unwanted magnetic or capacitive coupling between the cable and other adjacent cables should be negligible if mutual interference between them is to be avoided.

The following mechanical properties are important:

1. The cable should be flexible to allow ease of laying.
2. The cable should have sufficient tensile strength to allow for the stresses involved in laying, and in the case of overhead lines to allow for the tensions due to cable weight, the tugging of the wind, and the additional weight of snow.
3. The cable insulation between the conductors must be mechanically robust and should not deteriorate with the passing of time or be liable to an increased loss due to the penetration of moisture.
4. A submarine cable should also be able to withstand the corrosive effects of salt water, be proof against attack by fish and marine insect life and be armoured against accidental damage by ships' anchors or rocks moving about with the tidal waters.

2.3. Primary Constants

Before considering the construction of various types of cable, we shall review the applicable electrical constants. The primary constants of a cable may be listed as follows:

R, the ohmic resistance per mile

G, the conductance through the insulation between conductors per mile—often termed the *leakance*

L, the inductance per mile, and

C, the capacitance per mile run.

2.3.1. RESISTANCE

The resistance per unit length of the cable depends directly on the resistivity of the metal used (a factor which is temperature dependent) and varies inversely as the cross-sectional area of the conductors. At radio frequencies the resistance increases because the current tends to flow more on the surface of the conductors and thus uses less of the total cross-section. This tendency, called *skin effect*, becomes appreciable at frequencies of the order of 10 kc/s. At 100 kc/s, a conductor of 2 mm diameter may have a resistance of more than two and a half times its d.c. resistance value. At still higher frequencies, the radio frequency resistance increases as the square root of the frequency. R, the resistance per unit length, is taken to be the total resistance of the two conductors and not that of a single conductor. R may therefore be quoted in ohms per loop mile.

2.3.2. LEAKANCE

The loss conductance G, for unit length of cable for d.c., is the reciprocal of the resistance of the insulation between one conductor and the other. For a.c. however the loss in the cable insulation increases as alternating dielectric displacement currents occur. It is not easy to find by calculation the effective value of G for a given frequency and if the d.c. value cannot be assumed, then a measurement of the effective capacitance and loss angle of a specimen unit length of the cable should be made at the frequency concerned. (The loss angle of a capacitance is the complement of its phase angle. The phase angle is almost 90 degrees and the loss angle is very small. The tangent of the loss angle is nearly equal to the power factor of the capacitance and this should be as small as possible.) (*See also* Section 5.5.2.)

$$G = \omega C \operatorname{Tan} \delta \text{ mhos}$$

where $\omega = 2\pi f$ radians/sec,

C = capacitance measured at f c/s,

and $\operatorname{Tan} \delta$ = the tangent of the loss angle, a reading obtainable directly from certain types of a.c. measuring bridges.

2.3.3. INDUCTANCE

L, the inductance per unit length, depends on the inductance of both conductors and on the mutual inductance between them. The mutual inductance is in opposition to the self inductances of the conductors. As we should therefore suppose, the value of total inductance depends upon the spacing distance between the two conductors of a line and increases if they are moved farther apart. The line inductance depends also on the magnetic permeability of the medium separating the two conductors and inversely on the radius of the conductors used.

2.3.4. CAPACITANCE

The capacitance per unit length of cable, C, is that between the two conductors and depends upon their radius, the spacing between them and on the relative permittivity of the insulation. If the radius of the conductor is made larger, the capacitance is increased. If the spacing between conductors is increased, the capacitance decreases. The capacitance depends directly on the permittivity factor. The choice of insulation for the cable not only influences the capacitance per unit length but also determines the leakance loss. The insulation used should have sufficient dielectric strength, low permittivity and a dielectric hysteresis loss which is both small and constant over a wide range of frequencies. Air meets these requirements best but at the present time polythene comes first among the solid insulators. The relative permittivity of polythene is only 2·3.

2.4. Secondary Constants

2.4.1. CHARACTERISTIC IMPEDANCE

A knowledge of the primary constants of a line enables us to calculate two important secondary constants, namely the *characteristic impedance* and the *propagation constant.* The first of these is needed in order to decide the impedance to which any input generator or output load at the ends of the line should be matched to secure maximum transfer of power from generator to line, or from line to load. If proper matching is secured at the output end of the line, then current and voltage waves which reach the load transfer their power to the load without any reflected waves being sent back along the line. If, however, the characteristic impedance of the line has some reactive component (as it will if its losses are not negligible), then for maximum power transfer to the load, the load should have a reactance which is numerically equal but opposite in sign to that of the characteristic impedance, and the resistance of the load should be equal to the resistive part of the characteristic impedance.

If the line is matched by a terminative load equal to the characteristic impedance, it behaves as if it were of infinite length having no termination of any kind to reflect energy. Under such matched conditions, the ratio of the voltage between the conductors to the current in them (i.e. the ratio of line voltage to line current), at any point is equal to the characteristic impedance of the line. This value is independent of the length of the line and depends only upon the primary constants, R, L, G and C.

The characteristic impedance may be calculated from:

$$Z_0 = \sqrt{\frac{Z}{Y}} \text{ ohms} \tag{1}$$

Where Z is the impedance per unit length of line, i.e. the vector sum of R and the reactance of L, and Y is the admittance between the lines per unit length and is the vector sum of G and the susceptance (the reciprocal of reactance) of C. The calculation of Z_0 and also of the propagation constant is often complicated because these quantities contain resistance, reactance, conductance and susceptance. There are three examples however which are relatively simple, viz: when the current is, (a) direct, (b) at high frequency, (c) at very low frequency.

For direct current and for very low frequencies, the characteristic impedance is simply:

$$Z_0 = \sqrt{\frac{R}{G}} \text{ ohms} \tag{2}$$

where the reactance and susceptance are zero. For radio frequencies when the reactance of L is large compared with R and when the conductance G, is small compared with the susceptance of C

$$Z_0 = \sqrt{\frac{L}{C}} \text{ ohms} \tag{3}$$

Under these last conditions, the characteristic impedance is independent of frequency. This is also true if the ratio of $\frac{R}{L}$ can be adjusted to be equal to $\frac{G}{C}$, though this " ideal " transmission condition is hardly attainable in practice. Except for these special conditions Z_0 has both resistive and reactive parts.

2.4.2. PROPAGATION CONSTANT

The propagation constant is a quantity which reveals the loss of power per unit length of cable due to R and G and the phase change per unit length caused by the finite time required by the voltage and current waves to travel along the line. This secondary constant is given by

$$p = \sqrt{ZY} \tag{4}$$

where Z is the series line impedance per unit length and Y is the shunt admittance per unit length as before. Here too some simplifications may be made under certain conditions. For direct current,

$$p = \sqrt{RG} \text{ nepers/mile} \tag{5}$$

since L and C have no effect.

Or, if we neglect the line loss, an easy calculation of the phase change per unit length becomes possible,

$$p = \omega\sqrt{LC} \text{ radians/mile (or per metre according to length unit used),} \tag{6}$$

where $\omega = 2\pi f$ radians/sec.

At audio frequencies, the effects of L and G may be small compared with the effects of C and R. If the first two be neglected then cable loss, which is indicated by the real part of p (called the *attenuation constant*) is equal to

$$\alpha = \sqrt{\frac{\omega CR}{2}} \text{ nepers/mile} \quad (7)$$

where C is in farads/mile

and R in ohms/mile.

(For the meaning of nepers *see* Chapter 6.) Also if the effects of G and L are small compared with those of R and C, the reactive or unreal part of p, which indicates the phase change per unit length of cable, is given by

$$\beta = \sqrt{\frac{\omega CR}{2}} \text{ radians/mile} \quad (8)$$

where C and R have the same meanings as before.

2.4.3. ATTENUATION—FREQUENCY CURVES

The resistance of the wire of a line and the conductance through the insulation are the factors causing the loss of energy in a line, but

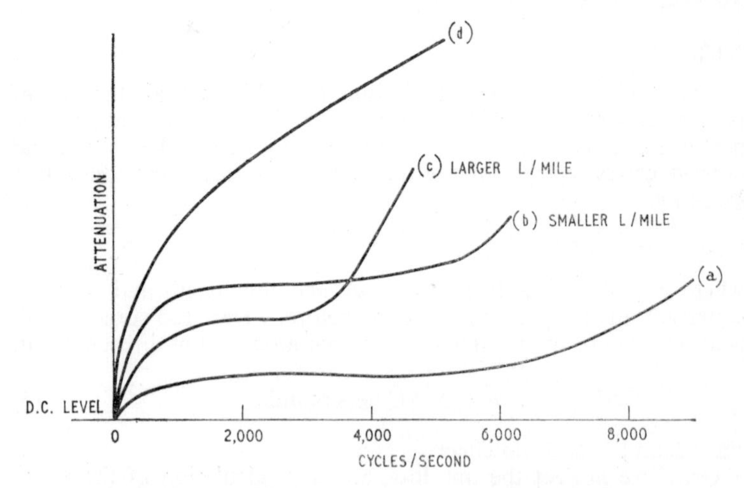

Fig. 2.1. Cable attenuation-frequency characteristics

the inductance and capacitance per unit length help to determine the size of the loss. If the inductance per unit length is increased, the characteristic impedance becomes bigger, which means that energy is transmitted at a higher voltage and thus the leakage through the insulation is increased. Should the capacitance per unit length be made larger, the additional shunt current through this susceptance gives added loss in the line resistance through which the current must pass to reach any particular part of the line capacitance. It follows that the attenuation varies differently with frequency for different types of cable. Fig. 2.1 shows attenuation-frequency curves for several different types of cable. An " ideal " cable would have constant attenuation for all frequencies of current, and curve (a) which applies to an open aerial line, has the most uniform attenuation. Curve (b) applies to an underground cable which has coils joined in series with it at regularly-spaced intervals. These are known as loading coils and the addition of the coils is an attempt to approach the ideal condition in which

$$\frac{R}{L} = \frac{G}{C}$$

and for which Equations 2 and 3 apply. Curve (c) also applies to an underground cable but with larger values of loading inductance. Curve (d) applies to an underground cable without loading coils. The construction and applications of various types of cables can now be considered.

2.5. The Open-wire Aerial Line

2.5.1. CONSTRUCTIONAL FEATURES

The use of overhead lines for multi-channel long-distance telephone trunk routes is obsolescent, but such lines are still used for short distance audio frequency routes and for local connection of subscribers to their nearest exchange. For several reasons, the chief being protection from weather, underground cable routes are more reliable than overhead lines but they are uneconomical to construct unless they can carry a large number of subscribers' lines or a large number of communication channels between exchanges. Thus the telephone pole is likely to remain a familiar sight in rural and suburban areas, where it carries connections from widely-distributed telephone users to the nearest underground cable duct, or, in some places, to the local exchange building.

Steel spindles pass vertically through the cross-pieces and carry a thread at their top end on to which is screwed a porcelain insulator designed to anchor the conductor. The insulator has a cross-section of the form shown in Fig. 2.2. It has a high polish to enable rain to wash off soot and dirt easily and its re-entrant form ensures that some part of the surface leakage path is always clean and dry whatever the weather.

Conductors used for overhead lines are of copper or cadmium copper. The latter has greater tensile strength although its resistivity is approximately 1·18 times that of copper. Subscribers' lines normally use cadmium copper of diameter approximately 0·05 in. although lines in places likely to be exposed to exceptionally high winds may use a heavier gauge of the same material of approximate diameter 0·06 in. (For trunk routes thicker conductors of copper have been used having diameters ranging from approximately 0·098 in. to 0·197 in., but these have been superseded by trunk routes using co-axial or other underground cables.) The spacing between a given pair of overhead conductors is nine inches.

2.5.2. BALANCE OF LINE IMPEDANCE TO EARTH

The conductors must be kept sufficiently taut to prevent the touching of adjacent conductors when they sway in high winds. It is

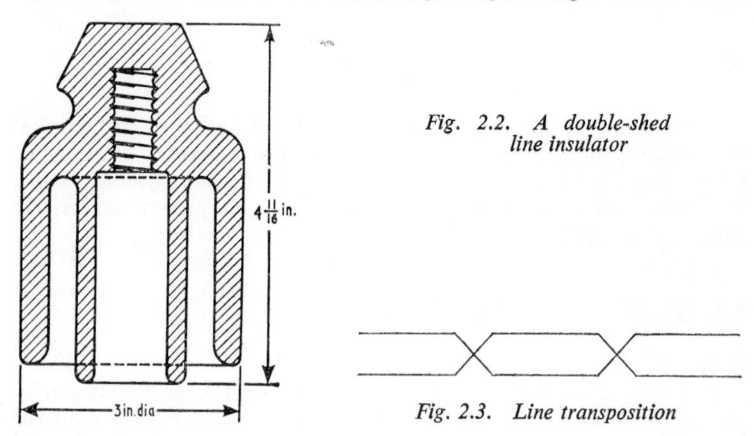

Fig. 2.2. A double-shed line insulator

Fig. 2.3. Line transposition

also necessary that the spacing remains as constant as possible so that both conductors of a given channel maintain an equal impedance to earth and to other lines. This balancing of the impedances of the two conductors is important because lack of balance causes a resultant coupling with other circuits so that conversations on adjacent lines can be heard, a phenomenon known as *cross-talk*. A perfectly-balanced line is also immune to interference from other stray fields such as those set up by lightning flashes or radiation from inadequately-suppressed electrical machinery. If, for example, an interfering field induces a voltage of V volts in a north–south direction in both conductors of a balanced line, these equal voltages balance and since they act in opposite directions round the loop circuit, produce no current. But this is only true if the impedances to earth of the two conductors are exactly similar. Equal voltages have equal effects only in equal impedances. To keep a balance between one pair of conductors and another, conductors are transposed at the insulators at regular intervals (*see* Fig. 2.3.).

2.5.3. CHARACTERISTIC IMPEDANCE

The ratio of line spacing to conductor diameter determines the inductance per unit length of line and the capacitance per unit length of line and therefore the characteristic impedance of the line. An expression from which the line impedance may be calculated is:

$$Z_0 = 276 \log_{10}\frac{S}{r} \text{ ohms} \tag{9}$$

where S is the line spacing and r is conductor radius in the same units.

By way of example, let us assume a line spacing of 9 in. with a conductor radius of 0·05 in.

$$\begin{aligned} Z_0 &= 276 \log_{10}\frac{9\cdot0}{0\cdot05} \\ &= 276 \log_{10}180 \\ &= 276 \times 2\cdot176 \\ &= 600\cdot6 \text{ ohms} \end{aligned}$$

Or if we assume the same spacing but with a larger conductor radius of 0·08 in., we find

$$\begin{aligned} Z_0 &= 276 \log_{10}\frac{9\cdot0}{0\cdot08} \\ Z_0 &= 276 \times \log_{10}112\cdot5 \\ &= 276 \times 2\cdot0512 \\ &= 566 \text{ ohms} \end{aligned}$$

These calculations show that the change of the conductor radius does not make a large alteration in the value of the characteristic impedance and 600 ohms is taken as the nominal impedance for a parallel open-wire line. Equation 9 above takes into account only the inductance and capacitance of the line and the value of Z_0 is modified if Equation 3 is not justifiable.

2.5.4. FREQUENCIES CARRIED

Where open-wire lines are used for connecting subscribers' equipment to an exchange, they carry frequencies in the speech band from 300 to 3,400 c/s in addition to the d.c. signalling and the 16-c/s current used for ringing the subscriber's bell. Limited use of aerial lines for carrier current telephony is practised. For example a (4 + 1)-system in which five channels are provided on one circuit is possible. One channel has audio speech frequencies but the other four channels consist of the lower sidebands produced when four different carrier frequencies are modulated each by one subscriber's line output. This involves a frequency range of 300 c/s to approximately 16 kc/s. A rather exceptional use of open-wire lines is made in the North West Highlands of Scotland where a 12-channel open-wire system links Inverness on the east coast at the northern end of Loch Ness, and Ullapool on the west coast and on the shores of Loch Broom. Transmission in one direction employs frequencies in the range 36 kc/s to 84 kc/s while in the reverse direction frequencies transmitted are in the range 92 to 140 kc/s. The distance from Inverness to Ullapool is about 60 miles.

2.5.5. OPEN-WIRE LINES USED AS TRANSMITTER AERIAL FEEDERS

Long-distance transmission of high radio frequency signals over open-wire lines would result in excessive radiation loss. This loss increases as the square of the frequency. It also increases as the square of the distance of separation between the two conductors of a line. Any unbalance results in a residual earth return current and an electric field between earth and the line which radiates electromagnetic waves strongly. Even in the high frequency band, however, open-wire lines are used for conveying radio-frequency current from a transmitter to its aerial over distances of the order of half a mile or a mile.

Radiation from a transmitter feeder reduces the efficiency of the aerial system and distorts the pattern of the radiation. Every precaution must thus be taken to ensure that any feeder is as accurately balanced to earth as possible. Currents in the two wires must be equal in amplitude and opposite in phase. The two lines should be symmetrically arranged with respect to earth with the height above earth at a maximum convenient level (often this is about 8 ft.). Unbalance to earth and to other objects can be minimised by the transposition of the conductors at the poles as in the case of telephone open lines (*see* Fig. 2.3). The supporting poles must be sufficiently near to each other to keep the spacing of the conductors constant.

A very short aerial feeder, where the power to be transmitted is only of the order of 1 or 2 kW, may form part of the resonant aerial circuit and may be a quarter, a half, three-quarters or even one wavelength long. It is then described as a *tuned feeder* and has standing waves along it, as does the aerial (*see* Section 4.4.1).

Normally, however, the feeder must be fed from a source impedance carefully matched to its characteristic impedance and matching is also necessary at the other end of the feeder where it is connected to the aerial. (Methods of matching a feeder to an aerial are referred to in Chapter 4.)

2.6. Underground Cables

2.6.1. STAR-QUAD CABLES—CONSTRUCTION

When a sufficient number of communication channels can be provided to justify the cost of providing cable ducts, then underground cables are to be preferred to open-wire lines because of their greater reliability and more efficient screening.

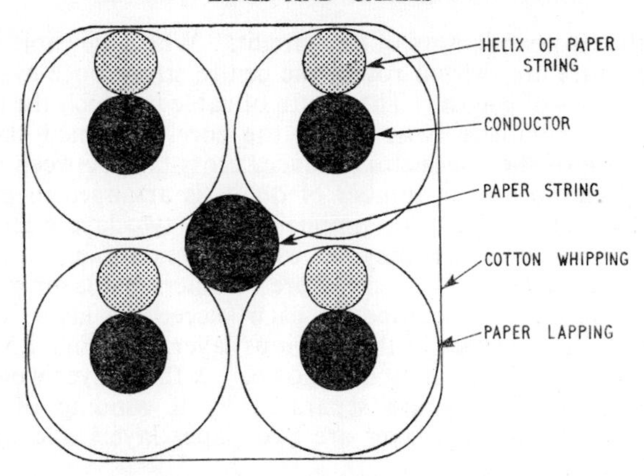

Fig. 2.4. The cross section of a paper-insulated quad

The majority of cables used in the United Kingdom on telephone circuits between exchanges are of the paper-core star-quad type whose construction is illustrated in Figs. 2.4 and 2.5.

Fig. 2.4 shows how the individual conductors are first arranged in groups of four. Each conductor may have a diameter as large as 0·05 in., but 0·036 in. and 0·025 in. diameter conductors may be

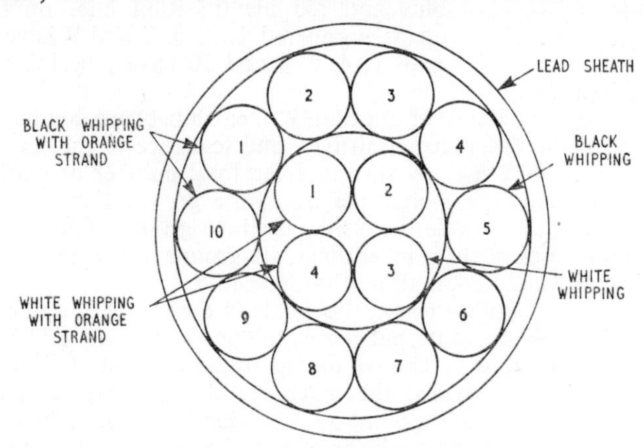

Fig. 2.5. A star-quad cable

used in lighter cables. Over each conductor is wound a helix of paper string. This acts as a spacer and holds the paper tape lapping which is wound on top, away from the surface of the conductor. As a result a large percentage of the volume of the insulation between conductors is air—the most efficient of dielectrics. A cotton whipping holds four conductor units together round a central paper string to form a *quad.* In one quad the two conductors, diagonally opposite to each other, form a pair, i.e. they are used as the two conductors of one circuit. When the cable is made the conductors are twisted round the centre string once every 8, 6, 5, 4 or 3 inches of length. The length of cable in which the conductors make a complete twist round the core is termed the *pitch*. This twisting of the conductors reduces cross-talk between circuits.

Fig. 2.5 shows how a number of quads is arranged in layers to form a complete cable. For simplicity, only two layers are shown but three, four or more may be used depending on how many circuits are required. The centre core or innermost layer may have 1, 3 or 4 quads. The number in each succeeding layer is then 6 greater than the number in the previous layer. In Fig. 2.5 a layer of 4 and a second layer of 10 are shown. A third layer would contain 16 quads. Layers are separated by a winding of cotton. Surrounding the outside layer are two paper layers and an outer sheath of lead.

2.6.2. STAR-QUAD CABLES—IDENTIFICATION OF CONDUCTORS IN A CABLE

Individual conductors in a particular quad are identified by groupings of red or blue inked lines printed on their paper lappings. One pair of conductors has paper with single lines and double lines respectively. The conductors of the other pair have paper, the lines of which are in groups of three, and groups of four respectively. Alternate quads have blue and red identification lines on their paper lappings. Thus quads numbered 1, 3, 5, 7 and 9 have red identification rings, if quads 2, 4, 6, 8 and 10 have blue ink rings on their lappings.

To maintain a balance of impedance to earth between conductors, the effect of the different permittivity and leakance of the marking ink, is annulled by making sure that the total number of marking lines per unit length of cable is the same for every conductor. Thus in a given length of cable there may be three groups of four inked lines, four groups of three inked lines, six groups of two inked lines or twelve individual lines, all uniformly spaced.

The cotton whippings round the quads in the centre and in even-numbered layers are white but the whippings round any odd numbered layers are black. The whippings round the first and the last quad in each layer (called the marker and the reference quads) have an additional orange strand for identification. These are Quads 1 and 4 in the centre layer of Fig. 2.5. and Quads 1 and 10 in the second layer.

2.6.3. LOADING COILS

In Fig. 2.1, curve (d) indicates the type of attenuation-frequency characteristic to be expected from an underground cable. Below about 300 c/s the effect of cable capacitance is negligible and the attenuation falls off rapidly towards the value it would have for d.c. (zero frequency). Above 300 c/s the attenuation rises more uniformly with frequency. The frequency range of commercial speech is 300 to 3,400 c/s and the rapid fall in attenuation below 300 c/s is unimportant. The attenuation for frequencies within the commercial speech range can be made more nearly constant by the use of loading coils. These coils have a toroidal core of a compressed powdered magnetic material. Each core carries two insulated coils, one in series with each conductor of a pair and the windings

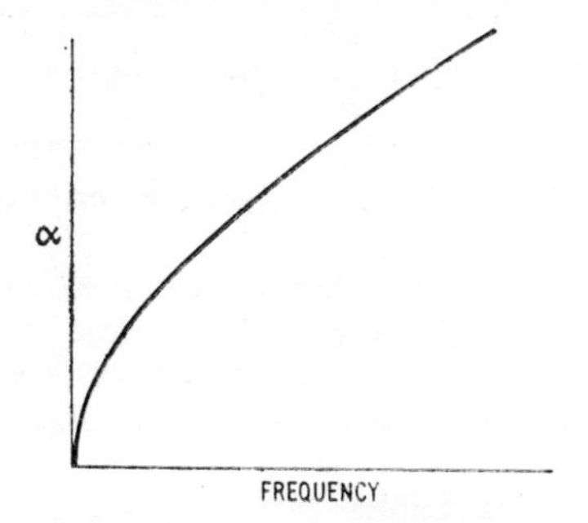

Fig. 2.6. Line attenuation-frequency characteristic

are so arranged that the mutual inductance between the two coils assists the self inductance. In this way a maximum inductance with a minimum of resistance is achieved. Loading coils for all the pairs in a particular cable are assembled in an iron case and installed along the cable duct at a manhole site. The spacing of the coils along a cable route must be regular. An effective inductance of 88 mH every 2000 yards is typical. Unfortunately, the application of loading coils in a line results in the line behaving as a low-pass filter which can pass freely only signals below a certain cut-off frequency. Curve (c) of Fig. 2.1 shows the attenuation variation with frequency as it might be with the loading suggested above. The cut-off frequency is of the order of 3·73 kc/s. Curve (b) shows a larger attenuation but with a higher cut-off frequency. This curve might represent a loading of 66 mH/2,000 yd. It is clear that the value of the loading coils and the cut-off frequency set limits to the frequency to be transmitted in such lines.

2.6.4. USE OF REPEATERS

The use of equalisers and repeaters (amplifiers), however, extends the frequency range of signals which can be transmitted along an underground cable route and enables 12- and 24-channel carrier current telephony to be employed. (The addition of equalisers and repeaters does not, however, overcome the limitations of the cut-off frequency caused by the use of loading coils.) The 24-channel carrier-current system requires a range of signal frequencies from 12 kc/s to 108 kc/s.

Fig. 2.6 may represent the attenuation characteristic of a section of the cable. If this is the case, an equalising attenuator E is inserted which has the inverse characteristic of Fig. 2.6 (*see* Fig. 2.8). Thus the cable and equaliser together have an output with the same loss at every frequency within the passband required. An amplifier *A*, with uniform gain (*see* Fig. 2.9) is then employed to restore the signal to the level it had at the input to the section of the line.

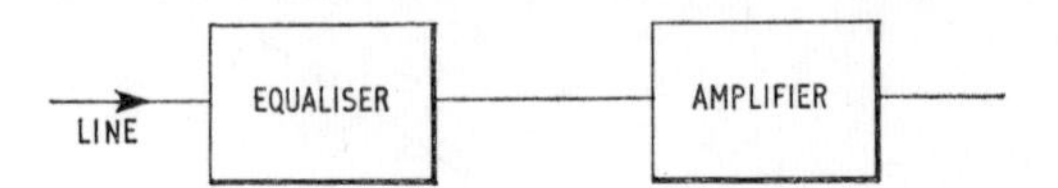

Fig. 2.7. Arrangement of equaliser and amplifier

Such an arrangement of equalisers and amplifiers is necessary approximately every 20 miles of cable. A cable containing 14 quads, each carrying 24 channels of carrier current telephony, could thus provide for 24 × 14, or 336 two-way conversations at the same time.

2.7. Concentric (coaxial) Cables

2.7.1. CONSTRUCTION FOR TELEPHONY

Fig. 2.10 (a) represents a coaxial line of a type widely used in carrier-current telephony. It is able to carry currents of much higher frequencies than can be used with a star-quad cable and this makes it possible to have a larger number of channels per line. A number of designs of coaxial line are available but the example illustrated has a centre conductor of copper wire whose diameter is 0·104 in. Circular polythene spacers fit over the central conductor and isolate it from the outer conductor which consists of a spiral of copper tape. The inside diameter of the latter is 0·375 in. Steel tape is wound over the copper tape and this helps to reduce unwanted coupling at low frequencies between adjacent coaxial lines. At high frequencies, the copper tape acts as an r.f. screen (*see* Section 5.4.1). The r.f. currents tend to flow on the inside surface of the outer conductor and on the outside surface of the inner conductor. It is usual to encase two or four of such coaxial lines in

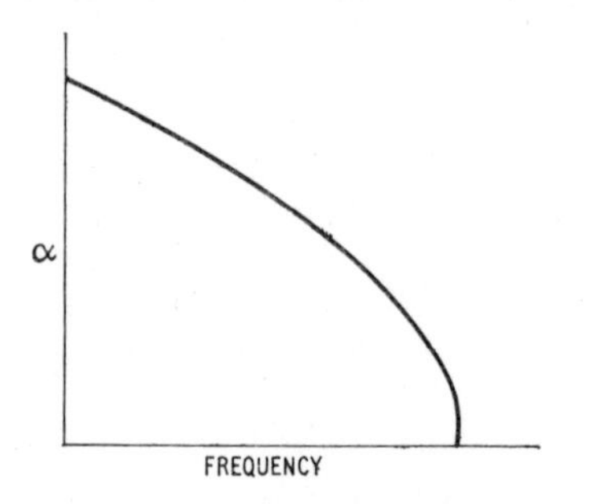

Fig. 2.8. An equaliser attenuation-frequency characteristic

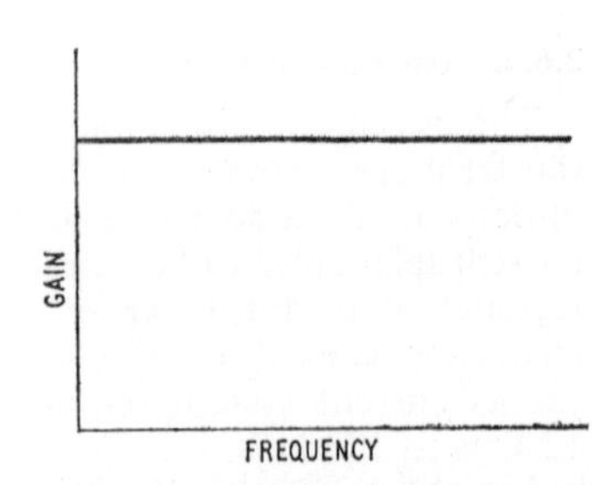

Fig. 2.9. Resultant characteristic of line and equaliser

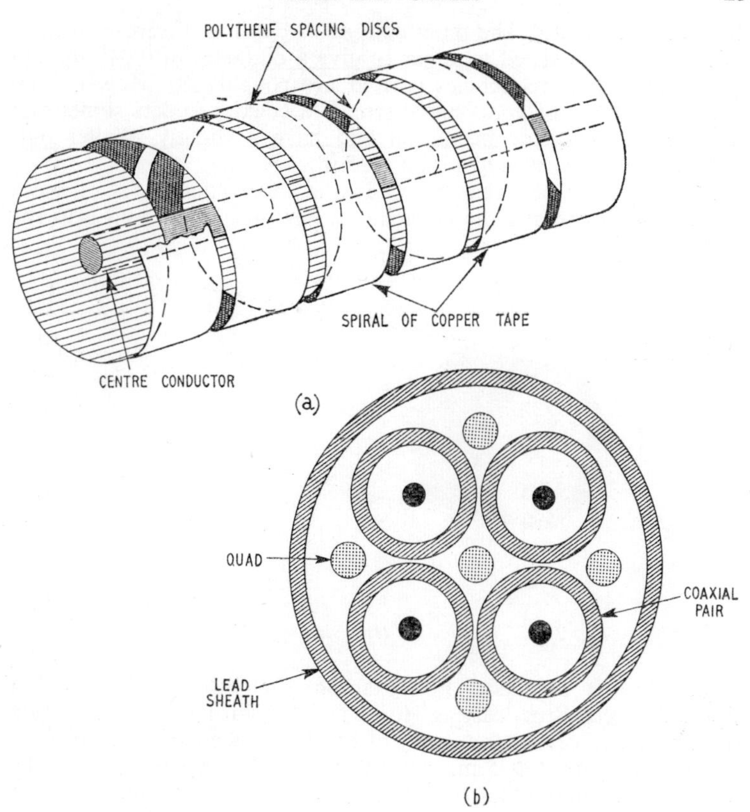

Fig. 2.10. (a) A coaxial pair, (b) cross section of a coaxial cable

one lead sheath. Two or four quads are also included in the interstices between the coaxial lines so that maximum possible use is made of the space within the lead sheath (*see* Fig. 2.10 (b)).

The frequency limit for the coaxial cable described above depends upon the total attenuation which can be tolerated and upon the distance which the signal must travel without amplification. The cable is, however, in common use for frequencies in the range 60 kc/s to 2·5 Mc/s and can be used without difficulty up to 3 Mc/s.

2.7.2. TELEVISION COAXIAL FEEDERS

Fig. 2.11 illustrates a design of coaxial feeder suitable for use as down-lead between a television or f.m. aerial and its receiver. In this type of cable the inner conductor is held in position by cellular polythene. This provides a large percentage of air in the volume of the dielectric and thus reduces the loss which the dielectric must cause. The outer conductor is a copper braid which is protected by a P.V.C. cover. The inner conductor consists of several strands of copper wire twisted together to give a diameter of 0·048 in. In such a cable the frequencies carried are roughly 40 Mc/s to 200 Mc/s. In areas of good signal strength, coaxial feeders sometimes use a solid polythene insulation and have a slightly smaller outside diameter—a little less than 0·25 in.

If the spacing between the outer and inner conductors of a coaxial line is increased the capacitance per unit length and also the

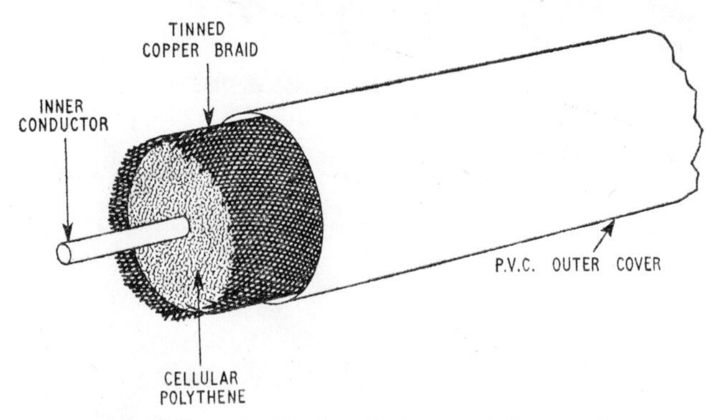

Fig. 2.11. A cable for television aerial connection

attenuation is reduced. For example the BBC send television signals between Broadcasting House and Crystal Palace transmitting station using a carrier frequency of 15 Mc/s and employing a coaxial cable of diameter 0·975 in., without the use of any repeaters in the course of the nine miles of cable. Had the smaller cable of 0·375 in. diameter been used, then intermediate repeater equipment would have been necessary.

Coaxial feeders between transmitters and their aerials can be used up to frequencies of about 3,000 Mc/s, because the transmission distance is relatively small. The mean circumference of the dielectric space indicates the minimum wavelength which can be transmitted without considerable loss of power. For frequencies higher than this a wave-guide system is required.

2.7.3. CHARACTERISTIC IMPEDANCE

The characteristic impedance of an air-dielectric coaxial feeder at high frequencies, for which Equation (3) applies, may be found from

$$Z_0 = 138 \log_{10}\frac{D}{d} \text{ ohms} \tag{10}$$

where D and d are the diameters of the external and internal conductors respectively. For a cable with a dielectric other than air the above impedance is divided by the square root of the relative permittivity of the insulation used. For example: To find the characteristic impedance of the coaxial feeder illustrated in Fig. 2.11, where $D = 0{\cdot}29$ and $d = 0{\cdot}048$ in. The relative permittivity of the cellular polythene is of the order of 1·45. Thus:

$$Z_0 = \frac{138}{\sqrt{1{\cdot}45}} \log_{10}\frac{0{\cdot}29}{0{\cdot}048}$$

$$= \frac{138}{\sqrt{1{\cdot}45}} \times 0{\cdot}7812$$

$$= 89{\cdot}54 \text{ ohms}$$

Or, considering a telephone coaxial cable for which $D = 0{\cdot}375$ and $d = 0{\cdot}104$ in., and assuming that the average permittivity is approximately that of air, then

$$Z_0 = 138 \log_{10}\frac{0{\cdot}375}{0{\cdot}104}$$

$$= 138 \times 0{\cdot}557$$

$$= 76{\cdot}88 \text{ ohms}$$

These values are typical of a concentric feeder impedance, though impedances as low as 40 or 50 ohms may sometimes be encountered.

2.8. Submarine Cables

2.8.1. ATLANTIC TELEPHONE CABLE

In 1866 the first telegraph cable across the Atlantic came into operation, but only in 1956 was a transatlantic cable suitable for telephone channels completed. Telegraph cables approximately 2,000 miles long, without repeaters, have a bandwidth of the order of 100 c/s. The total attenuation is excessive at all except d.c. and very low frequencies. Such cables are thus suitable for teleprinter or cable morse operation only. The transatlantic telephone cable has a bandwidth of 144 kc/s and can carry 36 telephone channels. The improvement in usable bandwidth has become possible partly as a result of the improved design of submarine repeaters which have to operate in some parts at a depth of two and a half miles, and partly because of the advent of polythene as an insulator for submarine cables instead of gutta percha. In the main cable from Oban in Scotland to Clarenville in Nova Scotia there are 102 repeaters, 51 in each cable. These amplifiers have an estimated life of 20 years and during this time must operate without attention, since the cost of breakdown and subsequent repair is excessive.

Fig. 2.12 shows the construction of the deep water cable. The centre conductor, which is approximately 0·132 in. diameter copper, has three copper tapes, each of thickness 0·0145 in., wrapped round

2

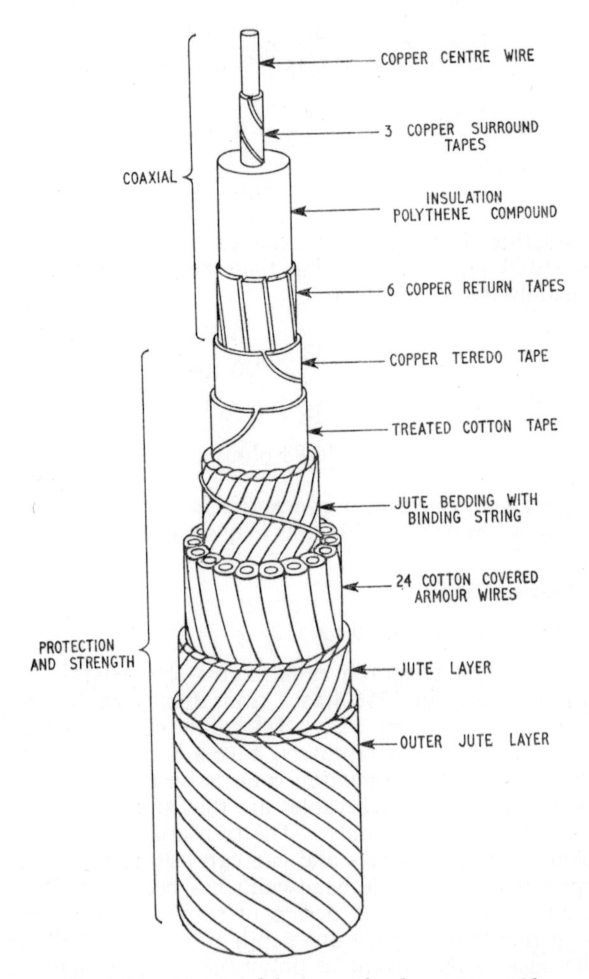

Fig. 2.12. Structural features of a deep water cable

it. This adds to the effective cross-sectional area of the centre conductor without reducing the flexibility of the cable too much. Next in order is the polythene dielectric whose external diameter is 0·62 in. The return conductor encases the polythene and consists of six copper tapes each 0·016 in. thick. A copper tape of 0·003 in. thickness then adds protection against the teredo worm (a marine worm which bores through submerged wood and attacks the insulation of cables). Then follow layers of cotton tape and jute bedding with binding string covered by 24 high-tensile steel armour wires each of 0·086 in. diameter. These are finally covered by two layers of impregnated jute. This is a type of concentric or coaxial cable. The exceptional design is necessitated, of course, by the severe physical conditions of strain and pressure to which it is subjected.

2.8.2. A LIGHT-WEIGHT TELEPHONE CABLE

Fig. 2.13 shows a further development in the design of submarine cables. By the omission of the armour wires, which in shallow water give protection from fishing trawls, etc., the deep water cable has the advantage of a much ligher weight. The tensile strength of the cable is increased by a core of steel strands of which there are 26 of diameter 0·033 in. and 4 of diameter 0·08 in. The inner conductor consists of a copper tape surrounding the core which has a box seam along the length of the cable. On top of the tape is the polythene insulation to an external diameter of 0·80 in. The outer conductor of six aluminium tapes each 0·015 in. thick is wound over the polythene. This external conductor is protected by a cotton tape and a polythene sheath to an outside diameter of 1·02 in. This design is now in use between Sweden and England from Göteborg to Marske (approximately 10 miles north of Middlesbrough). The submarine cable length is about 510 nautical miles. Frequencies carried are in the range 60 kc/s to 300 kc/s from Sweden to England and 360 kc/s to 608 kc/s in the reverse direction. This provides for 60 telephony circuits. The same cable design is being used in the Commonwealth round-the-world telephone cable scheme which starts with the " CANTAT " cable linking the U.K. with Canada. In the 2,100 miles of cable between the U.K. and Canada 90 repeaters and 7 equalisers are needed. Armoured cable is used at the shore ends but the same electrical characteristics as those of the light weight cable are maintained.

2.9. Comparison of Coaxial and Balanced Conductors

In this chapter we have considered several types of cable construction. All of them fall into two categories. The *go* and *return* conductors are either *balanced* in their impedance to earth, as they are in the open-wire lines or in the star-quad cable, or they are *unbalanced* in their impedance to earth as in the coaxial cables. A few points of comparison of these two categories will now be listed.

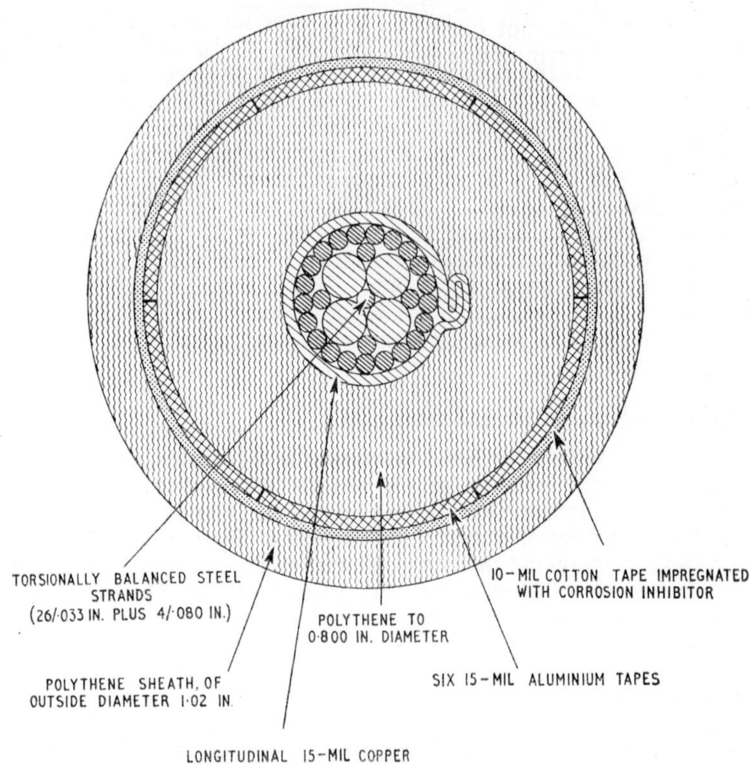

Fig. 2.13. A light weight submarine cable

Inductance per unit length is greater with the balanced pair especially when the spacing between the conductors is large, as in aerial lines.

Capacitance per unit length is smaller for the balanced pair of conductors than for the same length of coaxial conductors.

Characteristic impedance is larger for the balanced pair than for the coaxial cable because of the differences in L and C per unit length. When losses are neglected, the impedance for a balanced aerial line is approximately 600 ohms, and for a balanced pair of conductors in an underground cable it may be about 120 ohms. The characteristic impedance of a coaxial line is seldom more than 80 ohms and may be as low as 40 or 50 ohms.

Freedom from interference is greater for balanced conductors at low frequencies, say, below 60 kc/s, because the voltages induced in the opposite conductors balance out. But a coaxial cable has conductors with widely differing impedances to earth (the outer conductor is often at earth potential) and this cancellation of interfering voltages does not occur. But at high frequencies balance of impedance to earth of the two conductors of the balanced pair is very difficult to maintain and the advantage of the balanced system is not evident. Interference at high frequencies is not troublesome with coaxial cables because the outer conductor acts as a radio-frequency screen for the inner conductor, in the same way as the aluminium can round a coil in a radio receiver screens it from radio-frequency magnetic fields set up by currents in adjacent coils (*see* Chapter 5 for principles of r.f. screening). Further, if the outer conductor is firmly bonded to earth potential along its length it forms a perfect electro-static screen for the inner conductor. Thus balanced conductors are freer from interference at low frequencies but coaxial pairs suffer less interference at high frequencies.

In aerial feeders between a transmitter and its aerial, the balanced line is cheaper to construct and more suitable if the aerial itself is a balanced load with respect to earth. Coaxial aerial feeders are more suitable for an unbalanced load but are expensive for high-power transmission over considerable distances. At v.h.f. coaxial feeders suffer less radiation loss and receive less interfering signals when they are used for reception, as their pick-up is virtually zero.

Coaxial feeders to aerials can be buried in the earth. This helps to ensure that they are non-radiating and have zero pick-up when used for receiving aerials.

For submarine cables, coaxial conductors are more convenient and since they are screened by many cubic miles of sea water, are not subject to interfering signals.

Questions

1. Make careful sketches illustrating the construction of two of the following

(a) an aerial feeder for high power, m.f. broadcast transmitter,

(b) a submarine coaxial cable,

(c) a star-quad type trunk telephone cable.

Explain the reasons for the choice of insulating cables used and quote an approximate figure for typical characteristic impedance for each case you have described. (*C & G*, 1959.)

2. Tabulate the relative advantages and disadvantages of a balanced open-wire transmission line and an unbalanced coaxial feeder to connect a short-wave transmitter to its aerial.

Sketch and briefly describe one way in which a balanced line having a characteristic impedance of 600 ohms may be matched to feed a 75-ohm half-wave dipole aerial. (*C & G*, 1960.)

3. Sketch a multi-tube coaxial cable suitable for inland use. Give typical dimensions and indicate the materials used.

A coaxial pair has a loss of 20 dB at a frequency of 1 Mc/s. Calculate its loss at 9 Mc/s assuming that dielectric loss may be neglected. (*C & G*, 1961.) (The answer assumes that R is much greater than ωL so that attenuation is proportional to $\sqrt{f}$.)

4. A h.f. transmission line consists of open wires having an inductance of 0·004 H/mile and capacitance of 0·01 μF/mile. Neglecting losses, find the characteristic impedance, the phase coefficient and the wavelength along the line if the frequency is 100 kc/s.

5. A h.f. transmission line consists of a pair of open wires having a distributed capacitance of 0·01 μF/mile and a distributed inductance of 3 mH/mile. What is the characteristic impedance of the line? Describe a suitable method for connecting this line to an aerial having an impedance of 100 ohms.

6. What electrical and physical properties are desirable in cables? Discuss the factors which affect the attenuation constant of a cable.

7. Explain why a balanced aerial line may suffer less interference than a coaxial line at audio frequencies, but h.f. carrier current telephony is more effectively carried on the latter type of cable.

8. Explain the meaning of the terms: characteristic impedance of a line, attenuation constant and phase constant of a line. What do these constants depend on, (a) at v.l.f., (b) at h.f., (c) if leakance and line inductance are negligible?

9. What do you understand by the term "cross-talk" and how is this mitigated in different types of line transmission?

10. Describe the construction of a submarine telephone cable giving approximate dimensions and a value for its characteristic impedance. Why are equalisers associated with the repeaters used with submarine cables?

11. Sketch the arrangement of a four-tube coaxial cable for inland use. Show typical dimensions and mention the materials used.

A coaxial cable has a loss of 3·9 dB/mile at 1·2 Mc/s. If dielectric loss is negligible, what is the loss in 6 miles of cable at 4·34 Mc/s? (*C & G*, 1962.)

3

Communication Channels

3.1. Growth in Communications

With the constant development of rapid and efficient transport within countries and across continents, the parallel growth in international travel and the rapid expansion of industry and commerce it is not surprising that there is an ever-increasing demand for improved communication systems. The heavy demand on local and long-distance telephone systems and on all other forms of radio and line communication makes it necessary for engineers to find new methods of improving and expanding existing communication techniques. However, the introduction of new methods does not necessarily mean the disappearance of the old ones. Although artificial earth satellites may be used for reflecting or re-transmitting radio waves to distant parts of the earth and thus increase the number of channels available between continents, our common two-wire connection to local telephone exchanges will still be essential and the new Commonwealth telephone cable link will not fall into disuse. Methods which can be employed to provide a given communication link between two specific places must depend on the quantity of traffic between those places, the most economical way of satisfying the demand and the technical difficulties to be overcome. Future possible demand must also be considered. If a new technical advance can in some way make a particular path of communication cheaper, an increase in demand may result and thus justify the new method.

3.2. Telephony

3.2.1. A SUMMARY OF METHODS

Between a telephone subscriber and his nearest telephone exchange all that is required is a simple two-wire system, provided either by overhead wires or underground cable. The line loss should not exceed 3 dB. Telephone exchanges are normally placed somewhere near the centre of the area they serve, so that no subscriber lives at such a distance from the exchange that this 3-dB loss is exceeded. Dependent telephone exchanges are connected by two-wire systems to minor group exchanges and the latter in turn to zone exchanges. The line loss between a dependent exchange and a minor group

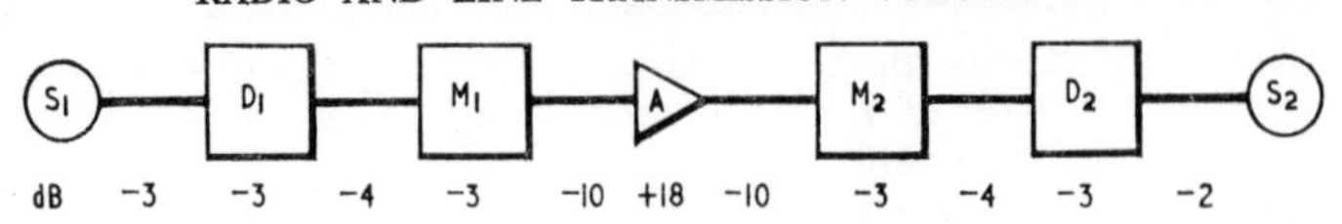

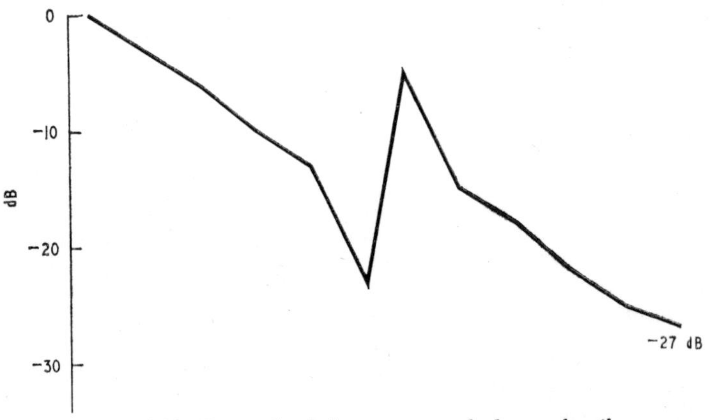

Fig. 3.1. Energy levels between two telephone subscribers

exchange should not be greater than 4·5 dB and between the minor exchange and the zone exchange it should be less than 3 dB. If the distance between the exchanges is such that despite the use of loading coils the total line loss is too big, amplifiers (repeaters) are inserted. Between zone exchanges four-wire systems are used and carrier-current telephony or telegraphy is needed to deal with the large volume of traffic carried over these main trunk routes. International communication may be by radio-telegraphy, radio-telephony, by cable telegraphy or, in a growing number of instances, by telephone cable. Where the destination or the origin of messages is mobile, as in aeroplanes, ships, taxis, ambulances, etc., a radio link with a frequency suited to the range is the only practical means of communication.

3.2.2. NEED FOR AMPLIFICATION

At each stage of its journey, the signal must be prevented from becoming so weak that it compares with small interference voltages from other lines (cross-talk) or from other external sources.

Fig. 3.1 shows how the energy level may vary between two telephone subscribers S_1 and S_2, who are interconnected via dependent exchanges D_1 and D_2 and minor group exchanges M_1 and M_2. The group exchanges are joined by a two-wire connection requiring (in this example), an amplifier. The distances between the various parts of this chain diagram do not, of course, represent the distances between the various parts of the equipment. Between M_1 and the amplifier A a drop of 10 dB is shown. If, by taking advantage of the gain of the amplifiers present, smaller gauge conductors are used, the cost of the repeaters can to some extent be off-set by the saving in copper. Assuming a line attenuation of 0·74 dB/mile (this being a typical loss for speech frequencies with the gauge of wire used), the distances between the minor exchanges and the amplifier are approximately 13½ miles. A is shown as having a gain of 18 dB. This gain is intended to take into account the losses in the balancing networks of the hybrid filters which must be used before and after each amplifier in a two-wire system. (For circuits of hybrid filters *see* Chapter 5 of *Radio and Line Transmission, Volume* 1.)

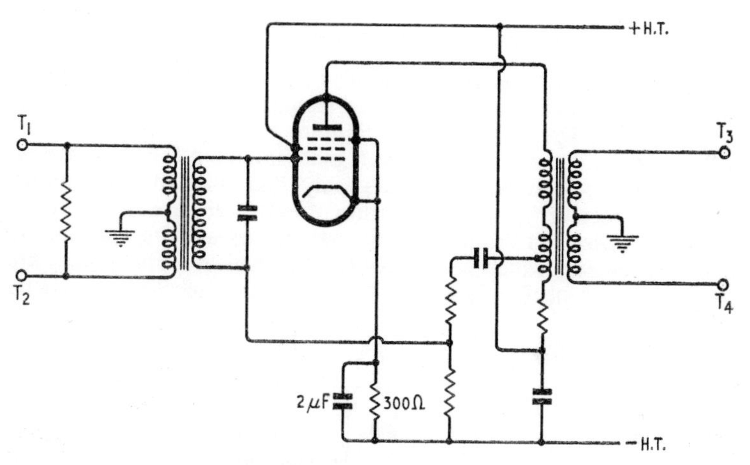

Fig. 3.2. A repeater circuit

Half the power arriving at the end of the line coupled to the input to an amplifier is dissipated in the matching network of the hybrid filter, and half the output power of the amplifier is wasted in the matching impedance of the hybrid filter on the output side of the amplifier. This means that a total loss of 6 dB due to hybrid filters is sustained at an amplifier. Hence the total gain of A must be 24 dB. A loss of 3 dB is shown at each exchange. The chief cause of this lies in the coupling impedance bridges by which the pair of lines entering the exchange is coupled to the correct pair of outgoing lines (*see* Chapter 1 of *Radio and Line Transmission, Volume* 1). Two couplings are needed at each exchange and each causes a loss of approximately 1·5 dB. The net loss between subscribers is thus 27 dB which is tolerable. (The maximum permissible loss is 30 dB.)

Fig. 3.2 shows the circuit of a suitable repeater. Negative feed-back (*see* Section 8.10), is used to improve the frequency response of the amplifier and also to reduce the effect on the gain of variations of the h.t. supply. The gain may be adjusted by altering the turns ratio of the input coupling transformer and may be between 24 and 30 dB.

In general the distance between group exchanges which are interconnected is not so great as to require the use of amplifiers. Amplifiers in audio circuits are avoided when possible because the provision of one amplifier for one audio channel, or even for a 1 plus 4 circuit (*see* Chapter 5 of *Radio and Line Transmission, Volume* 1) is expensive. On trunk circuits where a single amplifier can serve many channels simultaneously, the use of repeaters is far less extravagant. Trunk circuits using carrier telephony require the translation of audio bands to different and higher frequency levels. This principle will now be considered.

3.2.3. PRINCIPLE OF FREQUENCY TRANSLATION

The aerial of a radio receiver picks up small voltages from many different transmitters which radiate signals of many kinds: television signals, sound broadcasts, announcements from police cars, calls to taxi cabs and morse code signals intended for ships at sea. But from the loudspeaker comes only one selected programme. Within the receiver the selective action of inductive-capacitive circuits is used to select and amplify only that carrier and its appropriate side frequencies, which it is intended to receive. The others are attenuated to a negligible level. This separation of one radio signal

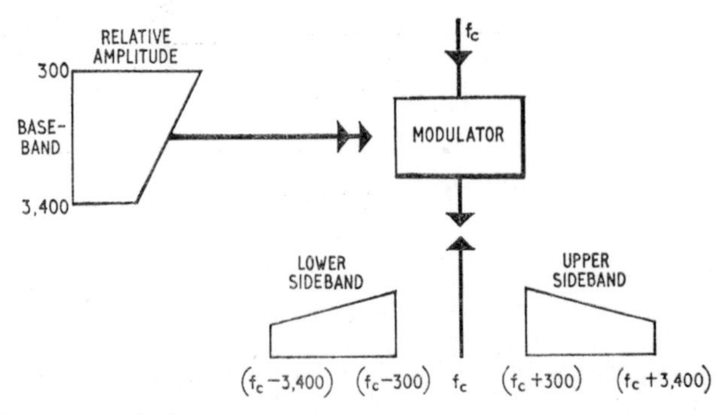

Fig. 3.3. Translation of base bands to side bands

from many others is only possible because the radiation from the many transmitters occurs at different frequencies derived by using different carrier frequencies. Similar principles are used in carrier telephony systems which operate between zone exchanges.

The frequencies which are used for modulation at the sending end and which are required at the end of a particular channel of communication after demodulation are termed the *base-band* of frequencies. For commercial speech the base-band is 300 to 3,400 c/s, for music it may be from 50 c/s to 10 kc/s or for television signals

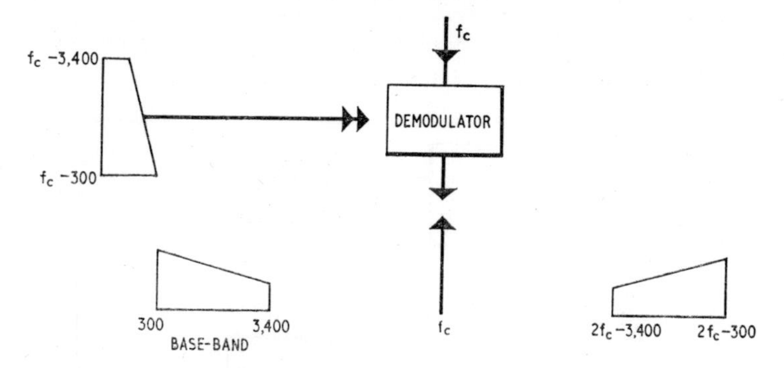

Fig. 3.4. Regaining the base band

from 0 to 3 Mc/s. When a carrier frequency is amplitude-modulated by a base-band of frequencies, a similar bandwidth of frequencies is generated above and below the carrier frequency as an upper and a lower sideband (*see* Fig. 3.3).

In Fig. 3.3 the relative amplitudes have been represented by the block figure which indicates that the 300-c/s signals have an amplitude of about 1·5 times the amplitude of the 3,400-c/s signals.

The lower sideband is inverted in the sense that the frequencies at the higher end of the band have the greater amplitude and those at the lower end of the band have the smaller amplitude, whereas the reverse amplitude distribution occurs in the original band of audio frequencies. However, if a filter circuit is used to select the lower sideband, and this only is transmitted along a line, then on mixing again at the receiving end with another oscillation of frequency f_c, we obtain, by modulation process, a further two sidebands. These range on the one hand from 300 c/s to 3,400 c/s and on the other hand from ($2f_c$–300) to ($2f_c$–3,400). The lower sideband of this second modulation is the original base-band and has been transposed again so that the signals of lowest frequencies once more have the highest amplitudes (*see* Fig. 3.4).

The process of re-obtaining the base-band at the receiving end is termed *demodulation* but it is essentially similar to modulation. The frequency level to which the base-band is translated by modulation at the sending end depends upon the carrier frequency f_c which is used. If f_0 is 108 kc/s the base-band is translated to two levels—(104·6–107·7) and (108·3–111·4) kc/s. The lower level is normally selected by filters.

Example 3.1

What carrier frequency should be used in order to translate a base-band of 300 c/s to 3·4 kc/s to a band 4 kc/s to 7·1 kc/s?

4 kc/s to 7·1 kc/s must represent the lower sideband obtained by modulation. Let the required carrier frequency be f_c kc/s. Then

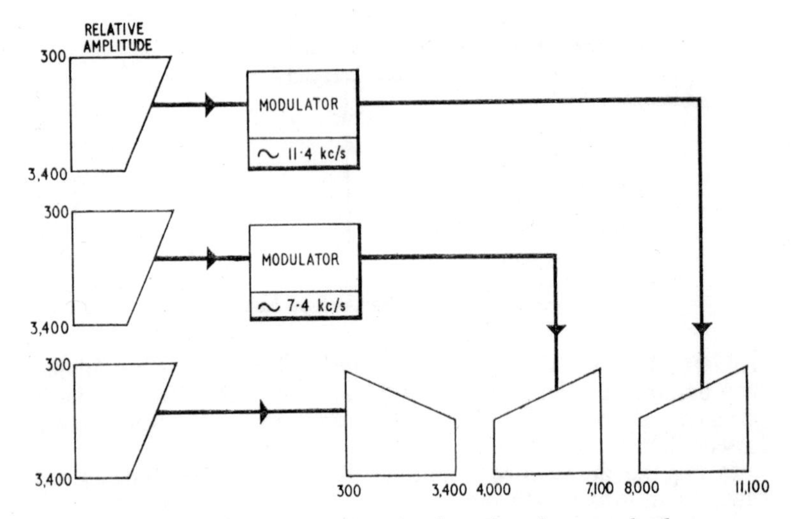

Fig. 3.5. Translation of base bands to three frequency levels

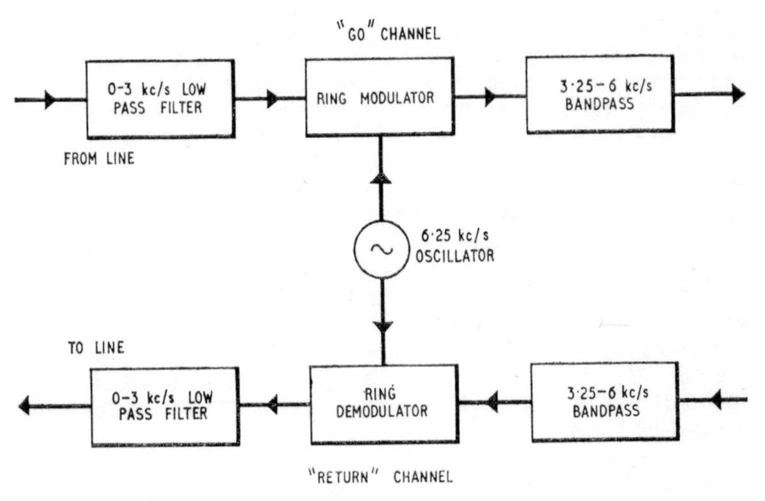

Fig. 3.6. Use of a common oscillator for " go " and " return " channels

$$f_c - 3{\cdot}4 = 4 \text{ kc/s}$$

and therefore, $$f_c = 7{\cdot}4 \text{ kc/s}.$$

If the example above had required the base-band to be translated to the band 8 kc/s to 11·1 kc/s, then a carrier frequency of 11·4 kc/s would have been required. If one base-band resulting from one telephone line output is translated to the band 4 kc/s to 7·1 kc/s and another telephone line output is translated by another modulating circuit to the band 8 kc/s to 11·1 kc/s, both of these, together with a third base-band from 300 c/s to 3·4 kc/s can be transmitted along one line and separated by filters at the receiving end. After separation by filters, carriers of 7·4 kc/s and 11·4 kc/s are required to restore the base-band frequencies for the two top channels. This is called a (1 + 2)-system. This arrangement of the three base-bands for transmission is shown diagrammatically in Fig. 3.5.

The manner in which one oscillator can supply the modulator for the " go " circuit and at the same time the demodulator of the " return " circuit of one particular channel of communication is illustrated by Fig. 3.6. The object of the modulator is to transpose a frequency band of 250 c/s to 3 kc/s to the band 3·25 kc/s to 6 kc/s, so that a similar base-band can be fed along with the transposed band for modulating a telephony transmitter. In this way two communication channels are provided in the frequency range 250 c/s to 6 kc/s. Before modulation, a low-pass filter must remove all audio frequencies above 3 kc/s. After modulation, the band-pass filter allows only the lower sideband of this modulation to pass forward to the next part of the circuit. In the return line a filter first selects the band of frequencies appropriate to the " return " channel which uses the 6·25 kc/s oscillator for demodulation. The demodulation process generates frequencies which are equal to the difference between the 6·25 kc/s and the selected line frequencies, and also frequencies which are equal to the sum of the 6·25 kc/s and the selected line frequencies. The filter which follows selects the difference frequencies produced in the demodulator and attenuates the sum frequencies.

3.2.4. THE MODULATOR

The circuit of the ring modulator normally used for both modulation and the complementary process of demodulation, is shown in Fig. 3.7 (a). Elements *1*, *2*, *3* and *4* are metal rectifiers which are rendered alternately conductive and non-conductive as the polarity of the carrier frequency reverses. When the polarity of the carrier oscillator is positive at X, elements *1* and *2* are conductive but the elements *3* and *4* are made non-conductive by the reverse polarity of voltage applied across them by the carrier oscillator. Under these conditions the effective circuit is that of Fig. 3.7 (b). If the secondary voltage of T_1 is as marked, then the current in the primary winding of the transformer T_2 due to the base-band input voltage,

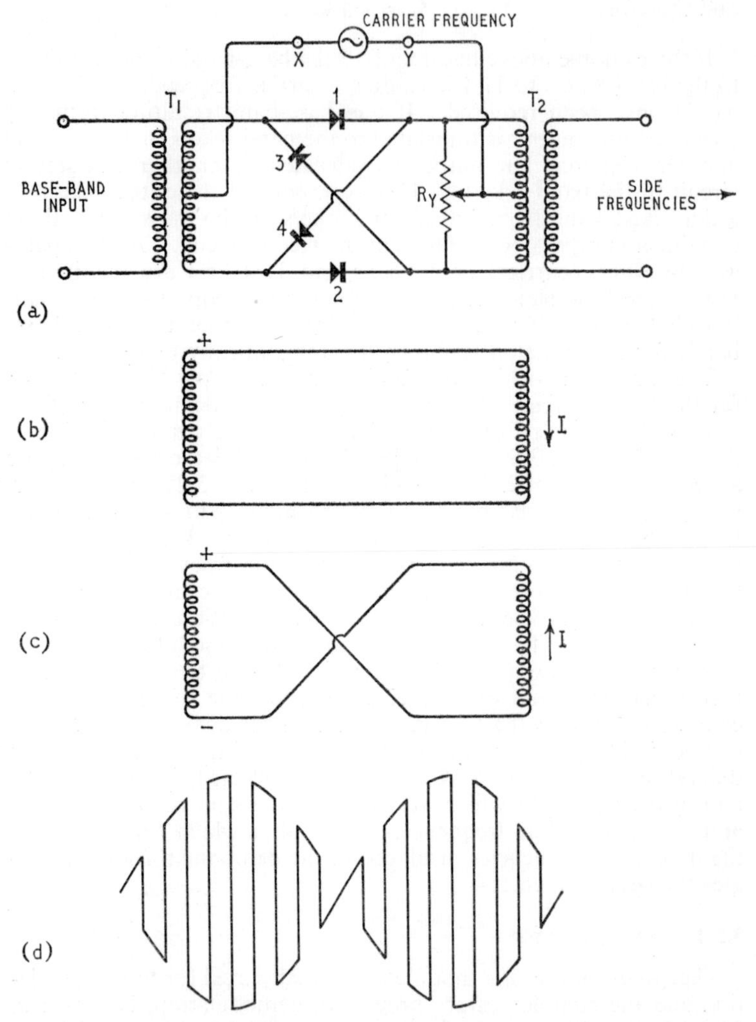

Fig. 3.7. (*a*) *A ring modulator,* (*b*) *effective circuit when rectifiers 1 and 2 conductive,* (*c*) *effective circuit when rectifiers 3 and 4 conductive,* (*d*) *output waveform*

is as indicated in Fig. 3.7 (b) by I. But if the terminal Y of the carrier output is positive to X, elements *3* and *4* are conductive while elements *1* and *2* are biased into non-conduction. The effective circuit is now as shown by Fig. 3.7 (c) and the direction of current due to the base-band input voltage is as marked by I in this diagram. The current I in Fig. 3.7 (c) is seen to be opposite in direction to the current I in Fig. 3.7 (b). The carrier voltage, which must be larger in amplitude than the base-band voltage, causes the current due to the base-band voltage to alternate in the primary winding of T_2, at the carrier frequency. The resultant current waveform in the output transformer T_2, is shown by Fig. 3.7 (d). The amplitude envelope assumes a sine waveform for the input to T_1, but whatever the input waveform to T_1 may be, the pulses of current in T_2 are at every instant proportional to the amplitude of the base-band voltage. The modulation envelope is therefore determined by the waveform of base-band voltage. The output of the carrier generator divides in opposite directions from the centre tap of T_2. Equal currents at the carrier frequency flowing in the two halves of T_2 produce no resultant magnetic flux at this frequency. Thus the carrier is suppressed and there is no output at carrier frequency from the secondary winding of T_2. A balance adjustment could be provided by a high resistor R_y with an adjustable centre slider. The output from T_2 contains both upper and lower side frequencies and also unwanted frequencies inherent in the clipped waveform of Fig. 3.7 (d) but a low-pass filter can easily separate the required lower sideband from the unwanted frequency components.

3.2.5. FILTERS

It will be evident from the foregoing paragraphs that filters play a prominent part in carrier-current telephony as well as in multi-channel radio transmission. It is therefore useful to give an elementary description of one or two simple electric wave filters. The mathematical treatment of filters is beyond the scope of this work but a qualitative approach is possible. Figs. 3.8, 3.9 and 3.10 show three different types of filter. The ideal filter offers zero attenuation within the pass-band and infinite attenuation for all frequencies outside the pass-band. To achieve this ideal would require reactive elements with zero loss. This is not possible but the higher the Q value of the components used the sharper the division between the pass-band and the attenuation band. The general principle is to offer to signals on unwanted frequencies a series impedance high compared with the series impedance offered to signals on the required frequencies, and to present at the unwanted frequencies a shunt impedance which is low compared with the shunt impedance at the frequencies to be passed forward. If it is remembered that:

(1) inductive reactance increases directly with frequency,
(2) capacitance admittance increases directly with frequency.

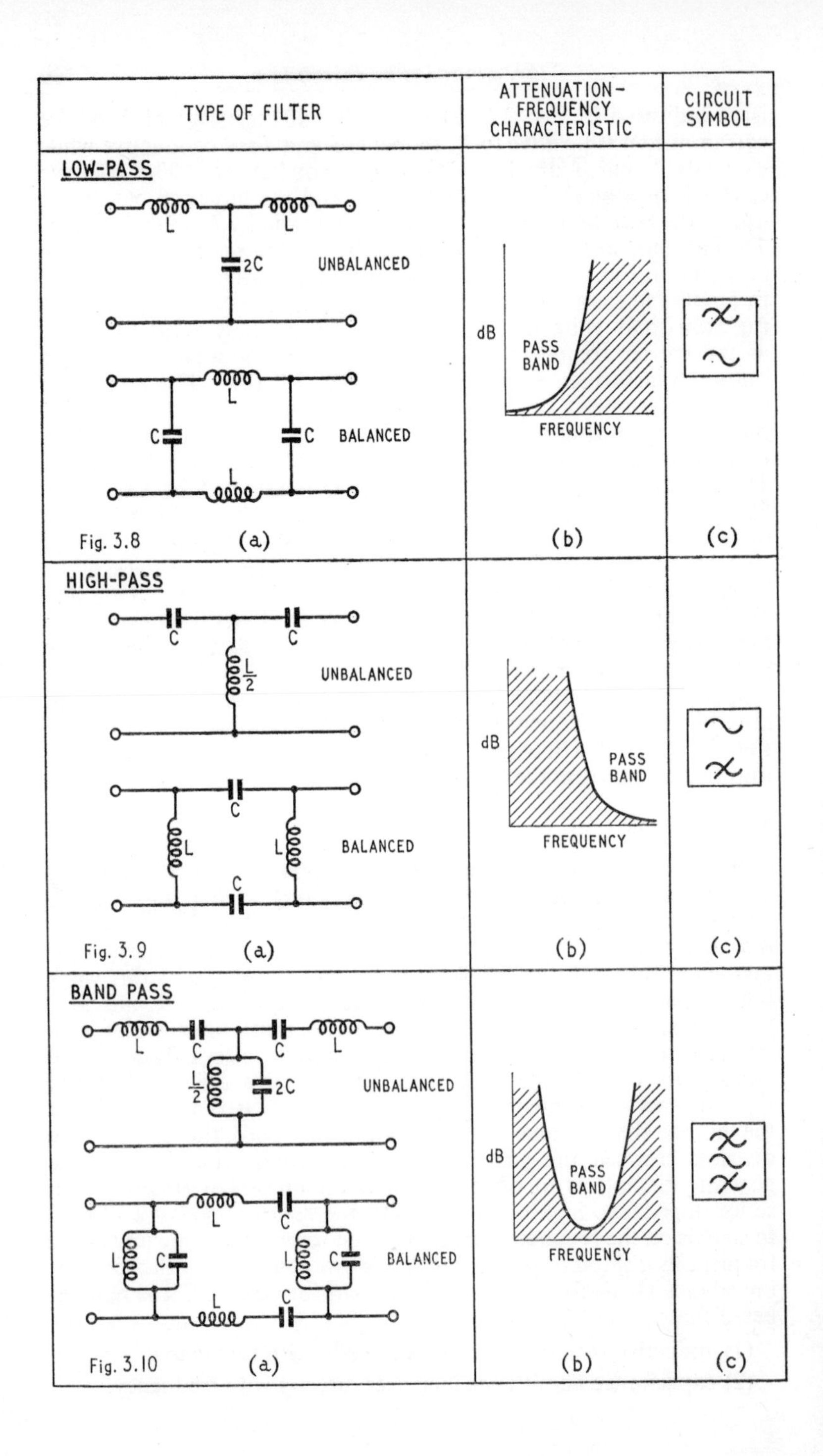

TYPE OF FILTER
ATTENUATION-FREQUENCY CHARACTERISTIC
CIRCUIT SYMBOL
LOW-PASS
L
L
2C
UNBALANCED
L
C
C
BALANCED
L
dB
PASS BAND
FREQUENCY
Fig. 3.8
(a)
(b)
(c)
HIGH-PASS
C
C
L/2
UNBALANCED
C
L
L
BALANCED
C
dB
PASS BAND
FREQUENCY
Fig. 3.9
(a)
(b)
(c)
BAND PASS
L
C
C
L
L/2
2C
UNBALANCED
L
C
L
C
L
C
BALANCED
L
C
dB
PASS BAND
FREQUENCY
Fig. 3.10
(a)
(b)
(c)

(3) a series circuit consisting of inductance and capacitance has a minimum impedance at resonance (and would have zero impedance were it not for the inevitable resistance of the circuit), and

(4) a parallel arrangement of inductance and capacitance has a maximum impedance at resonance,

then the diagrams of Figs. 3.8 (a), 3.9 (a) and 3.10 (a) will be seen to have a qualitative agreement with the attenuation-frequency characteristics shown opposite to them. In each of the circuits the value

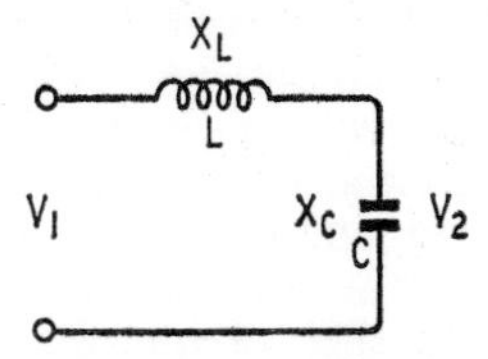

Fig. 3.11. A filter half section

of inductance represented by L and the value of capacitance represented by C are the values which a *half section* of filter would have. A *full section* of filter is then made by sliding two half sections together. To see how a half section might work, consider Fig. 3.11 in which V_1 is an applied alternating input voltage, and V_2 is the potential difference across the capacitive reactance and is the output voltage from the half section. The current due to V_1 is given by:

$$I = \frac{V_1}{(X_C - X_L)}$$

$$V_2 = IX_C$$

$$= \frac{V_1 X_C}{(X_C - X_L)}$$

or dividing top and bottom lines by X_c

$$V_2 = \frac{V_1}{\left(1 - \frac{X_L}{X_C}\right)}$$

When the frequency is low, then X_L is very small compared with X_C and the bottom line is approximately unity, so that $V_2 = V_1$. When the frequency is increased so that X_L increases and X_C decreases, then the whole denominator decreases and V_2 increases. At the resonance frequency at which X_L is equal to X_C, $\frac{X_L}{X_C}$ becomes unity. In theory the value of V_2 is then infinitely large since the bottom line has become zero. In practice, the line impedance and the filter resistance limit this growth of V_2 severely. A further increase of frequency will make X_L exceed X_C by an amount which continues to increase as the frequency rises further. The sign of the reactance has now reversed so that the phase of V_2 relative to V_1 has also reversed. The denominator rises continuously once the frequency at which resonance between L and C occurs has been exceeded so that V_2 must decrease after this particular frequency has been reached. This resonance frequency is the cut-off frequency for this simple filter.

3.2.6. THE USE OF CRYSTALS

Filter designs seek to attain the ideal of a sharp cut-off frequency with a minimum and constant attenuation within the pass-band. Although various improvements have been achieved in the design of filters, e.g. by adding inductance to the capacitive branches and vice versa, and by the use of components with lower loss and higher Q values, the most effective filters are those which use quartz crystals.

A thin slice of quartz with an e.m.f. applied between opposite faces contracts or expands very slightly according to the polarity of

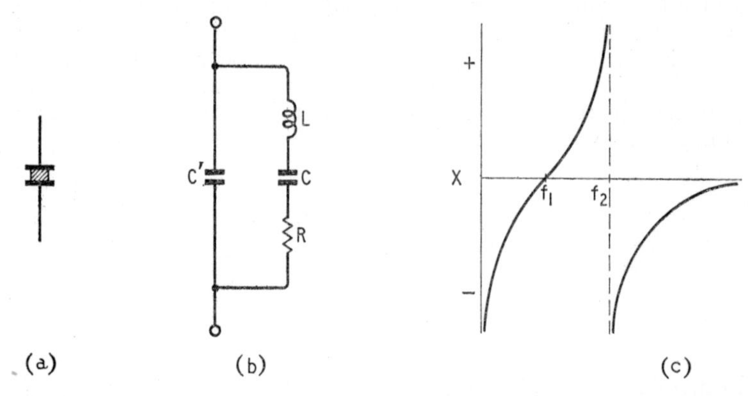

Fig. 3.12. (a) Quartz crystal—circuit symbol, (b) equivalent circuit of a quartz crystal, (c) a reactance-frequency graph for a quartz crystal

the applied voltage. If the voltage applied is alternating, the crystal alternately contracts and expands at the frequency of the applied signal. This effect is greatly magnified if the frequency of the applied e.m.f. is at or very near to the mechanical resonance frequency of the crystal. At this resonance frequency the vibrations are particularly strong and it is possible for a crystal supplied with too much electrical power to fracture itself by excessive vibration. Also a change in the mechanical dimensions of a crystal, brought about by varying mechanical pressure, results in opposite electrical charges appearing on opposite faces of the crystal and if the crystal is made to vibrate an alternating e.m.f. is generated at the frequency of vibration. This is called a *piezo-electric* e.m.f.

The crystal behaves in a similar way to an electrical circuit capable of exhibiting resonance except that the effective Q factor is much higher. The ratio of the energy stored in a mechanical vibrator to the energy supplied each cycle to maintain its vibration, is higher than that for an electrical oscillating circuit. This makes the quartz behave as an extremely selective circuit in which the mechanical resonance frequency of the crystal is the equivalent electrical resonance frequency. The piezo-electric e.m.f. can be considered to be the back e.m.f. of the circuit. Fig. 3.12 (a) gives the circuit symbol for a quartz crystal. Fig. 3.12 (b) is the equivalent electrical circuit. L is equivalent to the mechanical inertia of the quartz, C represents elastance or compliance, while R represents the energy loss due to vibration. This loss appears as heat in the crystal. C is the electrical capacitance of the crystal which is increased by the holder

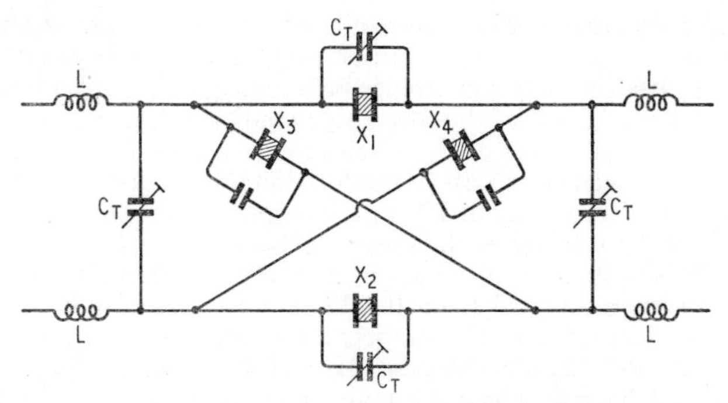

Fig. 3.13. A filter using quartz crystals

capacitance. Fig. 3.12 (c) is the graph of the reactance of this circuit plotted against frequency. The frequency f_1 is the frequency of series resonance while f_2 is a rejection frequency or frequency of anti-resonance. A variable trimming capacitor can be put in parallel with the crystal to enable small adjustments to be made to the resonance frequency of the parallel combination of crystal and trimmer. In this way the reactance-frequency graph of the filter can be modified within small limits.

Fig. 3.13 shows how four quartz crystals can be used to form a band-pass filter circuit. Crystals X_1 and X_2 in combination with their trimmers have equal frequencies of resonance and their rejection frequencies are also equal. The series resonance frequency of the crystals X_3 and X_4 is arranged to be equal to the frequency of anti-resonance of X_1 and X_2. The pass-band extends approximately from the series resonance frequency of X_1 and X_2 to the anti-resonance frequency of X_3 and X_4. The trimming capacitors marked C_T provide fine adjustment of the frequency range of the pass-band. One of the difficulties of using crystal elements in filter circuits is that of obtaining a sufficient bandwidth. The coils marked L have the effect of increasing the bandwidth which is brought up to

4 kc/s for use in carrier-current telephony circuits. The bandwidth obtained depends also on the frequency level at which the band-pass is centred. As the frequency is reduced it becomes increasingly difficult to obtain a given bandwidth, especially with crystal filters. To be able to take advantage of the sharp cut-off characteristics of a crystal filter, the frequencies at which these are used are generally between 60 and 120 kc/s and not lower.

3.3. Carrier Current—Multi-channel Systems

3.3.1. 24 CHANNELS

In *Radio and Line Transmission, Volume* 1, Chapter 5, a 12-channel trunk system such as might be used between zone telephone exchanges was described. For the sake of completeness we will summarise it here.

Twelve different audio base-bands of 0 to 4 kc/s are separately modulated with 12 different carrier frequencies which range from 64 kc/s to 108 kc/s with a difference between adjacent carrier frequencies of 4 kc/s. Each of the lower sidebands is selected by a crystal filter. The total range of the 12 sidebands is then from 60 kc/s to 108 kc/s. A further process of modulation using a group modulator with a carrier frequency of 120 kc/s transposes this range to a range of 12 to 60 kc/s, which is the lower sideband of this group modulation. This system as it stands is now almost obsolete but a 24-channel system adapted from the 12-channel system is in use. In the 24-channel scheme, 12 channels are treated as described above. A further set of 12 audio base-bands, each 4 kc/s wide, after the first modulation stage at which the range is 60 to 108 kc/s, is combined in the transmission line with the first set. Thus a total range of 12 to 108 kc/s is transmitted. The addition of the further 12 channels with an increase of bandwidth of 48 kc/s means the use of more equalisers and repeaters.

There is a limit to the increase of frequency in normal line transmission which is set by the increase in cable attenuation with frequency and by increased risk of cross-talk. Further multiplication of channels between zone exchanges has been met by a gradual conversion to coaxial cable circuits. Coaxial cables offer almost perfect screening at higher frequencies, and at such frequencies a wide band is available for the transmission of several hundred channels along one cable. Such a multi-channel scheme will now be considered.

3.3.2. 600 CHANNELS

The first stage in the assembling of the many channels at the correct frequency levels for transmission along the line is their arrangement into groups of 12 channels as in the 12- and 24-channel balanced line system. In Table 3.1 are set out the channels and the frequencies of the modulation carriers and lower sidebands. Each of these 12 channel groups is then modulated by a carrier frequency within

Table 3.1

GROUP 1: FREQUENCY RANGE 60 TO 108 KC/S

0–4 kc/s *Channel*	*Carrier frequency* (kc/s)	*Frequency range occupied by lower sideband* (kc/s)
1	64	60–64
2	68	64–68
3	72	68–72
4	76	72–76
5	80	76–80
6	84	80–84
7	88	84–88
8	92	88–92
9	96	92–96
10	100	96–100
11	104	100–104
12	108	104–108

the range 420 kc/s to 612 kc/s and again the lower sideband is selected after modulation. The groups and the frequencies of the modulating carriers and lower sidebands are set out in Table 3.2.

The super-group is then modulated with a super-group carrier to translate it to the frequency band which it is to occupy in the co-axial cable. Nine other super-groups, each formed as above and

Table 3.2

SUPER-GROUP 1: FREQUENCY RANGE 312 TO 552 KC/S

60–108 kc/s *Group*	*Carrier frequency* (kc/s)	*Lower sideband* (kc/s)
1	420	312–360
2	468	360–408
3	516	408–456
4	564	456–504
5	612	504–552

representing 60 channels each, are treated in a similar way but using a different super-group carrier frequency so that the line frequency range is different for each super-group. The super-group carriers and the line frequencies selected are set out in Table 3.3.

Fig. 3.14 shows the stages by which one group of 12 channels is assigned a place in the line frequency spectrum by the two processes of modulation described above. For each super-group five sets of group modulating equipment are required. The second super-group of 60 channels fits into the line frequency spectrum without a super-group frequency change. As indicated by Table 3.3 10 super-groups are assembled and frequencies range from 60 to 2,540 kc/s. However, a coaxial cable with suitable repeaters can transmit

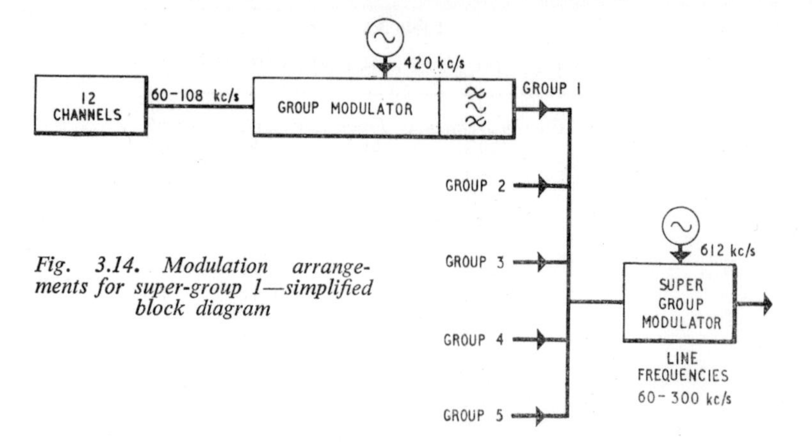

Fig. 3.14. Modulation arrangements for super-group 1—simplified block diagram

a frequency band 3 Mc/s or more wide so that all 600 channels can be accommodated. Wide-band repeaters may be required every six miles. Each repeater is preceded by an equaliser. Yet since only one equaliser and one amplifier is needed for each coaxial cable it is an economical system.

On trunk lines, each repeater is arranged to make good the loss on the section of the line which preceded it so that along the entire line the net loss is zero. This arrangement has the advantage that telephone calls can be routed by any available trunk line and always

Table 3.3

LINE FREQUENCIES 60 KC/S TO 2540 KC/S (600 CHANNELS)

312–552 kc/s *Super-group*	*Carrier frequency* (kc/s)	*Lower sideband* (kc/s)
1	612	60–300
2	Nil	312–552
3	1116	564–804
4	1364	812–1052
5	1612	1060–1300
6	1860	1308–1548
7	2108	1556–1796
8	2356	1804–2044
9	2604	2052–2292
10	2852	2300–2540

arrive at the ultimate zone exchange with the same energy level. The maximum attenuation is at the highest frequency and may be approximately 6·4 dB/mile. With a 6-mile spacing, amplifiers require a gain of 38·4 dB to replace this loss. A further 6-dB loss in the hybrid coils at the terminals must also be made good.

It is the function of the equaliser to increase the losses for all frequencies to that suffered by the highest frequency so that all the line frequencies enter each wide-band repeater at the same level. This principle is illustrated by the graphs of Fig. 3.16, in which curve *A* is the attenuation-frequency curve for a 6-mile length of cable, curve *B* is the attenuation curve for the equaliser used at the termination of this cable run and curve *C* is the graph of total attenuation against frequency due to both cable and equaliser.

At the receiving end reverse processes are needed to separate the many channels. The first requirements are 10 filter circuits to separate the various super-groups of line frequencies. Each separated super-group is then transposed to the band 312–552 kc/s by demodulators which use the oscillator frequencies listed in Column 2 of Table 3.3. Each of these ten different bands is then fed to a group of 5 different band-pass filters with pass-bands as listed in Column 3 of Table 3.2. Each of these 5 outputs is then demodulated using oscillator frequencies as listed in Column 2 of Table 3.2. This restores each of the separated five groups to the 60–108 kc/s level. At this level crystal filters can be used to divide each group into 12 bands each of 4 kc/s width, as listed in Column 3 of Table 3.1. Using the carriers of Column 2, Table 3.1, the 12 separated channels can then be restored to audio base-band level by demodulation. This chain of conversion is traced for one particular channel in

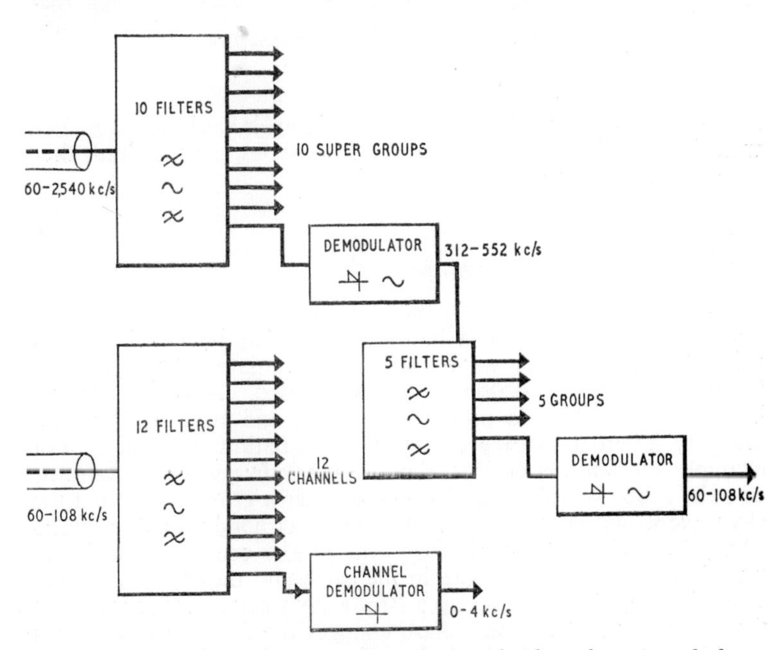

Fig. 3.15. Receiving end arrangements for a multi-channel carrier telephony system—a simplified block diagram

Fig. 3.15. In order that the correct frequency values shall be restored by demodulation it is essential that the carrier frequencies used at the demodulators shall be identical to those used by the modulators at the sending end. For this reason pilot carriers are transmitted at 60 kc/s and 2,604 kc/s.

One coaxial cable is needed for the go circuits and one for the return circuits. A cable enclosing four coaxial lines could thus provide two-way communication on 1,200 channels. In the above

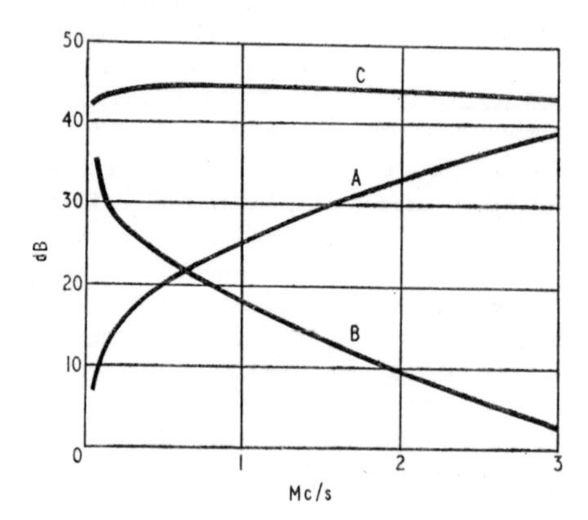

Fig. 3.16. Line and equaliser attenuation—frequency characteristics

multi-channel system after a group of 12 channels has been modulated with the second and the third carrier frequencies, the sidebands then represent a number of speech channels.

3.4. Radio Links

3.4.1. LIMITATIONS ON BANDWIDTH

Where cables of a suitable type do not link two places between which communication must be established, multi-channel schemes using electro-magnetic waves which are radiated through space instead of being guided along lines may be used. A difficulty arises, however, because radio propagation paths and attenuation values are very different at different levels of radiation frequency. For example, radiation at 60 kc/s, the lowest line frequency used on the co-axial cable, would be moderately good as a ground wave but the sky wave would be very weak. On the other hand radiation on 2·5 Mc/s, a frequency near the top end of the coaxial cable band, would be attenuated far more in the earth and thus have a weaker ground wave, but might set up a strong sky wave on reflection from the ionosphere. Also, the phase changes for different frequencies would be varied and indiscriminate, giving distortion. Therefore, when radio communication is used, the frequencies radiated must

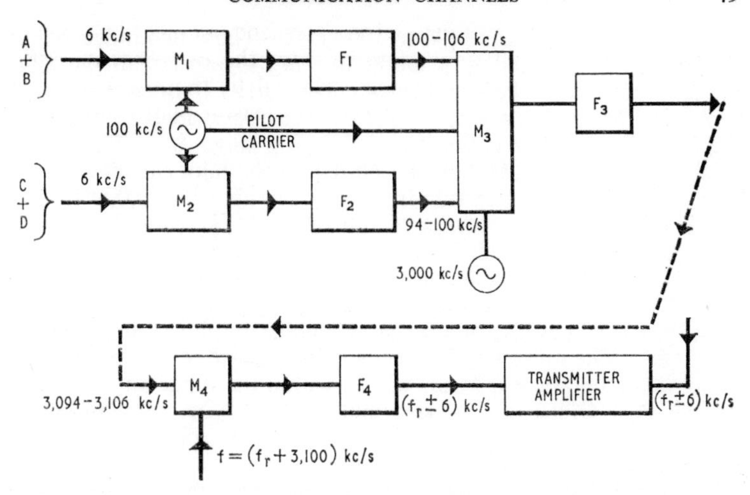

Fig. 3.17. An independent side band transmitter—simplified block diagram

all be of the same order. This restricts the total bandwidth transmitted to a value which is only a fraction of the mid-frequency radiated. To obtain a 3 Mc/s bandwidth it is necessary to radiate in the v.h.f. or even the u.h.f. bands. Further, while two cables may run side by side and carry a similar band of frequencies without interference between them, only one transmitter in the same area can use a given band of frequencies since similar frequencies radiated by two transmitters cannot be separated at the receiver. This means that the radiated bandwidth, especially in the lower frequency bands, must be very restricted. A transmitter working in the v.h.f. band 71·5 Mc/s to 100 Mc/s may provide only five speech circuits. In the band 1,700 to 2,300 Mc/s, 300 speech circuits can be accommodated while in the band 5,925 to 6,425 Mc/s, over 900 speech circuits can be included. In each example the bandwidth of radiation is small compared with the frequency level in use.

3.4.2. H.F. INDEPENDENT SIDEBAND WORKING

In the h.f. bands, independent sideband transmitters can provide four speech bands each of 3 kc/s width. Two are transmitted above the level of a pilot carrier frequency and two below. We have already seen in this chapter (*see* Fig. 3.6) how two speech bands can be accommodated in the band 0 to 6 kc/s. Fig. 3.17 illustrates the way in which two separate 0 to 6 kc/s bands can be arranged to provide separate and independent sidebands for the transmitter.

The two 6-kc/s audio bands are first applied to the modulators M_1 and M_2 together with a carrier frequency of 100 kc/s. For channels A and B, the filter F_1 attenuates all except the upper side-frequencies (100 to 106 kc/s). For channels C and D, the filter F_2

selects the lower sideband (94 to 100 kc/s). These two independent sidebands are then fed to a modulator M_3 together with the pilot carrier from the 100 kc/s oscillator. The carrier frequency supplied to M_3 is 3 Mc/s and, after modulation, the upper sideband is fed via the filter F_3 to the next frequency-changing stage M_4. This upper sideband (3,094 to 3,106 kc/s) is centred on the pilot carrier of 3,100 kc/s which is the upper side-frequency set up when 3,000 kc/s is modulated by the pilot carrier of 100 kc/s. M_4 is fed with a carrier frequency which is made equal to the radiation frequency carrier plus 3,100 kc/s. The difference frequencies produced by the modulator M_4 are between $(f_r - 6)$ kc/s and $(f_r + 6)$ kc/s. The transmitter proper then consists of a series of voltage amplifiers and a power amplifier. The radiation frequency f_r in the case of a long-distance radio link using the h.f. bands will be within the range 4 to 27 Mc/s. At the Post Office transmitting station at Rugby, 28 high-power transmitters operate within this range giving telephone services to most parts of the world.

The somewhat meagre bandwidth available for radio transmission is used more economically for telegraphy transmission. The bandwidth allowed for teleprinter working at 50 baud is only 120 c/s and many telegraphy channels can be accommodated on one sideband of a single transmitter. (*See Radio and Line Transmission, Volume* 1, Chapter 4.)

As for coaxial cable transmission, so with radio transmission; a pilot carrier must be transmitted to enable oscillation at the receiving end to be of correct frequency for demodulation. At high frequency only a small percentage change in the carrier frequency would make a large percentage change in the value of audio frequency set up on demodulation. If the oscillator frequency at the transmitter should change even by a small percentage then it is necessary that the oscillator used for demodulation at the receiving point should change by the same small percentage. This can be arranged by receiving the transmitted pilot carrier and amplifying it to the level needed by the demodulators in the receiver.

3.4.3. SINGLE SIDEBAND WORKING

Small transmitters with a radiated power of 60 watts or so can be used to provide communication in countries where telephone services have not been developed and where widely dispersed plant or sites belonging to such enterprises as oil companies, mining companies, railways, police forces, etc., require rapid communication to co-ordinate their efforts. Such transmitters may operate in the band 3 to 30 Mc/s and use single-sideband suppressed-carrier systems. A 60-watt transmitter of this type may be equally as effective as a double sideband transmitter using 450 watts. Furthermore, since the receiver bandwidth required is halved, the signal-to-noise ratio is, therefore, improved by the greater selectivity of the receiver.

Example 3.2

A double-sideband transmitter radiates a carrier power of 100 watts. Assuming sine-wave modulation to a depth of 70 per cent, calculate the power contained in the side-frequencies and state the power saving in decibels, if only one side-frequency is radiated.

Let the carrier current be I_c

$$\therefore \qquad \text{Each side-frequency current is } \frac{mI_c}{2}$$

where m is the depth of modulation.

Let R be the radiation resistance of the aerial.
The radiated power is then

$$I_c^2R + 2\left(\frac{mI_c}{2}\right)^2 R$$

$$= I_c^2R + \frac{m^2}{2}I_c^2R$$

But $I_c^2R = 100$ watts and $m = 0{\cdot}7$
Thus the total power is

$$100 + \frac{0{\cdot}7^2}{2}\cdot 100$$

$$= 100 + 24{\cdot}5$$

$$= 124{\cdot}5 \text{ watts.}$$

The power in one side-frequency is 12·25 watts. The saving is thus

$$10 \log\frac{124{\cdot}5}{12{\cdot}25} = 10 \log 10{\cdot}15$$

$$= 10{\cdot}064 \text{ dB}$$

$$= 10 \text{ dB (approx.)}$$

3.5. Summary

Telephone subscribers on the same exchange are linked by a transmission bridge across the 50-volt battery of their exchange. Loss in each subscriber line may be up to 3 dB. Loss in a transmission bridge approximately 1·5 dB.

Voice-frequency links can also be made between dependent exchanges and minor group exchanges, between minor group exchanges and group centres and between group centres and zone centres. Losses between minor exchanges should not exceed 4·5 dB.

If the distance between exchanges is greater than approximately 15 miles *repeaters* are used but loss is minimised generally by use of loading coils. Repeater gain is typically 24–30 dB.

Zone centres are linked by trunk systems using carrier-current telephony and telegraphy. Repeaters make transmission loss zero dB. A *24-channel system* uses line frequencies of 12 to 108 kc/s.

The multi-channel coaxial cable system uses line frequencies from 60 to 2,540 kc/s, and repeaters every 6 miles with a gain of approximately 40 dB.

Hybrid couplings (loss 3 dB) link 2-wire systems to 4-wire systems.

Equalisers precede amplifiers to equalise total loss for frequency range transmitted.

Ring Modulators translate base-bands to required place in line or radio-frequency spectrum.

Filters separate super-groups, groups and individual channels for demodulation at receiving end. Quartz crystal filters, by their sharp cut-off, make a close spacing of channels possible.

International Exchanges are linked by telegraph cables for cable morse and teleprinter use, by telephone cables using carrier-current working and repeaters or by h.f. radio links in range 4–27 Mc/s using independent-sideband pilot-carrier working.

Short-range radio links use multi-channel working in ranges 71·5 to 100 Mc/s, 1,700 to 2,300 Mc/s or 5,925 to 6,425 Mc/s.

Low-power transmitters in range 3 to 30 Mc/s use single-sideband working to save power and improve signal-to-noise ratio. (Effective gain over double sideband is 9 to 10 dB.)

Questions

1. Why is carrier transmission commonly used for long distance telephony circuits?

Explain, with the aid of a block diagram, how twelve audio-frequency telephone channels may be assembled to form a twelve-channel carrier group. Your diagram need include in detail only the equipment necessary for two channels. Show suitable carrier frequencies and filter-pass frequencies. (*C & G*, 1959.)

2. A carrier wave, $E_1 \sin\omega_c t$, is amplitude-modulated by the wave $E_2 \sin\omega_m t$, the depth of modulation being k. Calculate the amplitude of the carrier-frequency component and of each side frequency in the modulated wave.

Explain why it is common practice to use single-sideband suppressed-carrier working for multi-channel telephony working over co-axial cables. (*C & G*, 1960.)

3. Explain why single-sideband suppressed-carrier working is normally used for multi-channel telephony over cable circuits.

Draw a diagram of a balanced modulator using semi-conductors, suitable for use in such a system. Explain the purpose of the balance adjustment. (*C & G*, 1961.)

4. Explain the limitation on bandwidth which exists in radio-

communication and why line transmission tends to offer a greater number of communication channels.

5. What do you understand by an electric filter?

Explain some of the applications of filters in radio and line transmission circuits.

6. What do you understand by (a) independent sideband radiation, (b) single sideband radiation. Explain the purpose of the pilot carrier in the first of these systems.

What do you understand by the term " frequency translation "? Illustrate your answer by reference to a multi-channel system of line communication.

7. An audio line has an attenuation coefficient of 0·115 nepers/mile. A repeater is inserted into the line at 15 miles distance from the sending end and the output from this repeater enters the line at a level 5 dB above the level it had at the sending end of the line. Express the total gain of the repeater in dB.

8. What advantages does single sideband transmission provide over double sideband working, the carrier of the former being suppressed? If a single sideband suppressed carrier transmitter has an output power of 1 kW, what power would be radiated by a double sideband transmitter providing equivalent field strength assuming sinewave modulation of 50 per cent in the case of the latter?

9. Draw the equivalent electrical circuit of a quartz crystal and say what the various parts represent. What advantages do filters using quartz elements have over inductance-capacity filters and what disadvantage?

10. Two speech-frequency bands, each 300 c/s to 3 kc/s, are to be used for modulating an oscillator of 100 kc/s with a total modulation bandwidth of 6 kc/s. With the aid of a block diagram indicate how this may be done. If after modulation with the 100 kc/s r.f., the lower sideband is selected and the upper sideband discarded, indicate by diagrams which of the original speech bands is now inverted.

4

Aerials

4.1. Introduction

Although transmitting aerials can be used for reception, receiving aerials are not always sufficiently insulated and as conductors they do not always have a big enough cross-section to serve as aerials for transmitters. Further, aerials used for reception on m.f. and l.f. are not big enough to act as tuned radiators for those frequencies. In *Radio and Line Transmission, Volume* 1, aerials for receivers are described. This chapter is concerned with aerials for radiation.

4.2. The Dipole

4.2.1. STANDING WAVES

Fig. 4.1 (a) indicates a half-wavelength conductor with an alternating e.m.f. induced at its centre point from a transmitter. The vectors drawn above the line show the instantaneous phasing of a travelling wave of voltage which travels outwards to the ends of the aerial before being reflected at the terminals. The vectors indicate that the wave has a maximum value at the source of e.m.f. though the opposite terminals of the source have opposite polarity. The vectors showing instantaneous phase, rotate clockwise by 45° for each $\frac{1}{8}\lambda$ which the wave travels away from the centre. This represents the growing phase angle of lag which the wave has to the voltage at the centre as it travels farther towards the ends of the wire. At the ends of the aerial the wave is reflected and begins to travel back to the centre. The vectors drawn below the wire show the phasing at every $\frac{1}{8}\lambda$ of the reflected waves which are seen to arrive back at the generator in opposite phase to the outgoing wave. At whatever point in the wave cycle the relative phase of the returned voltage to that of the outgoing wave is considered, the reflected wave is always opposite in phase to the generator voltage when it arrives back after reflection at the ends. This must be so because the wave takes a quarter of a period to travel from the centre of the aerial to the end and a further quarter of a cycle to return. If no losses occurred either as heat or as electromagnetic radiation, the reflected wave amplitude would be equal to the outgoing wave amplitude at the generator which would apparently make the terminal voltage of the e.m.f. zero. This impossible condition would amount to a short circuit on the source of e.m.f. but does not have practical significance It is true, however, that the centre-fed dipole behaves as a low impedance load and the transmitter final stage must be matched to an impedance of the order of 70 or 80 ohms.

Fig. 4.1 (b) shows vectors for travelling waves of current in the same dipole. The outgoing current wave vectors are seen to be in phase with the voltage vectors at the generator. This implies that the transmitter " sees " a resistive or resonant load. The vectors of reflected current are shown reversed as compared with the direction of the vectors for the current wave arriving at the ends. The polarity of the charge at the ends of the aerial does not reverse but the direction of flow of charge does. In consequence the reflected current wave returns to the generator in phase with the outgoing current wave and the resultant current at the generator terminals is high. This agrees with the conclusion of the previous paragraph that the generator supplies a low-impedance load.

Fig. 4.1 (c) shows the distribution of the r.m.s. result current (I) and the r.m.s. resultant voltage (V). The current antinode occurs where the incident and reflected current travelling waves are in phase. The voltage antinodes are where the two voltage travelling waves reinforce one another. This diagram is arrived at in Chapter 14 of *Radio and Line Transmission, Volume* 1, by a different approach.

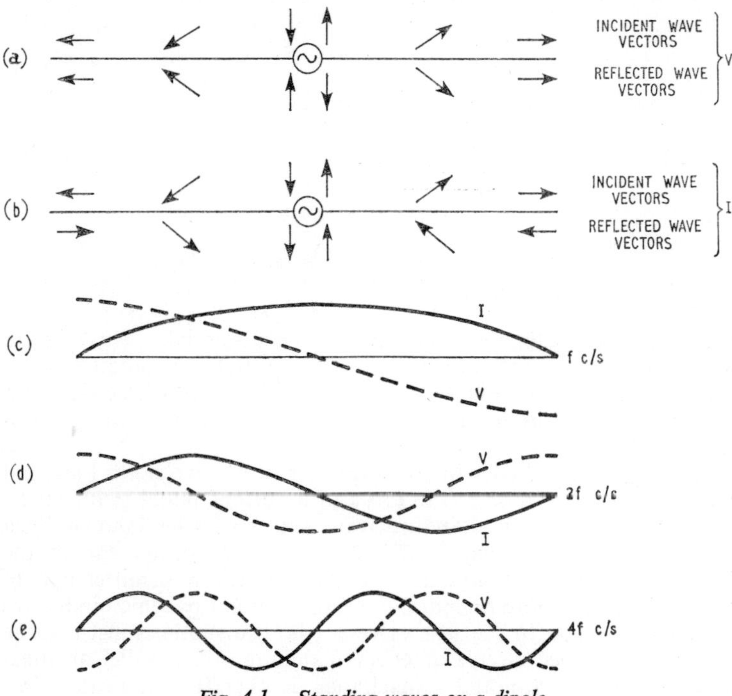

Fig. 4.1. Standing waves on a dipole

If a conductor is a resonant dipole for a given frequency it is also resonant at twice this frequency because the length is that of a wavelength for this higher frequency (*see* Fig. 4.1 (d)). For the fourth harmonic of the same frequency the aerial is two wavelengths long and standing waves form as in Fig. 4.1 (e). This feature makes it possible to use one aerial for several frequencies which are harmonically related. It may be possible to use one aerial for frequencies in the 4-, 8- and 16-Mc/s transmission bands and another dipole for the 3-, 6-, 12- and 24-Mc/s bands.

4.2.2. STANDING WAVES ON FEEDERS

Fig. 4.2 and Fig. 4.3 illustrate two methods of connecting a dipole aerial to a transmitter by tuned feeders. In these examples the feeder line forms part of the resonant aerial circuit and standing

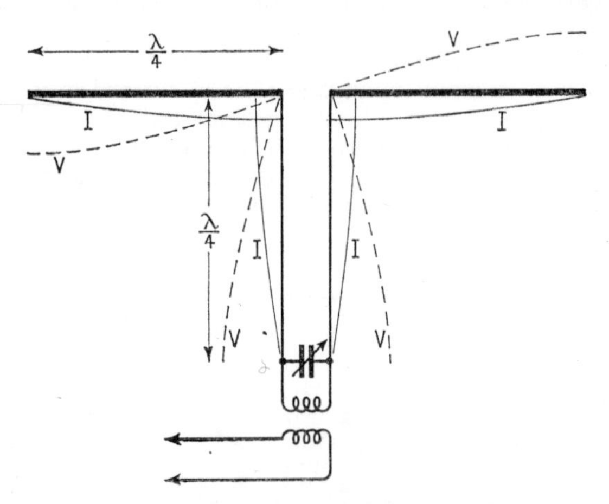

Fig. 4.2. A quarter wave tuned feeder and dipole —standing waves

waves form along the feeder as well as along the aerial. This method is only used when the transmitter is very near to the aerial because long feeders with standing waves on them have high dielectric losses at the voltage antinodes and high copper losses at the current antinodal points.

In Fig. 4.2, the feeder line is a quarter of a wavelength long and has a current antinode at the top and a current nodal point at the bottom. The voltage standing wave on the feeder (shown by a dotted line) has a voltage antinode at the transmitter end of the feeder. The ratio of voltage to current at the transmitter end of the feeder is therefore high and the tuning circuit by which inductive coupling is made to the last valve of the transmitter must also be a high impedance circuit to effect matching. A parallel arrangement of inductance and capacitance is therefore indicated (*see*

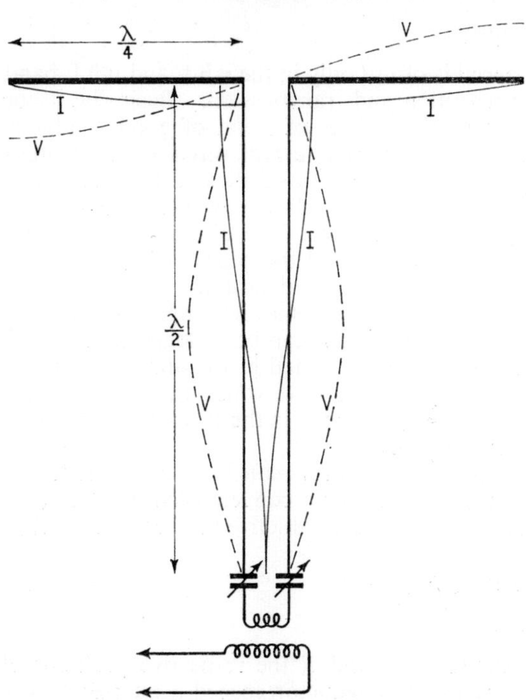

Fig. 4.3. A half wave tuned feeder and dipole—standing waves

Section 9.4.1). Transformer matching to the output valve is then possible. This arrangement can be used whenever the feeder length is an odd number of quarter wavelengths long and the dipole is centre fed.

Fig. 4.3 applies to a tuned feeder which is a half wavelength long. This gives a voltage nodal point and a current antinode at the base. The ratio of voltage to current at the transmitter end of the feeder is now low and a low input impedance to the feeder exists. Consequently a low-impedance coupling circuit is necessary and this is provided by a series tuned circuit. Mutual coupling to the last transmitter valve must now match the low impedance coupling of the circuit to the higher output impedance of the valve. Any even number of quarter wavelengths of feeder require this type of feeder tuning for a centre-fed dipole.

A longer feeder line must be matched at the transmitter and at the aerial to avoid reflection of energy at either end of the feeder and the standing waves which this would cause. Matching between the feeder line which may have an impedance of about 600 ohms for a twin line or 60 to 70 ohms for a coaxial line, and the aerial can utilise an r.f. transformer or the delta match (*see Radio and Line Transmission, Volume* 1, Fig. 14.5).

4.2.3. SLOT DIPOLES

The slot aerial is of use only in the v.h.f. and u.h.f. bands and can be used for reception and transmission. Basically, it consists of a narrow slot in a plane sheet of metal of good conductivity. The length of the slot is just over a half wavelength. Connections are made to the edges of the slot as indicated by Fig. 4.4. A vertical slot radiates a horizontally-polarised wave. The electric field flux is shaded in across the slot. The alternating current producing magnetic flux circulates round the edges of the slot as indicated by the curved arrows on the same diagram.

Centre-fed, the slot aerial has a high impedance of approximately 500 ohms but a coaxial feeder can be matched to the slot by a coupling system such as that indicated in Fig. 4.5. This is really a type of u.h.f. transformer coupling in which the centre conductor (held in the slot by insulating strips) is the primary, and the walls of the slot are the secondary.

Arrays of slots and reflectors can be used to produce directional radiation or reception as for a conventional dipole. (The reflector for a slot aerial is a plane conductive surface or mesh.)

4.3. Use of Parasitic Elements

4.3.1. REFLECTOR

Fig. 4.6 shows by A and R the respective positions of a dipole aerial and a reflecting conductor spaced a quarter of a wavelength farther away from the intended destination of the radiation. The point P is a receiving point at a distance from the aerials which is large compared with the spacing between the aerial and its reflecting element. Given this condition, the directions AP and RP are practically parallel and the angle ADR approximates to a right angle. The distance DR is the extra distance which a re-radiated signal from R will have to travel to reach P, compared with the distance AP travelled by the signal from the dipole A. The reflected signal reaching P from R is out of phase with that reaching P from A for two reasons. Firstly, because the re-radiated signal from R is leading 90° on the radiation from A and secondly because the extra distance of wave travel causes a phase lag of $\left(2\pi\frac{DR}{\lambda}\right)$ radians.

Let α be the phase difference at P between the two signals. Then,

$$\alpha = \frac{\pi}{2} - \frac{DR}{\lambda}.2\pi \quad (1)$$

From Fig. 4.6

$$DR = \frac{\lambda}{4}.\cos\theta \quad (2)$$

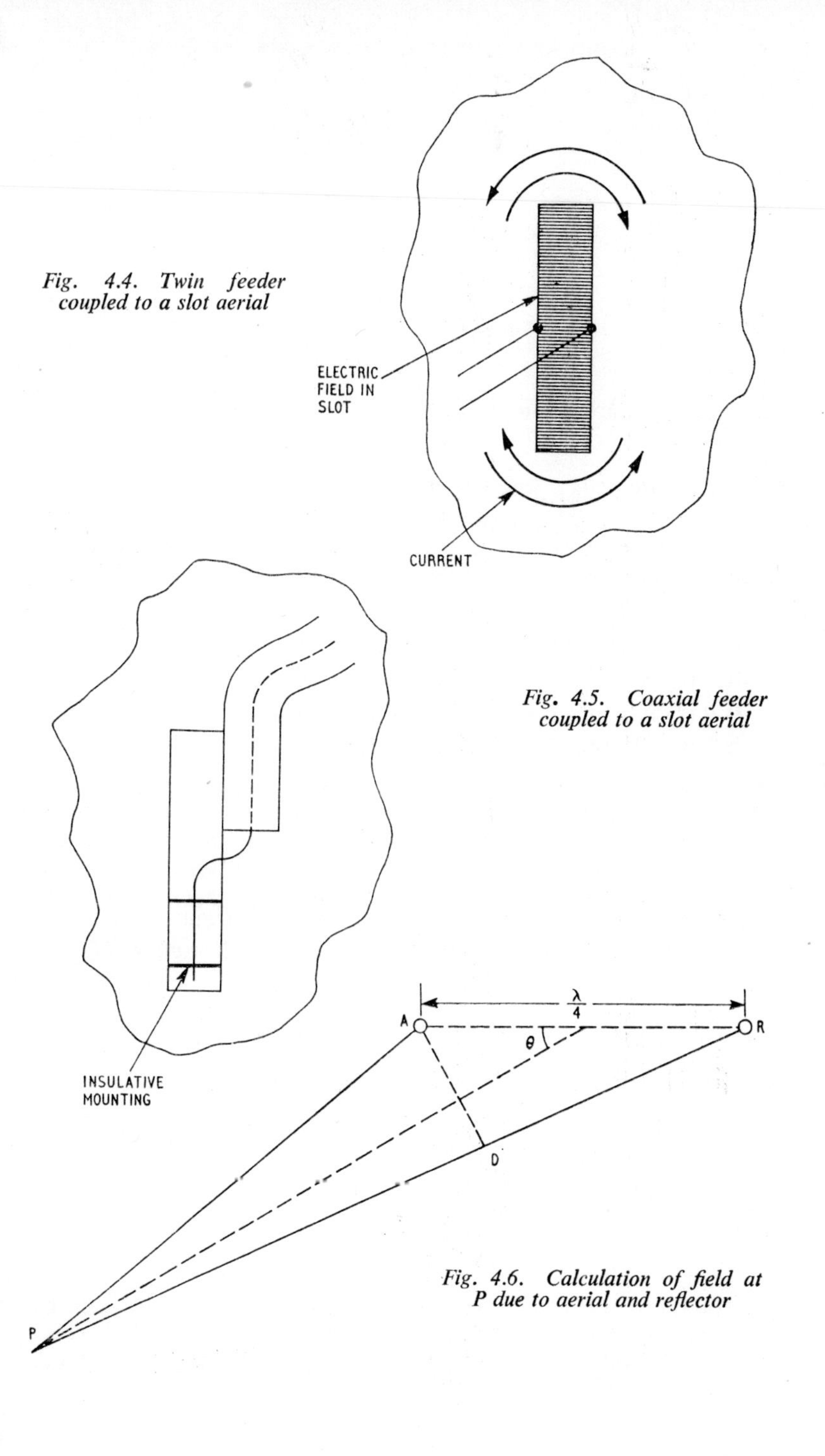

Fig. 4.4. Twin feeder coupled to a slot aerial

Fig. 4.5. Coaxial feeder coupled to a slot aerial

Fig. 4.6. Calculation of field at P due to aerial and reflector

Substituting (2) in (1)

$$\alpha = \frac{\pi}{2} - \frac{\lambda}{4} \cos \theta \cdot \frac{2\pi}{\lambda}$$

$$= \frac{\pi}{2} - \frac{\pi}{2} \cos \theta$$

$$= \frac{\pi}{2} [1 - \cos \theta] \qquad (3)$$

Let the signal field strength at P due to A

$$= X_m \sin \omega t$$

then the field strength at P due to R,

$$= X_m \sin (\omega t + \alpha)$$

where α is the phase difference indicated by equation 3.

By inserting different values for the angle θ in equation 3, the phase difference between the two signals reaching P can be calculated. The total signal at P is the vector sum of the radiation from A and the radiation from R.

If θ is 0°, then $\cos\theta = 1$ and $\alpha = 0°$
If θ is 180°, then $\cos\theta = -1$ and $\alpha = 180°$
If θ is 90°, then $\cos\theta = 0$ and $\alpha = 90°$
If θ is 60°, then $\cos\theta = 0{\cdot}5$ and $\alpha = 45°$

By evaluating angle α for a large number of values of angle θ, and then finding the vector sum of the radiations from the aerial and its reflector, the resultant signal in different directions relative to the base line AR can be found. A polar diagram of radiation plotted from the results will be as in Fig. 4.7. Alternatively if α is the phase difference between the radiations reaching P from A and that reaching P from R, the sum of the two signals at P is given by

$$X_m \sin\omega t + X_m \sin(\omega t + \alpha)$$

$$= 2X_m \left[\sin\left(\omega t + \frac{\alpha}{2}\right)\cos\frac{\alpha}{2}\right]$$

This indicates an alternating signal of *peak value*: $2X_m \cos\frac{\alpha}{2}$ which is leading in phase by $\frac{\alpha}{2}$ radians on the signal from the aerial A. Thus the peak signal in any direction is found once the phase difference between the two signals is known for the direction considered.

The reflector is made slightly longer than a half wavelength causing it to have an inductive reactance at the signal frequency. The reflector current lags by approximately a quarter of a cycle on the e.m.f.

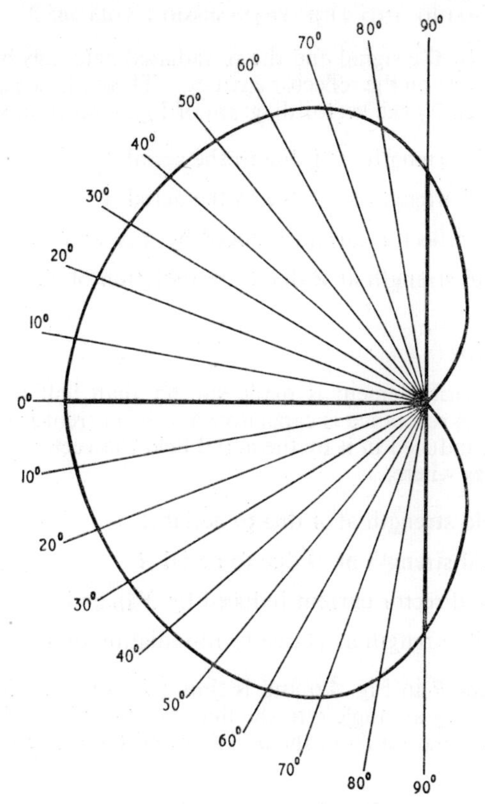

Fig. 4.7. A polar diagram for an aerial and reflector

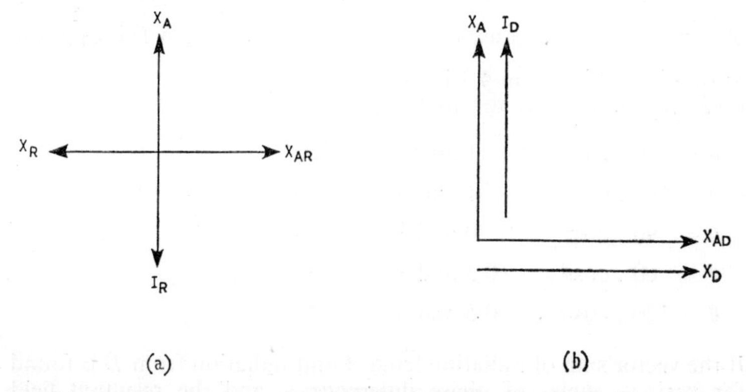

Fig. 4.8. (a) Vectors for aerial and reflector fields, (b) vectors for aerial and director fields

induced in it by the signal and the re-radiated field lags by a further quarter of a cycle on the reflector current. These phase relationships are summarised by the vector diagram of Fig. 4.8 (a), in which:

X_A = field strength at A due to the aerial A

X_{AR} = field strength at R due to the aerial A

I_R = the reflector current induced by X_{AR} and

X_R = field strength at R due to re-radiation of R.

4.3.2. DIRECTOR

If the parasitic element is made shorter than half a wavelength long so that its reactance is capacitive and its current leading by 90° on the e.m.f. induced in it by the aerial field the vector diagram Fig. 4.8 (b) applies where,

X_A = field strength at A due to aerial A

X_{AD} = field strength at D due to aerial A

I_D = the director current induced by X_{AD} and

X_D = field strength at D due to re-radiation of D

which replaces R in Fig. 4.6 and is thus $\frac{1}{4}\lambda$ from A. In any given direction making an angle θ to the line AD, the phase difference α, between the radiation from the aerial A and the element D is given by:

$$\alpha = -\frac{\pi}{2} - \frac{\pi}{2}\cos\theta^\circ = -\frac{\pi}{2}(1 + \cos\theta^\circ) \qquad (4)$$

which is comparable with Equation (3) excepting the change of sign of $\frac{\pi}{2}$ which occurs because the phasing of the current in D is opposite to that in A of Section 4.3.1.

Inserting some values for angle θ gives

$\theta = 0°$, $\cos\theta = 1$ and $\alpha = -180°$

$\theta = 180°$, $\cos\theta = -1$ and $\alpha = 0°$

$\theta = 90°$, $\cos\theta = 0$ and $\alpha = -90°$

$\theta = 60°$, $\cos\theta = 0{\cdot}5$ and $\alpha = -135°$

$\theta = 120°$, $\cos\theta = -0{\cdot}5$ and $\alpha = -45°$

If the vector sum of radiation from A and radiation from D is found for various angles of phase difference α, and the resultant field strength plotted as a polar diagram against the angle of direction θ, the polar curve will be found to be a mirror image of Fig. 4.7.

This indicates that the maximum radiation is in the direction from *A* towards *D*. The parasitic element *D* placed between the aerial and the intended reception point is called a *director*.

The above treatment of the two-element aerial is simplified by neglecting the resistance of the parasitic element which in each example has been regarded as a pure reactance. Moreover we have neglected the effects of mutual inductance between the two elements. The best effective length and spacing for a director or reflector is determined in practice by experiment rather than by calculation. In typical aerials a reflector of length 0·526 of a wavelength may be spaced about a quarter of a wavelength behind the aerial while a director of length approx. 0·47 of a wavelength may be spaced 0·157 of a wavelength on the other side of the aerial.

The calculation of the resultant field strength in a given direction due to more than two aerial elements is more difficult than the example considered above but the same basic principles apply.

4.3.3. AERIAL GAIN

The directivity of an aerial is often expressed as the angular width between the directions on the polar diagram where the field strength has a value of 0·707 $\left(\text{i.e. } \frac{1}{\sqrt{2}}\right)$ of the maximum value, and the radiated power is thus half the maximum power. This is called the *beam width.*

The gain obtained in consequence of the directional properties of an aerial can be expressed in various ways; one is to divide the power which must be supplied to a simple dipole by the power which must be supplied to the directional aerial to give the same field strength in the direction of best radiation and then to assess this in decibels.

Example 4.1

A dipole aerial is fed with 25 kW of power to produce a required signal strength in a certain service area. The addition of a reflector element makes the same field strength available in the service area with only 10 kW of input power to the aerial. What is the gain obtained by the addition of the reflector?

The gain in decibels

$$= 10 \log \frac{25}{10}$$

$$= 10 \log 2{\cdot}5$$

$$= 10 \times 0{\cdot}3979$$

$$= 3{\cdot}979 \text{ dB or}$$

4 dB to the nearest whole number.

4.4. Travelling Wave Aerials

4.4.1. RHOMBIC

A rhombic aerial is used to give uni-directional transmission or reception where a wide-band untuned aerial is required. It is usually supported on poles as a horizontal diamond shape, as shown by Fig. 4.9. Travelling waves of current travel outwards from the transmitter *T*, and any power not wasted or radiated on reaching the

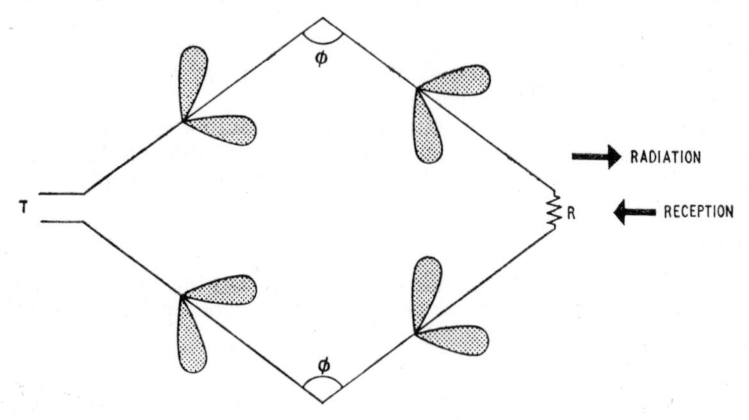

Fig. 4.9. The rhombic aerial

termination is absorbed by the matching resistor *R*. This prevents reflection, standing waves and back radiation. Each arm of the rhombic aerial is long compared with a wavelength at the frequency radiated and may be two or more wavelengths long. The radiation pattern for a straight conductor of this length has two main lobes as indicated in Fig. 4.9. The direction of these main lobes varies only a little for as much as a 2 to 1 variation of frequency. The angle is arranged so that each limb has one main lobe of the polar diagram pointing in the required direction of radiation. A large number of rhombic aerials are in use at the G.P.O. transmitting station at Rugby for the frequency bands 4–27·5 Mc/s. Smaller types used at this station have a pole height of 75 feet with a side length of 315 feet, and an angle ϕ of 140°. A larger type has a pole height of 150 feet, a side length of 540 feet and an angle ϕ of 135°.

The greater the side length of the rhombic in terms of a wavelength the narrower the beam width of radiation. The polar diagrams of Fig. 4.10 are drawn for an angle ϕ of 140°. The widest lobe applies to a side length of two wavelengths, the middle lobe to a side length of four wavelengths and the narrowest lobe to a side length of six wavelengths.

Rhombics may use double wires for their sides, the spacing between the two wires of a side being adjusted to give a required

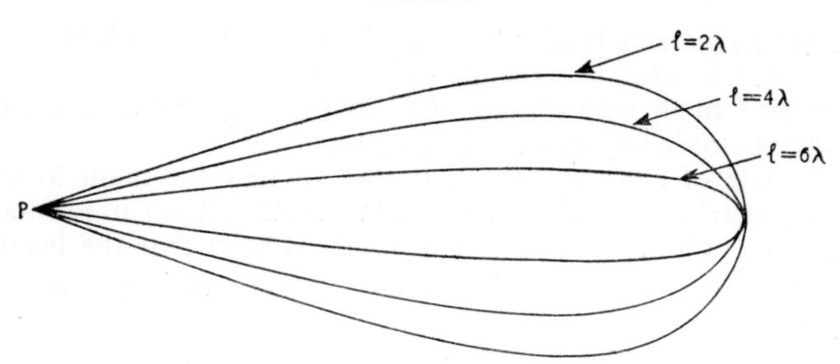

Fig. 4.10. Showing increase of directivity with increase of rhombic dimensions

impedance value for the aerial. The use of multiple wire rhombics can give a gain of 0·5 to 1·5 dB over similarly dimensioned single-wire types.

The radiation from rhombic aerials is horizontally polarised. This type of aerial has application to frequencies approximately in the range 3 to 200 Mc/s.

4.4.2. THE INVERTED-V AERIAL

The inverted-V aerial can be regarded as half of a rhombic aerial with its plane turned through 90°. The other half of the rhombic aerial is in effect replaced by an image aerial in the earth beneath the inverted V. The inverted-V aerial has the main characteristics of a rhombic aerial in that it is an untuned, wide-band travelling-wave aerial giving a directional radiation whose beam width

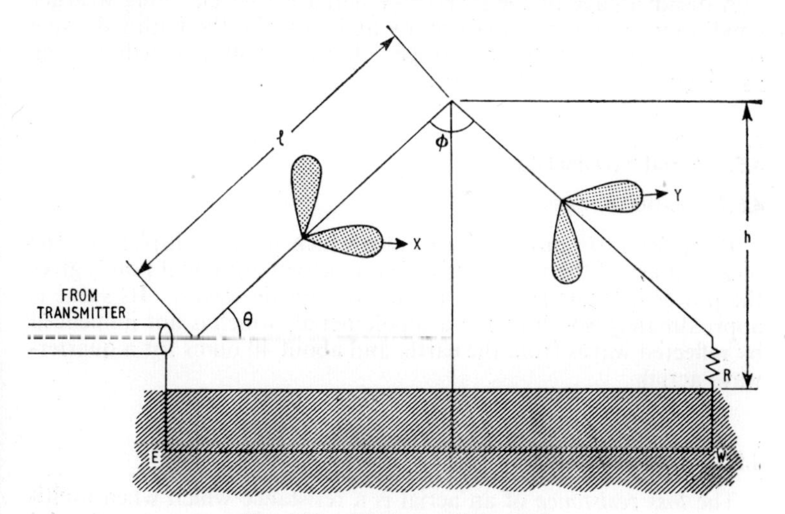

Fig. 4.11. The inverted-V aerial

decreases as the side length l is made longer, but the radiation from the inverted-V aerial is vertically polarised.

A buried earth wire marked EW in Fig. 4.11 helps to increase efficiency by reducing earth resistance loss.

The angle of elevation θ is adjusted so that the main lobes of radiation marked X and Y in Fig. 4.11 mutually assist in producing forward radiation. An optimum design is such that the height h is approximately equal to $\sqrt{l\lambda}$, where l is the side length in wavelengths.

Example 4.2

Let the side length be 4λ

Then
$$\sin\theta = h/l$$
$$= \frac{\sqrt{4\lambda.\lambda}}{4\lambda}$$
$$= \tfrac{1}{2} = 0{\cdot}5$$
$$\therefore \angle\theta^\circ = 30^\circ$$
and
$$\angle\phi = 2 \times (90 - 30)$$
$$= 120^\circ$$

A disadvantage of the inverted-V aerial is that changing weather conditions vary earth conductivity and so make the required value of terminating resistance change. Buried artificial earth netting can reduce this variable factor.

4.5. Aerial Terminology

4.5.1. RADIATION RESISTANCE

Radiation resistance is a resistance which when multiplied by the mean square of the aerial current at a current antinodal point, gives the power in watts radiated as electromagnetic waves. Its value is approximately 80 ohms for a dipole aerial, which is not influenced by reflected waves from the earth, and about 40 ohms for a quarter-wave aerial.

4.5.2. LOSS RESISTANCE

The *loss resistance* of an aerial is a resistance which when multiplied by the mean square of the aerial current at a current antinodal point gives the power lost by the aerial other than by radiation. The loss resistance may be due to

1. The radio-frequency resistance of the aerial.
2. Induction by the aerial of currents into adjacent conductors whose resistance dissipates energy as heat.
3. Dielectrics in which alternating dielectric displacement currents are set up by the intense electric field near to the aerial.
4. Leakage across the surface of aerial suspension insulators which increases if the insulators become covered with soot or other atmospheric deposits.
5. The resistance of the earth round the aerial which is high if the ground is dry.
6. Corona discharge which may occur at voltage antinodal points if the electric field gradient is sufficiently steep to ionise the air round the aerial and cause a convection current discharge through the air to the nearest earthed point.

4.5.3. TOTAL AERIAL RESISTANCE

The *total aerial resistance* is a value in ohms which when multiplied by the mean square of the aerial current measured at an antinodal point gives the total power supplied to the aerial. It is equal to the sum of the loss resistance and the radiation resistance.

4.5.4. AERIAL EFFICIENCY

The *aerial efficiency* is the ratio of the power radiated as radio waves to the power supplied to the aerial by the transmitter. It is the ratio of the radiation resistance to the total resistance of the aerial. The efficiency can be increased by increasing the effective height (*see* Section 4.5.5) and radiation resistance and by minimising all losses.

4.5.5. EFFECTIVE HEIGHT

The *effective height*, also known as the radiation height, is the length of aerial conductor, which, if it carried the value of current measured at a current antinodal point uniformly along its length, would produce the same radiated power as the actual aerial. For a vertical quarter-wave aerial the effective height is approximately two-thirds of the physical height. The effective height is increased by any arrangements which increase the aerial top capacitance.

4.5.6. RADIO-FREQUENCY RESISTANCE

The *radio-frequency resistance* is a value which when multiplied by the mean square of the aerial current measured at a current antinodal point gives the power wasted as heat in the aerial conductor itself. Because of skin effect, it is considerably greater than the d.c. resistance of the conductor and is reduced by the use of tubular conductors where possible and by the use of stranded non-magnetic aerial conductors.

4.5.7. HERTZIAN DIPOLE

A *Hertzian dipole* is the concept of a short linear conductor carrying the same r.m.s. value of current along its whole length. Because the magnitude of radiation from this can be calculated, it is useful to regard some practical aerials as consisting of a chain of such dipoles joined in series.

4.5.8. ISOTROPIC RADIATOR

An *isotropic radiator* is the concept of a radiator which radiates equally well in all directions in all planes. Although no such radiator exists in practice this fictitious aerial is sometimes used as a reference standard with which to compare other aerials.

Questions

1. Assuming that the velocity of a radio wave in a dipole aerial is 98 per cent of its free-space velocity, what length of conductor should be cut to provide a dipole for 98 Mc/s?

2. A dipole aerial is resonant to a frequency of 10 Mc/s. It is centre-fed via a tuned feeder of length 22·5 metres. Draw a sketch to show the standing waves of current and voltage on the feeder and state whether the feeder should couple to a transmitter via a series or a parallel tuned circuit.

3. Explain the principle of the reflector element used with a transmitting dipole. Assuming that the field re-radiated by a reflector is leading by $\pi/2$ radians on the field radiated from a dipole spaced a quarter of a wavelength from it, what is the field strength radiated at 80 degrees to a line joining the aerial to the reflector, relative to the field strength in the direction of best radiation?

4. The addition of a reflector to a dipole produces an aerial gain of 3 dB. If the power needed by the simple dipole is 2 kW for a required field strength, what power will the dipole need after the reflector has been added in order to provide similar field strength in the direction of best radiation?

5. Explain what is meant by a radiation polar diagram and say what information can be obtained from it.

6. Describe the construction of a rhombic aerial and sketch its polar diagram in a horizontal plane. Over what frequency band would you expect a rhombic aerial to be used?

7. Describe an untuned broad-band aerial suitable for the radiation of vertically polarised waves.

8. What is meant by the efficiency of an aerial and what factors affect its value?

9. What is meant by the term isotropic radiator? Draw radiation patterns (polar diagrams) in the horizontal plane for the following:

(a) a vertical half-wave dipole aerial,

(b) a horizontal half-wave dipole,

(c) a vertical half-wave aerial with a reflector,

(d) a horizontal half-wave dipole with reflector.

Calculate suitable dimensions for an array comprising a half-wave dipole and reflector for use at 180 Mc/s. (*C & G*, 1962.)

10. What is meant by the term *Hertzian dipole*? In what way does the practical dipole differ from the Hertzian dipole?

5

Components

(In this chapter we extend the introductory treatment of components given in *Radio and Line Transmission, Volume* 1.)

5.1. Resistors

5.1.1. SELF INDUCTANCE AND CAPACITANCE

As a generalisation it may be said that carbon compound resistors, whether insulated or non-insulated, are used for low wattage applications, while those for high-power dissipation above (say) 5 or 10 watts, are wire-wound using a nickel–chromium or other high-resistance alloy. Apart from their relative cheapness carbon resistors have the advantage of having a very small self-capacitance and inductance. The coils of a wire-wound resistor have inevitably a larger inductance than a straight carbon rod, and capacitance between each turn of wire and the next provides a shunt circuit across the resistance which becomes effective when the frequency is sufficiently high.

Bifilar winding of resistance coils is used if a very low inductance value for the resistor is required. (In this type of winding the resistance wire is doubled back on itself at its mid-point so that the current in adjacent parallel conductors travels in opposite directions.) If the bifilar winding is in several layers on an insulated core the self-capacitance of the resistor is high. An alternative method of winding, which reduces capacitance is to wind the resistor in sections on a porcelain former, with a small space between the sections, and then to join the sections in series. The capacitance of each section is much smaller than that of a single large coil and when the sections are joined in series, the total capacitance becomes approximately equal to that of one section divided by the number of sections joined in series.

Another way of making a low-inductance resistor is to wind the resistance wire in an even number of insulated layers one on top of the other changing the direction of winding on alternate layers. The magnetic fields due to alternate layers are thus in mutual opposition.

Mat resistors (often used in battery-charging circuits) are made by weaving resistance wire into an asbestos mat, with the wire zigzagging back and forth in the mat so that adjacent conductors carry the current in opposite directions so that the resultant magnetic field is very small. (Non-inductive charging resistors help to reduce the inductive flash which tends to pit switch contacts when the charging current is switched off.)

5.1.2. TEMPERATURE COEFFICIENT

A further important difference between carbon and wire-wound resistors lies in the effect of a change in temperature on resistance. The value of a carbon resistor decreases if it gets hot, whereas the resistance of a wire-wound resistor increases with temperature. This may be alternatively expressed by saying that carbon has a negative temperature coefficient while metals have a positive temperature coefficient. The temperature coefficient of resistivity of a material is the factor or coefficient by which its resistivity at 0° C must be multiplied to give the resistivity increase for each degree (C) rise of temperature. This assumes that the resistivity increases linearly with temperature and this is true up to approximately 100° C.

Example 5.1

The resistivity of pure iron at 0° C is 11·5 $\mu\Omega$ cm. What is the resistivity of iron at 40° C if the temperature coefficient of resistivity is 0·0055?

The change in resistivity for 1° C

$$= 0{\cdot}0055 \times 11{\cdot}5$$

The change in resistivity for 40° C

$$= 40 \times 0{\cdot}0055 \times 11{\cdot}5$$
$$= 2{\cdot}53$$

The resistivity at 40° is thus

$$11{\cdot}5 + 2{\cdot}53 = 14{\cdot}03\ \mu\Omega\,\text{cm}.$$

In general terms, we may write

$$\begin{aligned}\rho_t &= \rho_0 + \alpha\rho_0 t\\ &= \rho_0(1 + \alpha t)\end{aligned}$$

where ρ_t = resistivity at t° C

ρ_0 = resistivity at 0° C

t = rise of temperature above 0° C

and α = temperature coefficient of resistivity.

Or, since the resistance value depends directly on the resistivity of the material used, we may write

$$R_t = R_0 + \alpha R_0 t$$
$$= R_0 (1 + \alpha t)$$

where R_t = resistance of resistor at $t°$ C

and R_0 = resistance of resistor at 0° C

The temperature coefficient of carbon is approximately −0·0005.

5.1.3. THERMISTOR

The change of resistance of materials with a change of temperature can lead to a very large change in valve heater circuit resistance, especially when several valves are joined in series, as the valve heaters rise from room temperature to almost white heat. A heater resistance may be 21 ohms when hot but only an ohm or less when cold. When valves are in series, this can lead, on first switching on, to high surge currents which may overheat or even burn out those valves which reach their correct operating cathode temperatures first. Only when all the valves in the chain have reached the correct temperature can the current settle down to its right value. A special type of resistor has been developed to counteract this effect. It is called a *thermistor* or thermal-sensitive resistor. In appearance it is often like a rather short, thick carbon resistor, black in colour, being unpainted and having brass end-caps to which connecting leads are soldered. It is made by firing metal oxides with a ceramic base and bonding material. The temperature coefficient is negative and tends to vary inversely as the square of the absolute temperature ($x°$ C = $(x + 273)°$ absolute). The resistance of a thermistor of value 760 ohms at 20° C may fall to 18 ohms when carrying a heater current of 0·6 amps. One with a room-temperature resistance of two kilohms may fall to only 300 ohms at 80° C. A thermistor connected in series with a chain of valve heaters limits the initial surge of current because of its high initial resistance. By the time the valve cathode and heater temperatures have approached their normal level, the thermistor temperature has risen and its resistance has fallen, allowing the heater current to reach its full rated value. Valve heaters in such circuits take longer to warm up, but their life is longer. The circuit symbol for a thermistor is as in Fig. 5.1. Thermistors are also used in automatic amplitude limiting circuits in oscillators and have an application in the measurement of temperature.

5.1.4. OXIDE FILM RESISTORS

A recent development has been the tin oxide film resistor. A film of tin oxide is formed on a small ceramic tube while the tube is at a very high temperature. On cooling, the oxide film is exceedingly hard and tough. A helical track is ground from the film, leaving a spiral-like surface path of high resistivity from one end of the tube to the other. The broader the track is ground, the narrower is the path of tin oxide film left, and the greater the resistance value. End-caps and leads are attached, making contact with the ends of the tin oxide film resistance path. A range of miniature film resistors has been prepared and it is claimed that they are of good stability, are smaller in size than carbon resistors of similar wattage rating and

Fig. 5.1. The thermistor symbol

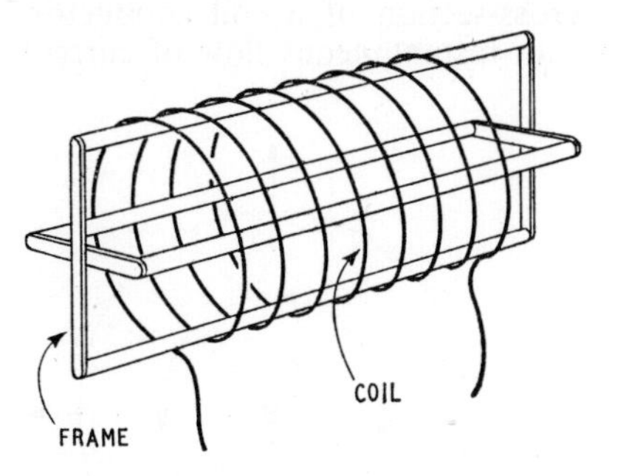

Fig. 5.2. A frame supported inductor

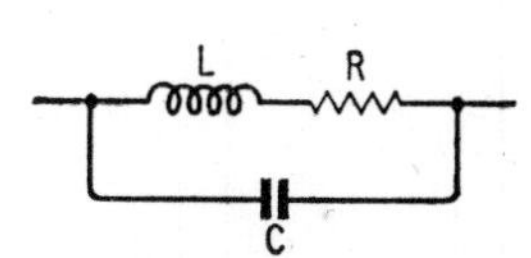

Fig. 5.3. Equivalent circuit of a coil

that they have a low temperature coefficient, are little affected by moisture, can operate at high temperatures and are mechanically very robust.

5.2. Inductors

5.2.1. EQUIVALENT CIRCUIT AND LOSSES

Coils in which no ferro-magnetic materials are used for any part of their intentional magnetic flux circuit are said to be air-cored. The turns of these coils may be air-spaced or, alternatively, they may be supported on a non-magnetic former of porcelain, polystyrene or resin-bonded pressed paper. Air-spaced coils in receivers are found in high-frequency circuits and are single-layer coils constructed of a sufficiently heavy gauge conductor to be self supporting. They may be seen in use as chokes in television and v.h.f. equipment. In transmitters, where coils are larger in size, the coils may be formed round a skeleton former, as illustrated in Fig. 5.2, which is a compromise between air spacing and the use of a complete former.

An equivalent circuit of an inductor is shown in Fig. 5.3. L represents the inductance, R the losses and C the winding capacitance. The inductance value depends directly on the square of the number of turns used, the area surrounded by one of the coil turns and inversely on the winding length and on a flux-leakage factor which depends on the ratio of winding-length to radius. Coils are generally designed to secure the required value of inductance with a minimum amount of loss of power to the circuit of which they form a part.

The loss of the coil depends to a large extent on the radio-frequency resistance. This could be defined as the resistance which, when multiplied by the mean square of the current in the coil, gives the power wasted as heat in the coil conductor. The value of the radio-frequency resistance is affected by any factor which influences the d.c. resistance, and also by skin and proximity effect. These last two factors will now be described.

Fig. 5.4 shows the enlarged cross-section of a coil conductor. The cross in the centre indicates an instantaneous flow of current

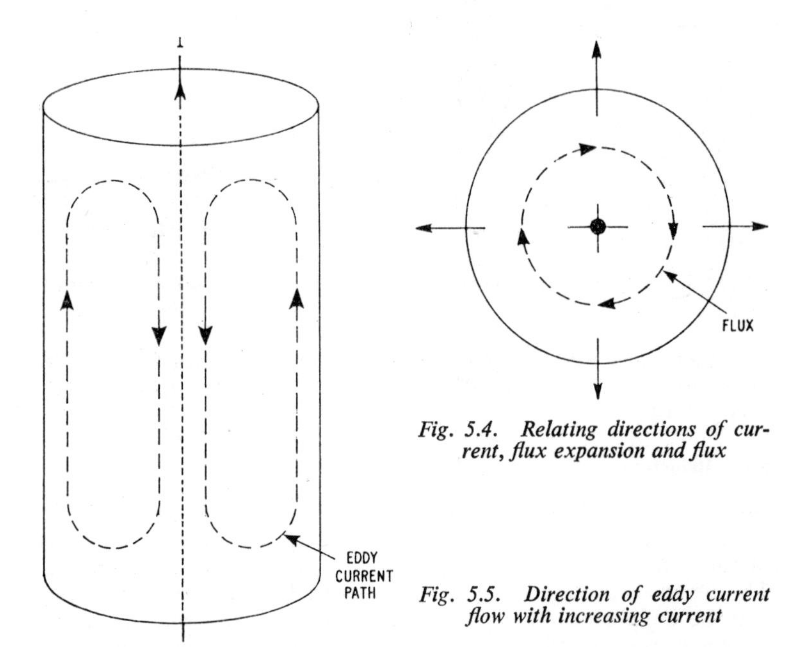

Fig. 5.4. Relating directions of current, flux expansion and flux

Fig. 5.5. Direction of eddy current flow with increasing current

into the plane of the paper. The dotted circle carrying arrows shows one line of magnetic flux and its appropriate field direction can be determined by applying Maxwell's " cork-screw " rule. The radial arrows indicate an outward expansion of the field energy as the current is assumed to be growing in value. If at any point round the dotted circle we assume a relative motion of that part of the conductor in the opposite direction to the radial arrow (i.e. inwards towards the centre) and apply Fleming's Right Hand Rule, we discover a direction of induced e.m.f. out of the plane of the paper and in opposition to the assumed current flow. If the current is growing in the opposite direction, then the e.m.f. induced in the conductor again opposes it. The induced e.m.f. is in the same direc-

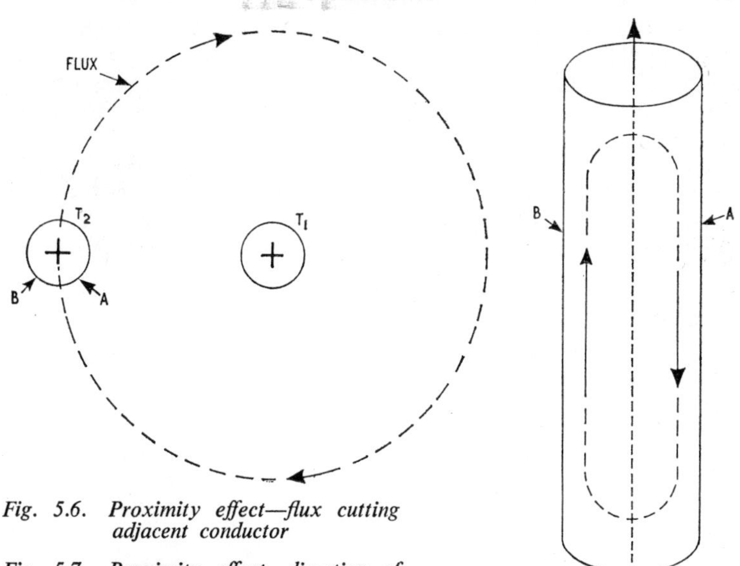

Fig. 5.6. Proximity effect—flux cutting adjacent conductor

Fig. 5.7. Proximity effect—direction of induced eddy current

tion as the current when the current value is decreasing. This back e.m.f. of self induction is greatest at the centre of the conductor where the flux linkage is greatest, and least on the surface where only the flux external to the conductor can be regarded as linking the conductor. This causes an unbalance of induced e.m.f. as between centre and surface so that eddy currents circulate as indicated in Fig. 5.5, flowing in opposition to the main current near the centre of the wire and in the same direction as the main current near and on the surface. The resultant current is very small at the centre and large at the surface. The eddy currents thus have the effect of causing loss by their own heat dissipation and by pushing the resultant current out to the surface of the conductor they reduce the effective cross-section and the conductor resistance increases in consequence. If a tubular conductor is used, the cross-section available to the current is almost as big as the cross-section through which the current would flow if it were presented with a solid conductor of equal diameter, and the eddy current heat loss is avoided if the conductor is tubular.

Eddy current loss is also caused by the field of one turn of a coil cutting across the turns next to it. Fig. 5.6 shows an expanding flux line due to turn T_1 cutting across the cross-section of a turn T_2. Both turns carry current into the plane of the paper at the instant shown. The expanding flux induces an e.m.f. in T_2 which acts out of the plane of the paper. This is greater in the half of the cross-section which is nearer to T_1 than in the other half. Thus an unbalance exists and eddy currents tend to flow against the main current in side *A* and with the main current in side *B* (*see* Fig. 5.7). The eddy currents alternate at the frequency of the coil current and dissipate heat in the resistance of the material. This is proximity effect and is one of the factors which calls for adequate spacing of the turns of an inductor at high frequencies.

Both eddy-current and proximity-effect losses can be reduced by the use of interwoven stranded wire. The first loss is reduced because each conductor strand in turn occupies both centre and surface positions in the complete conductor. Thus each strand has the

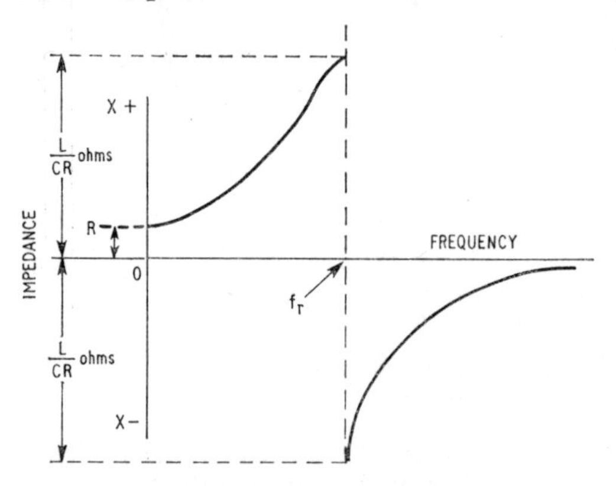

Fig. 5.8. The impedance-frequency characteristic of a coil

same total resistance and carries an equal share of the total current. This ensures a uniform distribution of the current through the cross-section. The proximity effect is reduced because in the weaving of the strands each is given a complete twist at regular intervals so that each half of the cross-section of each strand is alternately on the nearer side and then on the farther side of the next coil turn, for equal lengths of conductor.

There is a further loss in a coil if it has an insulating covering on the conductor or if it is wound on a coil former. In such coils the alternating electric field which must exist between the ends of the coil when the coil carries an alternating current, sets up a dielectric hysteresis loss in the insulation. This loss is avoided by using self-supporting conductors with no insulation other than the air space between the turns. When a coil former is essential the material is chosen to have a power factor as low as possible.

The capacitance *C* which shunts the coil must be small so that its reactance is high compared with the reactance of the coil inductance at the frequency for which the coil is to be used. The coil has maximum impedance when parallel resonance occurs. For frequencies above this, it behaves as a capacitive reactance which approaches zero as the frequency of voltage applied approaches infinity. The impedance-frequency graph for a coil has the form of

Fig. 5.8. Coil capacitance is minimised by using a former whose relative permittivity is low (an air-spaced coil is best in this respect) and by increasing the spacing between turns as far as is necessary. The higher the frequency for which a coil is required, the wider should the turns spacing be.

5.2.2. IRON-DUST CORES

Iron-dust cores have for many years been used in low- and medium-frequency coils to increase the average permeability of the magnetic circuit and so reduce the number of turns (and thus the winding resistance) needed for a given inductance. The core is made of finely-divided ferro-magnetic particles of an alloy such as permalloy, coated and compressed in an insulating binder. The magnetic flux has to pass alternately through the binder (equivalent to a minute air gap) and the magnetic particles. The relative permeability of this composite material may be of the order of 100 and depends on the closeness of packing of the particles. The eddy-current loss and the magnetic hysteresis loss add to the total loss of the coil and may be represented by added resistance in series with the coil. Above a certain critical frequency the eddy currents begin to reduce the effective permeability. Eddy currents induced in the core particles by the alternating magnetic field due to the coil current, set up fields in antiphase to the inducing field which reduce the

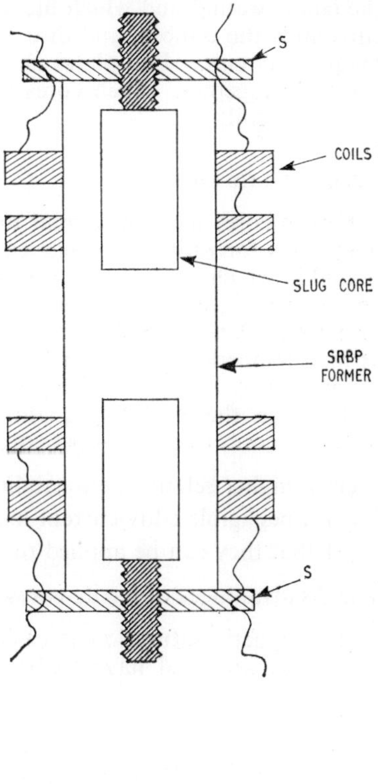

Fig. 5.9. An i.f. transformer

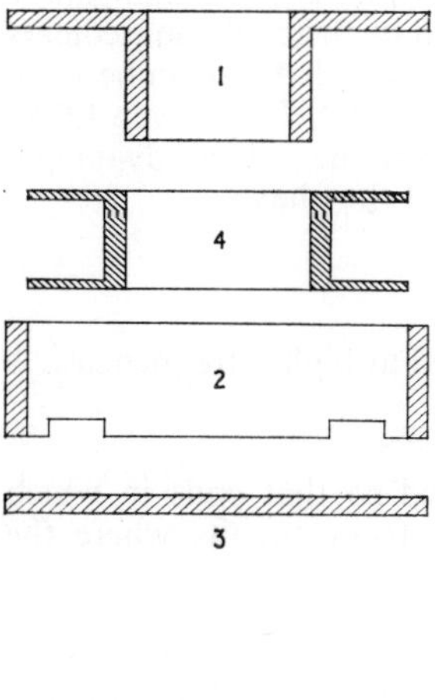

Fig. 5.10. A potted coil

resultant field strength in the particles. This limits the useful frequency range of iron-dust cores. The shielding effect of eddy-currents restricts the use of laminated cores to low or audio frequencies and it is only with extreme lamination in the dust particles that ferro-magnetic metal cores are at all practicable at radio frequencies. Dust cores shaped as small cylinders or slugs can be used at frequencies up to 100 Mc/s. In Fig. 5.9 two coils are shown each with an adjustable dust core. The coils form part of a radio-frequency transformer. Each coil is wound in two banks which are connected in series to reduce the winding capacitance. The wire is silk-covered and wound in layers which are held in position by wax. The wax also serves to exclude damp from the insulation. The former is of paxolin. A brass strip, *S*, is fitted across the former at each end to carry the screw to which is attached the dust core slug. Movements of the slug into the coil to increase the inductance or out of the coil to reduce the inductance are then possible. This device allows permeability-tuning of the resonant circuit of which the coil forms a part.

An alternative shape of core known as the " pot " core may be used for lower radio frequencies for which larger inductance values are needed. Fig. 5.10 shows the four main parts of the " pot " core inductor. Parts *1*, *2* and *3* are made of the core material. Part *1* is a thick disc with a central pillar. Part *2* is in the shape of a ring and part *3* is a disc. Part *4* is an insulating bobbin on which the coil is wound and which fits over the pillar of part *1*. The ring surrounds the bobbin, and disc *3* fits over the end to complete the magnetic circuit. End-plates and clamping nuts and bolts hold the assembly together. Such cores serve also as magnetic screens for their own coils.

5.2.3. FERRITE CORES

Ferrite cores have in a great many instances now replaced iron-dust cores for radio-frequency coils. A ferrite is a chemical compound in which one atom of iron in the iron oxide molecule (Fe_3O_4) has been replaced by an atom of another metal such as manganese or nickel. A mixture of ferrites is produced by fusing the ingredients at high temperatures. Such a material, of which Ferroxcube is an example, may have a relative permeability of 500, 750 or even 1,000, and yet has the insulating qualities of a ceramic. The advantages of ferrite cores over iron-dust cores are that they have

(1) a higher relative permeability,
(2) a negligible eddy-current loss, and
(3) that they can be applied to coils used at higher frequencies.

The disadvantages of ferrite cores are

(1) magnetic saturation is easily reached so that coils in which they are used have to be excluded from circuits where the

radio-frequency current is superimposed on a large d.c. component,

(2) a low value of maximum operating temperature (the Curie point at which the material temporarily loses its ferro-magnetic property is lower than for iron-dust cores),

(3) the ferrites have a high relative permittivity which tends to increase coil capacities.

5.2.4. LAMINATED IRON CORES

Laminated iron-cored coils are used for smoothing chokes in power supply units. The shape of the magnetic circuit is as for a small power transformer (*see* Fig. 5.11). The core is built up of

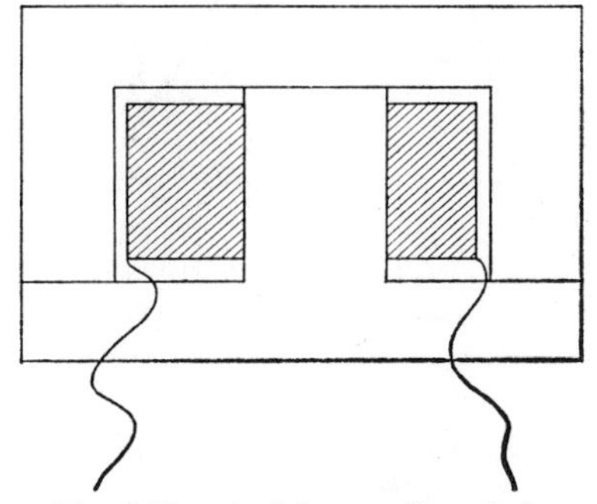

Fig. 5.11. An l.f. smoothing choke

high-permeability alloy laminations of thickness of the order of 0·015 in. each. The stacked laminations form T- and inverted U-pieces which fit together as indicated by the diagram. The windings must be of suitable gauge wire to carry both the output direct current of the power unit and its ripple component of current. Neglecting flux leakage, which is small for a completely closed magnetic circuit, the inductance is given by

$$\frac{\mu_0 \mu_r A N^2}{L} \text{ H}$$

where

μ_0 is the magnetic space constant equal to $4\pi \times 10^{-7}$,

μ_r is the relative permeability of the core laminations,

A is the cross-sectional area of the core in square metres,

N is the number of turns, and

L is the length of the magnetic circuit in metres.

The relative permeability of the core is affected, however, by the direct current flowing in the choke and the steady magnetisation produced. Fig. 5.12 shows how the relative permeability of a magnetic alloy varies with flux density. If the core area and the number of turns are so chosen that the minimum direct current expected to be carried by the choke produces maximum permeability in the core, peak swings of current and increased load current values cause a fall in inductance. The fall in inductance is permissible for some

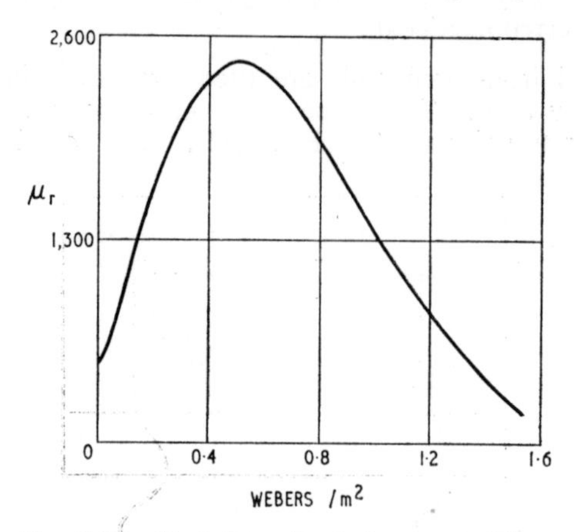

Fig. 5.12. Variation of relative permeability with flux density

applications, in particular for a choke-input smoothing filter, because it is desirable that the inductance should become less when the load current increases and greater when the load current decreases. (*See also Radio and Line Transmission, Volume* 1, Chapter 16.) When this system of design is used the smoothing choke is termed a *swinging choke.* If it is required that the value of the inductance of a smoothing choke shall remain practically constant over a range of load current values, then an air gap must be included in the magnetic circuit. The ratio of the air gap length to the total length of the magnetic circuit is called the *gap ratio.* The value of this ratio is likely to be between 0·0002 and 0·002. Since the air gap reluctance is large compared with that of the same length of magnetic circuit through the core, the change of core permeability with a change of d.c. magnetisation becomes a less significant factor of the total reluctance of the magnetic circuit. The inclusion of the air gap has the unwanted effect of reducing the inductance available, possibly to a third of that of a similar choke with no air gap.

5.2.5. VARIABLE INDUCTORS

Variable inductors are used in radio transmitter circuits to obtain resonance in conjunction with fixed or variable capacitors over different frequency ranges. The adjustment of the slug-core position has already been cited as one method of controlling the inductance of a coil. Alternatively, a small inductance value may be adjusted at high frequencies by the movement of a brass slug within the coil. Eddy currents are induced in the brass slug which are almost in antiphase to the coil currents, as indicated by the vector diagram of Fig. 5.13. In this it is assumed that the impedance of the brass cylinder, or slug, has very little resistive component so that the impedance is mainly that of an inductive reactance with the slug currents lagging by almost 90 degrees on the e.m.f.'s induced in the slug. The resultant field is thus weakened by the opposing magnetic field set up by the eddy currents in the slug. As the slug is moved farther into the coil the inductance value becomes smaller. This contrasts with permeability tuning where the magnetic slug increases the inductance value when it is moved farther into the coil.

Low-power transmitters make use of variometers for aerial tuning. These variable inductors consist of a rotatable coil on an insulating

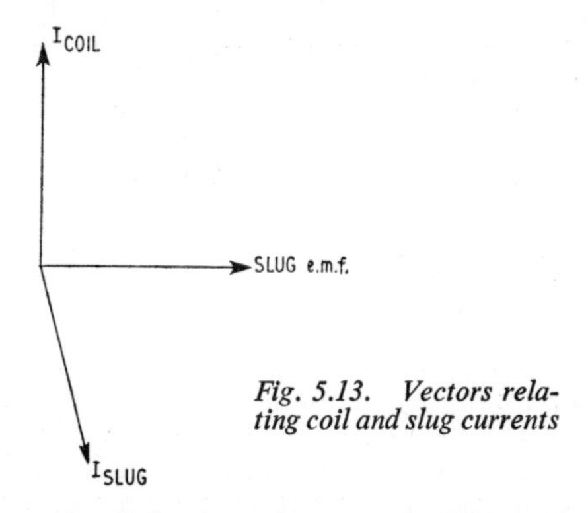

Fig. 5.13. Vectors relating coil and slug currents

former situated inside the hollow former of a larger coil. The two coils are connected in series. If the inductance values of the coils are represented by L_1 and L_2 and their maximum possible mutual inductance by M, then the maximum inductance value of $(L_1 + L_2 + 2M)$ exists if the two coils are parallel and their magnetic fields are at every instant in the same direction, i.e. the coils are in series-aiding configuration (*see* Fig. 5.14 (a)). When the two coils are mutually at right angles their mutual inductance is zero and the total inductance is $(L_1 + L_2)$ (*see* Fig. 5.14 (b)). With the movable coil turned so that its field is in line with, but opposing the field of the larger coil, the total inductance is at its minimum value of $(L_1 + L_2 - 2M)$ (*see* Fig. 5.14 (c)).

A tapped inductor can be used with a multi-point switch if the required changes in inductance value are large. The inductance, however, changes by large increments since a change in the number of turns alters both the field flux per unit current and the number of turns with which it can link.

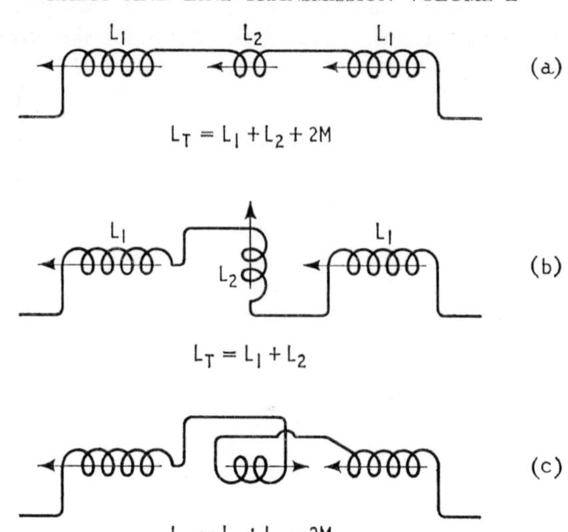

Fig. 5.14. Inductance values of a variometer

In receivers, waveband switches are used to select the coils needed for the various ranges of tuning.

Example 5.2

A variable tuning capacitor is used with an inductor to tune a receiver over the medium frequency range of 525 to 1,605 kc/s. If the capacitance has a minimum value of 20 pF what is the value of the inductor used and what maximum value of capacitance is needed, assuming a fixed circuit stray capacitance of 30 pF in parallel with the coil?

For resonance the inductor reactance must be equal to the capacitor reactance. The smallest capacitance will tune the coil to the highest frequency end of the range.

$$2\pi fL = \frac{1}{2\pi fC}\,\Omega$$

$$L = \frac{1}{4\pi^2 f^2 C}\text{ H}$$

where $$C = 20 + 30\text{ pF}$$

and substituting, $$L = \frac{10^{12} \times 10^{6}}{4\pi^2(1{,}605)^2 10^6 50}\,\mu\text{H}$$

$$= 196{\cdot}7\,\mu\text{H}$$

The capacitance required to tune this same inductor to the lower frequency can then be found by proportion. The resonant frequency of a circuit being

$$f = \frac{1}{2\pi\sqrt{LC}}$$

it is seen to be inversely proportional to $\sqrt{C}$ if L is constant.

$$\therefore \qquad \sqrt{\frac{C_{max}}{50}} = \frac{1{,}605}{525}$$

$$\therefore \qquad C_{max} = 50 \times \left(\frac{1{,}605}{525}\right)^2$$

$$= 467 \text{ pF}$$

Thus 467 pF is the total capacitance. Deduction of the stray capacitance of 30 pF gives us the final answer of 437 pF.

5.3. Transformers

5.3.1. OUTPUT TRANSFORMERS

In *Radio and Line Transmission, Volume* 1, Chapter 6, a transformer was represented as two coils with the symbol for an iron core drawn between them. Fig. 5.15, however, indicates more accurately the features of a transformer which affect its behaviour. The parts of this equivalent circuit are as follows. The dotted rectangle encloses

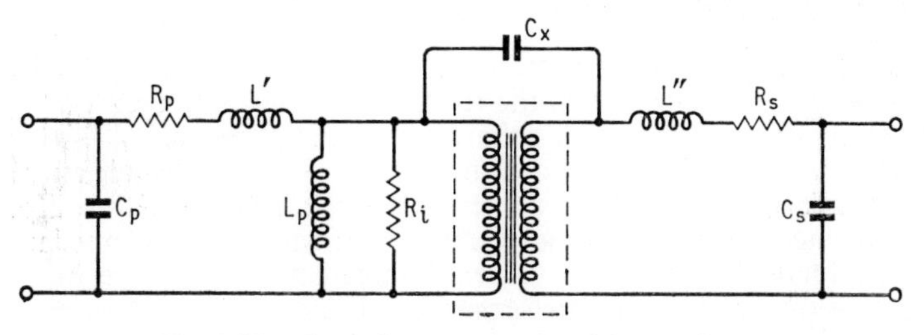

Fig. 5.15. Equivalent circuit of an l.f. transformer

what is called the *ideal transformer*. This is the transformer without losses, flux leakage and winding capacitances. The ideal transformer has the following properties:

1. On open-circuit secondary, no primary current flows.
2. If T is the ratio of the number of secondary turns to the number of primary turns, then on load, the primary current is precisely T times bigger than the secondary current and the secondary voltage is exactly T times bigger than the primary voltage.

3. The ampere-turns product of the secondary winding on load is equal in value to the ampere-turns product of the primary winding, e.g. if the turns ratio T is 20, with 25 turns on the primary and 500 turns on the secondary, then with 1 ampere flowing in the primary winding, the secondary current is 0·05 amperes. Thus $1 \times 25 = 0{\cdot}05 \times 500$. But the magnetising forces due to the primary and secondary currents are in opposite directions at every instant so that the resultant magnetic flux of the ideal transformer is non-existent.

L_p = inductance of the primary winding,

R_i = resistance representing the loss in the core of the transformer (usually called the iron loss),

R_p = primary winding resistance,

R_s = secondary winding resistance,

L' = primary flux leakage inductance which represents the effect of that part of the magnetic flux due to the primary current which fails to link with all the turns of the secondary winding,

L'' = secondary flux leakage inductance which shows the effect of that part of the flux due to the secondary current which fails to link with all the turns of the primary winding,

C_p = distributed capacitance of the primary winding,

C_s = distributed capacitance of the secondary winding,

C_x = mutual capacitance between the two windings.

The separation of the aspects of a transformer which in practice make it less than ideal, from the remaining ideal component clarifies the study of the design and behaviour of transformers.

By way of example some design features of a loudspeaker matching transformer are now considered. It is the function of the output transformer of a receiver or an amplifier to match the low impedance of the loudspeaker to the value of impedance which the output valve needs for optimum performance. For example a valve whose optimum load is 8 kilohms requires a step-down transformer of 40 : 1 to match it to a 5-ohm load. The number of primary turns used is chiefly dependent upon the primary inductance required. (This is shown by L_p in Fig. 5.15.) The larger the primary inductance provided, the better is the low-frequency response of the transformer. The response at the low frequency end of the scale falls 3 dB below the mid-frequency response at the frequency for which the reactance of the primary winding is equal to the effective load resistance (in our example 8 kilohms) in parallel with the primary resistance R_p added to the a.c. resistance of the output valve. The number of primary turns necessary for a given value of inductance depends inversely on the flux density level at which the core material is used and the cross-sectional area of the core. The number of secondary turns depends on the number of primary turns selected and on the required turns ratio. A gauge of wire is used for the primary winding which can carry, without over-heating, the direct current of the output valve, e.g. S.W.G. 34 (diameter 0·0092 in.) which carries up to 66 mA safely.

The shape of the core is usually similar to that of the choke core illustrated in Fig. 5.11. A size is chosen which will accommodate in its window-space the desired number of primary and secondary turns of sufficient cross-sectional area. The winding space allowed for the secondary turns is approximately equal to that allowed for the primary turns. Although the number of secondary turns may be only a fortieth of the number of primary turns the wire used is a thicker gauge, e.g. S.W.G. 18 (diameter 0·048 in.). The core is made up of stacked silicon–iron laminations. The lamination of the core minimises eddy current loss and this material has a relatively low hysteresis loss. Flux leakage loss may be minimised by winding half the secondary winding underneath the primary winding and half on top of it. The high-frequency response limit of the transformer is chiefly set by the value of flux leakage inductance. The response falls 3 dB below the mid-frequency response at approximately the frequency for which the total flux leakage inductance (calculated as a total equivalent value in the primary circuit) has a reactance equal to the sum of the equivalent valve load resistance (e.g. 8 kilohms) in series with winding resistance and the a.c. resistance of the output valve if C_p and C_s are negligible. For such considerations, the total equivalent flux leakage inductance is given by $(L' + T^2L'')$, and the total winding resistance is calculated from $(R_p + T^2R_s)$ (*see* Fig. 5.15).

Paper insulation between layers of turns may be used in addition to a coating of insulating shellac on the conductors. The thicker the paper used between layers, the smaller are the capacitances C_p, C_x and C_s, but to economise in space the paper can only be thin. The resistance R_p is typically between 400 and 500 ohms but the secondary resistance R_s is of the order of 0·1 ohms.

Some output transformers have a tapping about a tenth of the number of the primary turns from one end of the primary winding to which tapping the H.T. supply positive lead is connected. Supply current to other valves in the equipment flows through this tenth section of the primary winding and causes a d.c. magnetising force in opposition to that due to the output valve d.c. which flows through the other nine-tenths of the primary winding in the opposite direction. This not only helps to reduce d.c. magnetisation of the core and the danger of core saturation but also facilitates removal of mains hum from the output. Hum-voltage induced in the transformer secondary winding by unsmoothed ripple in the power amplifier d.c. is partially balanced by an opposing hum-voltage induced into the secondary by the ripple in the other valve supply currents flowing through the tapped-off section of the primary.

The interwinding capacitance marked C_x in Fig. 5.15 can be largely eliminated by interposing a copper or aluminium foil screen between the primary and secondary windings. The foil can be wound spirally as a strip in a single layer, but successive turns must not be in contact and the ends of the strip must be open-circuited. The screen must not form a closed electrical circuit round the core or it will behave as a short-circuited turn and effectively reduce the primary inductance to a very low value by a large circulating current in anti-phase to the primary magnetising current. The foil is connected to earth via the transformer casing so that the two windings have little or no capacitance to each other though each has a capacitance to the foil.

5.3.2. TOROIDAL CORES

Toroidal-cored transformers in which the core is a doughnut shape, offer some advantages over the conventional type of transformer and are sometimes used in high-quality amplifiers. The speedy winding of uniform coils over toroidal cores has been facilitated by the development of new machines for this purpose. The nickel alloy used for the core is wound up in rings from thin tape and after heat treatment is fitted into a plastic case which protects the core and acts as a smooth former on to which the windings can be wound. A tough lacquer wire insulating coating is used without silk, rayon or cotton covering, so that a large number of turns per unit of winding space can be used. This provides a very high primary inductance for a relatively small winding resistance. The uniform distribution of the winding round the core assists the dissipation of the inevitable, if small, heat loss. In a conventional transformer the windings are piled up on top of about a third to a half of the length of the magnetic circuit and this discourages the dissipation of heat due to winding resistance. Further advantages arising from a toroidal-shaped core are a reduced winding capacitance and a smaller leakage flux. In power transformers the smaller leakage flux enables the transformer to be mounted nearer to other pieces of equipment without stray mutual induction causing unwanted mains-hum in the output of the receiver or amplifier.

A toroidal output transformer is advertised by its manufacturer as having the following specifications

Ratio:
to match 6,400 ohms to load of 16 ohms or 4 ohms.

Primary inductance at 400 c/s:
20 H.

D.C. resistance:
primary 16 ohms, secondary 1 ohm.

Primary resonant frequency with the secondary short-circuited:
50 kc/s.

Frequency response:

—0·5 dB down from mid-frequency response at 10 c/s and 17 kc/s with level response over the intervening range.

Toroidal windings on permalloy powder cores have also been developed to give inductance coils with higher Q factors, greater stability of inductance with time and temperature change, lower leakage flux and self-capacitance, lower hysteresis loss and more constant permeability in spite of field strength or frequency change, than are realised with conventional coil core shapes.

5.4. Screening

5.4.1. RADIO-FREQUENCY SCREENING

The effects of using a conducting can to screen a coil which is used in a radio-frequency circuit, are to increase its total loss and to reduce the effective value of its inductance. Fig. 5.13 which shows vectorially the relative phasing of a coil current and the eddy current induced by the coil field in a brass slug, may equally well indicate the approximate phasing of the currents set circulating round the screening can of an r.f. coil or i.f. transformer. The eddy-current field opposes the coil field outside the can and the two

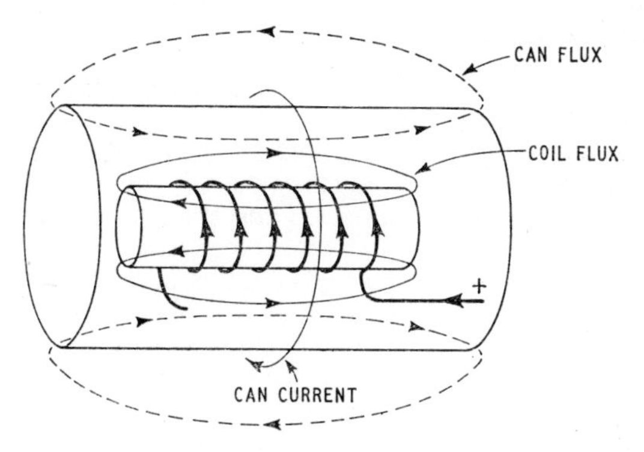

Fig. 5.16. The principle of eddy current screening

tend to cancel mutually. This is further illustrated in Fig. 5.16 in which the can is considered as a single extended turn surrounding the inner coil. The directions of current flow in the coil and in the can are shown as opposite in accordance with Lenz's Law. The field due to the coil current is shown by a full line while the field due to the can current is marked in dotted lines. Inspection of the figure shows that the two fields have opposite directions *outside* the can although they have the same direction in the space between the inside of the can and the outside of the coil. The cancellation of the external fields is the end sought in the eddy-current screening process, but nevertheless, it is not achieved without a reduction in the resultant total flux linkage per ampere with the coil. The coil inductance is thus reduced. Further, the screening can is not without resistance and even though it is kept as small as possible it dissipates some power obtained by inductive process from the coil. The loss due to the can is an added loss to the coil and can be represented as an increase in the resistance of the coil in an equivalent circuit diagram. Since the coil inductance is reduced while its losses are increased, the coil Q factor is also reduced. This, of course, has to be accepted as an inevitable by-product of the screening of r.f. components.

The effectiveness of the screen depends on a path of good conductivity round the screen for the flow of eddy current. Any longitudinal break in the screen would render it useless. A radio direction finder loop aerial has a brass tubular screen round it to make the capacitance to earth of the loop evenly distributed, but here there is a deliberate break in the continuity of the screen to prevent its acting as an r.f. magnetic screen because this would prevent the signal reaching the enclosed loop aerial. To minimise the reduction in Q value of the coil, a screen must not fit too closely round the coil. A spacing equal to the coil radius between the coil surface and the inside of the can is adequate.

5.4.2. LOW-FREQUENCY SCREENING

When a coil which carries audio-frequency currents is to be screened, the eddy-current principle described above is unsuitable since the eddy-current amplitude could not be made sufficiently big. The screen surrounding low-frequency coils is therefore made of a highly permeable magnetic alloy. Thus the flux external to the coil passes within the very low reluctance path offered by the enclosing screen instead of taking the higher reluctance path via the surrounding air space. The leakage flux which does not accept the path through the screen is negligible. The total reluctance of the coil magnetic circuit is reduced by the presence of the screen and more flux per ampere-turn of magnetising force is set up, causing the value of the coil inductance to be increased. A screen of ferromagnetic material for use at radio frequencies would increase the coil loss excessively by its hysteresis loss and this is the reason why the eddy-current process is adopted for radio-frequency screening.

Whether the screen be for radio-frequency or audio-frequency signals it must be effectively connected to earth or to the chassis of the equipment so that it can also provide electrostatic screening. The can being earthed, any potential difference which the screened component may have to earth results in an electric field which ends on the nearest earthed point, which is the inside of the screening can. The capacitance to earth of the screened component is slightly increased by the screen but the stray capacitance to other components is removed.

5.5. Capacitors

5.5.1. ELECTROLYTIC CAPACITORS

Having already considered the inductor, the reader will not be surprised to find that the capacitor is not the simple circuit element which its usual circuit diagram symbol suggests. For example an electrolytic capacitor may be represented by one of the equivalent circuits shown in Figs. 5.17, 5.18, 5.19, 5.20. The first of these may

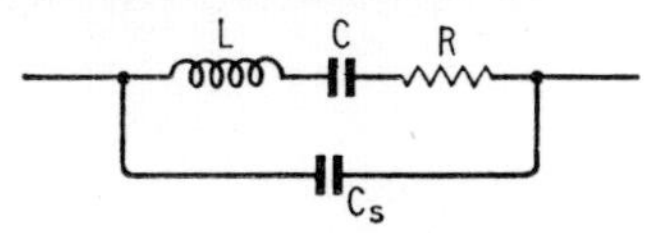

Fig. 5.17. An equivalent circuit of an electrolytic capacitor for low frequencies

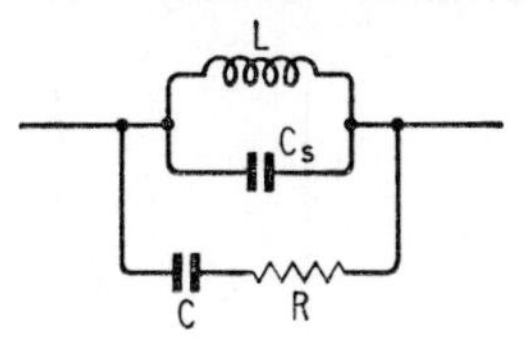

Fig. 5.18. An equivalent circuit of an electrolytic capacitor for high frequencies

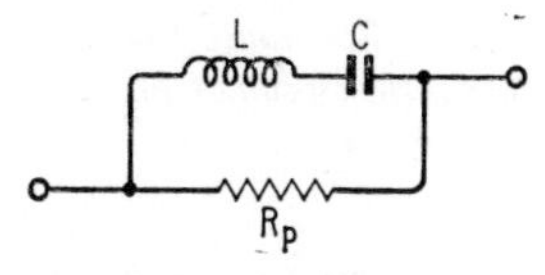

Fig. 5.19. An equivalent circuit for a rolled paper capacitor

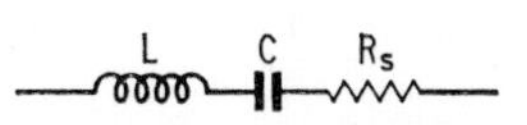

Fig. 5.20. An alternative circuit for a rolled paper capacitor

represent an electrolytic capacitor at low frequencies. C is the effective capacitance required across the oxide film on the anode plate and the capacitance which tends to exist across the naturally-formed oxide film on the cathode plate. These two capacitances are in series. The effect of the unwanted film on the cathode plate is worst in capacitors designed for low voltage working and in which, therefore, the anode film is almost as thin as the unwanted cathode film. C_s is a shunt capacitance which would exist between the two capacitor electrodes if both aluminium oxide films were taken away and the electrolyte were absent. This shunt capacitance is small and is significant only at high frequencies. L shows the inductance of the capacitor which results from the winding of the foil electrodes. It is kept to a minimum by the manufacturer but may be of the order of 0·01 to 1·0 μH. R is an equivalent resistance representing all the power losses of the capacitor. The losses are relatively high for an electrolytic component compared with those for other types. The reactance-frequency graph for the circuit of Fig. 5.17 is that of Fig. 5.21. This shows a series resonance frequency f_1 at which the capacitor has minimum impedance. This frequency is probably in the range 5 to 50 kc/s. As a decoupling capacitor across a cathode bias resistor, the capacitor functions best at this resonance frequency. The anti-resonance or rejection frequency f_2 may occur somewhere in the range between 1 and 10

Mc/s and at this frequency an electrolytic capacitor has a high impedance and as a decoupling capacitor is ineffective. Such high frequencies do not occur in audio amplifiers but this anti-resonance could be troublesome in video amplifiers. The difficulty might be avoided by using two capacitors in parallel, each having half the total capacitance needed and about half the inductance of the larger capacitor. This would present about the same total capacitance but only a quarter of the inductance of the single large

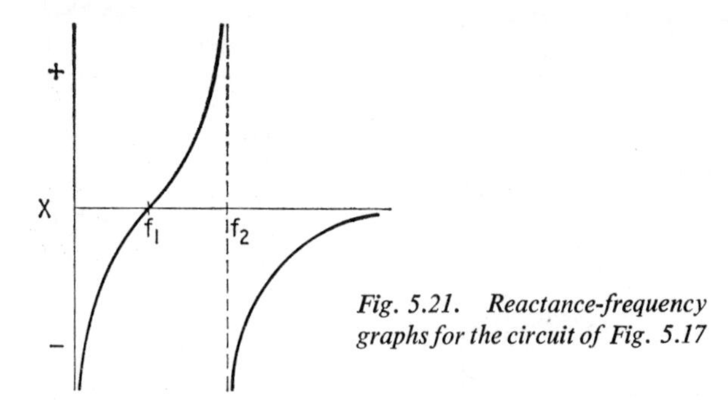

Fig. 5.21. Reactance-frequency graphs for the circuit of Fig. 5.17

capacitor so that the rejection frequency would be nearly doubled and thus outside the range to be decoupled. For high frequencies Fig. 5.18 is a more suitable equivalent circuit.

Example 5.3

An electrolytic capacitor has a maximum impedance at 4 Mc/s and an effective inductance due to foil windings and leads of 0·5 μH. What is the effective value of capacitance in shunt with the inductance at this frequency?
Assuming the formula for resonance

$$f = \frac{1}{2\pi\sqrt{LC}} \text{ c/s to apply,}$$

on re-arranging to find C, we have

$$C = \frac{1}{4\pi^2 f^2 L} \times 10^6 \ \mu\text{F}$$

$$C = \frac{10^6 \times 10^6}{4\pi^2 16 \times 10^{12} \times 0{\cdot}5} \ \mu\text{F}$$

$$= \frac{1}{32\pi^2} \ \mu\text{F}$$

$$\text{F} = 0{\cdot}00317 \ \mu$$

5.5.2. OTHER CAPACITORS

A rolled-paper capacitor also has some inductance and might be represented either by Fig. 5.19 in which the losses are shown by a high value of parallel resistance, or by Fig. 5.20 where the losses are shown by a small value of equivalent series resistance.

In mica capacitors or plate capacitors of other types, the inductance factor is much smaller and can be neglected. Only the loss and the required capacitance are then significant elements and the capacitor can be represented by a loss-free capacitor in series with a small loss resistance or in parallel with a high value of loss resistance. The relationship between the small series loss resistance (R_s) and the high parallel loss resistance (R_p) is given by

$$R_s = \frac{1}{\omega^2 C^2 R_p} \text{ ohms,}$$

the capacitance being approximately the same in each equivalent circuit. The parallel resistance shows more clearly the effect of

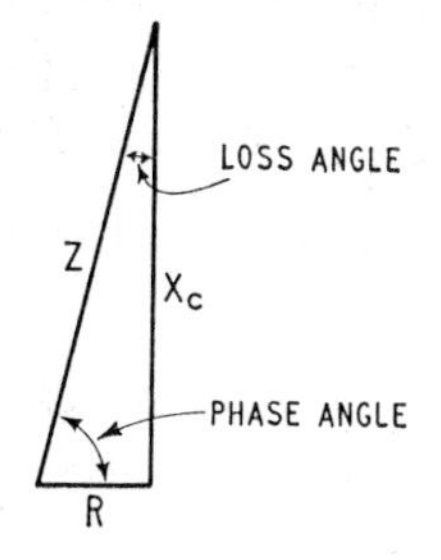

Fig. 5.22. The impedance triangle of a capacitor

leakance through the capacitance but a series equivalent value may be more convenient for purposes of calculation.

The capacitor may be represented by an impedance triangle as in Fig. 5.22.

In the ideal capacitor, the phase angle is 90 degrees and the loss angle is zero. The impedance triangle with its very small loss angle shows that Z, the impedance approximates to the reactance X_C. Thus the power factor R/Z is approximately equal to R/X_C. This last ratio is also the tangent of the loss angle and this measurement, which should be as small as possible, is usually taken as a goodness factor of a capacitor. Losses in capacitors can be due to the following causes:

1. Loss due to conductance (leakance) through an imperfect insulation used for the dielectric.
2. Leakance across the surface of a dielectric.
3. Energy loss within a dielectric due to alternating dielectric displacement currents—dielectric hysteresis loss.

4. Resistance at r.f. of plates and leads.
5. Additional losses if damp and other impurities are present in the dielectric.

The losses are a function of frequency so that both the equivalent resistance and the reactance are variables and the power factor for a capacitor must be stated for a particular frequency. Losses also vary with temperature and makers usually specify a working temperature range for their components, and often a temperature coefficient to indicate possible variation of capacitance with tempera-

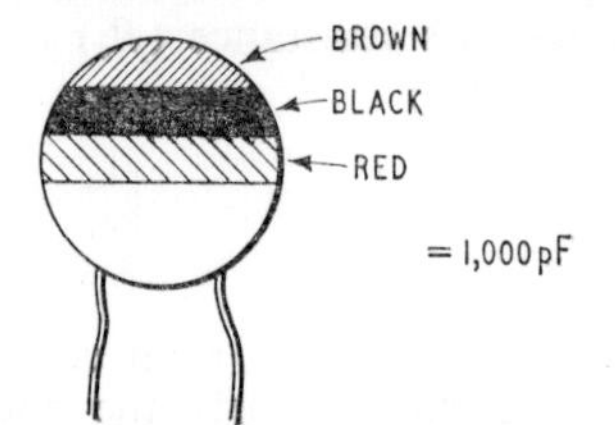

Fig. 5.23. A ceramic capacitor

ture. In Table 5.1 values which are typical of several types of capacitors are given for comparison.

Note.—Ceramic capacitors use thin flakes of a ceramic material on to which the plates are deposited as an extremely fine silvered layer. They are coated with plastic and then painted according to a colour code similar to that used for resistors but which gives their capacitance in picofarads (*see* Fig. 5.23). The ceramic may be chosen for its high permittivity as in the high-K type or for its low loss as in the high-Q type.

5.6. Gas Discharge Tubes

In a valve data manual there may be a section headed *voltage reference and stabiliser tubes.* Such tubes are intended to assist in

Table 5.1

Type of Capacitor	*Power factor and frequency*	*S.W.V. (d.c.) Volts*	*Typical capacitance*	*Temperature range (°C)*
Aluminium oxide electrolytic	0·1 at 50 c/s	6–500	1,000–2 μF	20 to 85
Tantalum electrolytic	0·05 at 50 c/s 0·15 at 10 kc/s	6–70	100–1·5 μF	−40 to 85
Rolled paper	0·01 at 1 kc/s	225–1,000	0·001–0·5 μF	−100 to 70
Silvered mica	0·001 at 1 Mc/s	1,000	5–50 pF	−30 to 70
Ceramic high-K types	0·03 at 100 kc/s	1,500	470–0·02 pF	−20 to 85
Ceramic high-Q types	0·002 at 1 Mc/s	500	1·5–260 pF	−20 to 85

the provision of steady voltages from sources of power whose output may tend to fluctuate. Stabiliser tubes are intended to supply a load with direct current which may vary in value but at a voltage which does not change by more than one or two volts. A reference tube is used to provide a constant voltage with no more than, say, 0·1 per cent variation over its useful life, although it is generally arranged that the load current is practically constant. The voltage is required as a reference standard rather than a source of power.

Both stabilisers and reference tubes resemble radio valves in appearance and have normal valve bases. Yet although a 7-pin base may be used, inside the envelope there are only two electrodes which may be called cathode and anode. The envelope contains an inert gas which may be neon or argon, at a pressure which is approximately a hundredth of that of the atmosphere. (Normal atmospheric pressure = 10^5 newtons/sq. m.) The cathode may be

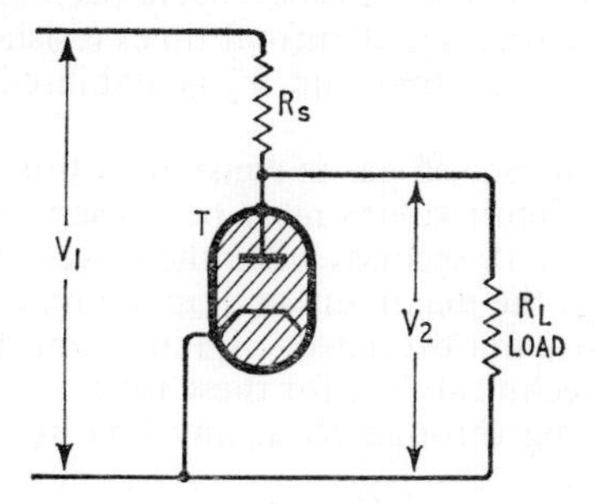

Fig. 5.24. A gas filled stabiliser tube circuit

in the form of a small rod, with the anode as a surrounding cylinder. Both electrodes are of molybdenum of a high degree of purity.

If the voltage applied between the two electrodes of a gas-filled tube is low (say, below 100 volts) there is practically no current flow. If the voltage is increased, a value will be reached at which the few free electrons which are always present in any gas, are given sufficient kinetic energy by the electric field between the anode and cathode to dislodge by impact an electron from a gas molecule with which they may come into collision. In this way further free electrons and also positive gas ions are produced. These, too, produce ionisation by collision and thus the number of current carriers and the current value increases sharply. The voltage at which this occurs is called the *ignition voltage*. Fig. 5.24 shows how the tube is connected in circuit. V_1 is the unstabilised supply and V_2 is the voltage to be connected across the load and which is to be stabilised. (In practice the load is often an oscillator valve whose frequency can only remain stable if its H.T. supply is stable.) V_1 must exceed the ignition voltage of the tube. Before the gas ionises as described above, the current through R_S and the p.d. across R_S is small so that the voltage across the tube is high enough to strike the tube. Once the gas has ionised, the additional tube current flowing through R_s reduces the p.d. across the tube to the stabilised level which is called the *burning voltage*. The load is in parallel with the tube and shares this voltage. Should the supply voltage, V_1, increase, there will be a momentary increase in voltage across the tube which gives rise to a further increase in ionisation and a reduction in the tube resistance. Since the tube resistance decreases at the same time as the supply voltage increases, the current increase in R_s can be such that the increased p.d. across R_s almost equals the increase in supply voltage. Alternatively, should the supply voltage fall, the tube de-ionises slightly and its resistance increases. The reduction of current in R_s is such that the fall in p.d. across R_s practically compensates for the reduction in supply volts. It will be seen that the tube behaves as an automatically varying resistance. When the current increases its resistance decreases. When the tube current decreases the resistance increases. Thus the product of current times resistance for the tube remains substantially constant and V_2 is stabilised at the burning voltage.

The p.d. across an ionised gas is constant within certain limits of burning current. Circuit values must be chosen so that except for a brief period of some 10 seconds when the valves of the load circuit may be warming up, the maximum and minimum current values for the stabiliser tube are not exceeded. Ignition and burning voltages are also quoted in technical data for these tubes.

The current flowing through R_s at any time is

$$\left(\frac{V_1 - V_2}{R_s}\right)$$

The current through the tube is therefore

$$I_T = \left(\frac{V_1 - V_2}{R_s}\right) - I_L$$

where I_L is the load current.

Or $$I_T + I_L = \left(\frac{V_1 - V_2}{R_s}\right)$$

whence $$R_s = \left(\frac{V_1 - V_2}{I_T + I_L}\right)$$

From this, limits for R_s can be calculated.

Example 5.4

A stabiliser tube has an ignition voltage of 110 volts, a burning voltage of 75 volts and is to be used to stabilise an H.T. supply to a valve requiring a current of 10 mA from a source liable to fluctuate between 220 volts maximum and 200 volts minimum. If the maximum and minimum current values for the tube are stated to be 22 mA and 2 mA respectively, what are the possible limits of value for a series regulating resistance?

The minimum value is found by assuming the maximum supply voltage and the maximum tube current

$$R_s = \frac{220 - 75}{22 + 10}$$

$$= \frac{145}{32}$$

$$= 4{\cdot}54 \text{ k}\Omega \text{ (since current is in mA).}$$

The maximum value is found by assuming the smaller value of the supply voltage and the smaller value of tube current. (We are assuming in this case that the load current remains at 10 mA and the stabilisation remains perfect giving a constant value of burning voltage of 75.)

$$R_s = \frac{200 - 75}{2 + 10}$$

$$= \frac{125}{12}$$

$$= 10{\cdot}4 \text{ k}\Omega$$

However, a third condition must be satisfied. That is that the ignition voltage must be attained across the tube on switching on. If we assume that the tube current before ionisation is zero and that the supply voltage has its minimum value of 200 volts, then the voltage across the tube on switching on is:

$$\left(\frac{R_L}{R_L + R_s}\right) \times 200 \text{ volts}$$

This must exceed the ignition voltage, in this case 110 volts. The load which takes 10 mA from 75 volts is 7·5 kΩ. If we assume that the above tube voltage is just equal to the ignition voltage, we can find the maximum value of R_s set by the ignition voltage factor. Thus

$$\frac{7{\cdot}5}{7{\cdot}5 + R_s} \times 200 = 110 \text{ volts}$$

whence $$110\, R_s = 1{,}500 - 825$$

and $$R_s = \frac{675}{110}$$

$$= 6{\cdot}14 \text{ k}\Omega$$

Our final conclusion is that our series resistor should be between 4·54 and 6·14 kΩ.

The maximum current which the load can draw without the tube current falling below the minimum limit is approximately equal to the difference between the maximum and the minimum tube current values. If the supply voltage (V_1) were to stay constant but the load current demand were to vary, then with perfect stabilisation, the sum of the load current and the tube current would remain constant. The tube current would fall by the same amount as the load current increased and vice versa. Perfect stabilisation is not possible and either a change in supply voltage or a change in load current demand will cause a small change in the load voltage. The difference between the burning voltage at maximum current and the burning voltage at minimum tube current is termed the *regulation voltage*. It should not be more than 5 or 6 per cent of the average burning voltage.

5.7. Silicon Voltage Regulators

An alternative voltage stabiliser is a semi-conductor device often called the *zener diode*. (For an introduction to semiconductors *see Radio and Line Transmission, Volume* 1, Sections 7.2.1, 7.2.2 and 7.2.3.) The semiconductor used in the manufacture of these diodes

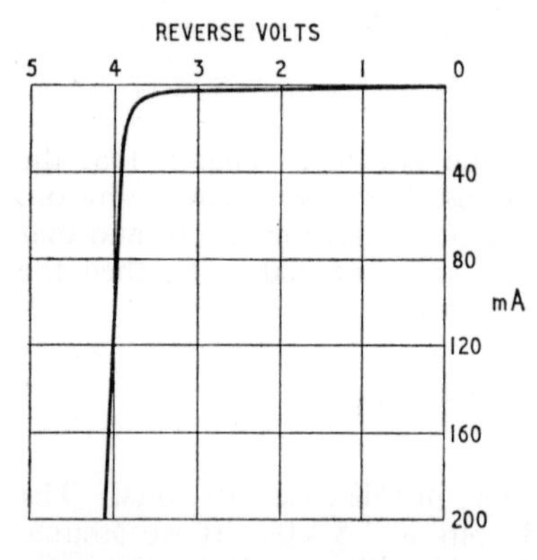

Fig. 5.25 The reversed voltage conductance characteristic of a silicon diode

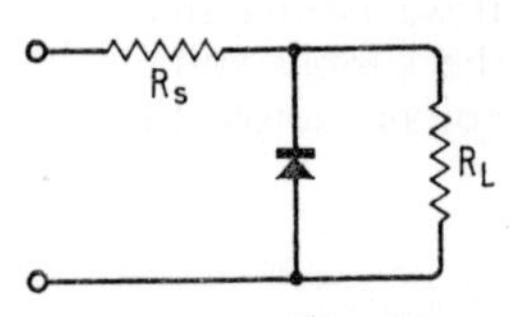

Fig. 5.26. The silicon diode stabiliser circuit

is silicon. A junction diode is prepared of *p*-type and *n*-type silicon whose reverse break-down voltage is approximately equal to the voltage to be stabilised. The reverse break-down voltage can be determined during manufacture to be between 2 volts and several hundred volts according to operating requirements.

Break-down of a reverse-biased silicon diode when it occurs at voltages below 5 volts approximately, is attributed to the electric field between the *p* and *n* regions becoming sufficiently intense to pull valency electrons out of their orbits so that they become free charges and leave behind positive ions. Both ions and free electrons are then available as current carriers within the depletion layer.

(The depletion layer is the region of the electric field existing across the semiconductor junction and caused by the migration of electrons from the *n*-type material into the *p*-type material and of holes from the *p*- to the *n*-type material.) This field emission effect was first suggested as the reason for semiconductor break-down by C. Zener in 1934, hence the name *Zener effect.*

Higher voltages produce break-down by a different mechanism. At high reverse voltages the electric field between *p* and *n* regions accelerates the few current carriers normally present to such a velocity that they free electrons by impact with atoms. The charges released in this way are in turn accelerated to velocities at which they also cause ionisation of other atoms. Thus carrier charges multiply rapidly and a very large increase of current results. This is called *avalanche effect.* Diodes involving either Zener effect or avalanche effect may be loosely called Zener diodes.

The shape of the reverse-current characteristic curve of the silicon diode is similar to that of current-voltage curve of a gas-filled voltage regulator valve, although it is usually drawn in the inverse direction (*see* Fig. 5.25). When the reverse voltage of the silicon diode exceeds the break-down value then even a small increase of applied voltage causes a large increase in current. Conversely, for a large change in current the voltage across it remains almost unchanged. This behaviour compares with that of the gas-filled diode after it has ionised.

Fig. 5.26 shows the circuit which may be employed using a semiconductor diode to maintain a stable voltage across a load resistor R_L which may vary. Apart from the substitution of the silicon diode for the gas-filled tube *T*, Fig. 5.26 is seen to be similar to Fig. 5.24. Resistance R_s limits the current through the silicon diode after break-down has occurred. If the maximum working temperature of the diode is not exceeded no damage is suffered by the diode as a result of its operation at reverse voltages just greater than break-down value. The principle of stabilisation is similar to that described for the gas-filled tube in Section 5.6.

Advantages of silicon voltage regulator diodes over gas-filled regulating tubes are

1. They are smaller in size.
2. They have a better voltage regulation.
3. Their life expectancy is longer.
4. They can be designed for a wider range of stabilised voltages.
5. No voltage higher than the stabilised voltage is needed for " striking " as in the instance of the gas-filled device.

Questions

1. It is often necessary to restrict the effects of the magnetic and electrostatic fields associated with radio components by the use of

shields or screens. What materials would be used for the screen surrounding the following items?

(a) a microphone transformer,
(b) a 450-kc/s intermediate-frequency transformer,
(c) a medium-frequency loop aerial.

Briefly explain the reasons for the choice of material in each case. (*C & G*, 1959.)

2. Explain the operation of a neon tube voltage stabiliser.

A neon tube and resistor are connected in series to provide a 180-volt stabilised output from a 220-volt d.c. supply. If the load resistance is 15 kilohms and the current through the neon stabiliser is 12 mA, determine the value of the series resistor and the power dissipated in it.

If the load resistance is reduced to 10 kilohms, determine the new current which flows through the neon tube.

Assume that the operation is ideal. (*C & G*, 1960.)

3. Explain briefly but clearly the reasons for the following design features in radio and line transmission components

(a) the use of stranded conductors in an inductor for use at 200 kc/s,
(b) the use of non-magnetic conducting screens surrounding an i.f. 450-kc/s transformer,
(c) the small air gap in the magnetic circuit of an audio-frequency output transformer. (*C & G*, 1961.)

4. What do you understand by a " thermistor "? What is its characteristic property and how is this utilised in practice? Draw a simple circuit to illustrate its use.

5. What do you understand by the " Q factor " of an inductor? What features of the design of inductors for radio frequencies ensure a maximum Q factor?

6. Explain the principle of a variometer. The maximum inductance of a variometer is 400 μH and its minimum value is 300 μH. If the fixed coil has an inductance of 200 μH, what inductance has the rotatable coil and what is the maximum mutual inductance between the two coils used?

7. Explain the principle of action of a silicon voltage regulator diode and draw a simple circuit to show the connection to a load of such a regulator. What advantages has this diode over the gas diode regulator?

8. Draw the equivalent circuit diagram for an electrolytic capacitor for low frequencies and sketch the reactance-frequency graph for the circuit. Indicate on your graph the frequency for which the capacitor will be most effective as a decoupling device and that for which it will be least effective for this purpose.

9. A paper capacitor of 0·05 μF capacitance has a power factor of

0·01 at 1 kc/s. Represent the loss of the capacitor (a) by a series resistance value and (b) by a shunt resistance value, giving the value of each resistance.

10. Explain the effect of a screening can used with an i.f. coil on the inductance value, the loss, and the Q factor of the coil. Why is it essential that the can should have no high resistance joins? Do you expect the relative permeability of the magnetic screen used for this purpose to be high or approaching unity? Give reasons for your answer.

6

Logarithmic Units and Scales

6.1. Introduction

In the Appendix of *Radio and Line Transmission, Volume* 1, the decibel unit (dB) is introduced as a logarithmic unit for the comparison of power, current, voltage and field strength ratios. This earlier work is now recapitulated in the following four simple examples.

(a) The radiated power of a transmitter is 1 kW when radiating on C.W. but the power radiated increases to 1·4 kW when modulation is applied. Express the increase in power in dB.

$$\text{The increase} = 10 \log \frac{1{\cdot}4}{1{\cdot}0}$$

$$= 10 \times 0{\cdot}146$$

$$= 1{\cdot}46 \text{ dB}$$

(b) The r.m.s. current in a 600-ohm resistor is 10 mA when supplied direct from an audio source but the current falls to 10 μA when an attenuator is inserted between the source and the 600-ohm load. Express the fall in power in the 600-ohm resistor in dB˙

$$\text{The fall} = 20 \log \frac{10}{10 \times 10^{-3}}$$

$$= 20 \log 10^3$$

$$= 60 \text{ dB}$$

(c) A transistor amplifier stage is found to give a voltage gain of 35. Express this gain in dB.

$$\text{The gain} = 20 \log 35$$

$$= 20 \times 1{\cdot}544$$

$$= 31 \text{ dB approx.}$$

(d) The field strength of an m.f. transmitter at night fluctuates between levels of 180 and 20 μV/m. Express this fluctuation in dB.

$$\text{The variation} = 20 \log \frac{180}{20}$$
$$= 20 \log 9$$
$$= 19 \text{ dB.}$$

6.2. Amplifiers in Tandem

The usefulness of the decibel unit becomes more apparent when more than one loss or gain is to be considered. This is exemplified in the consideration of amplifiers in tandem which follows.

Amplifiers are said to be connected in *tandem* or *cascade* when the

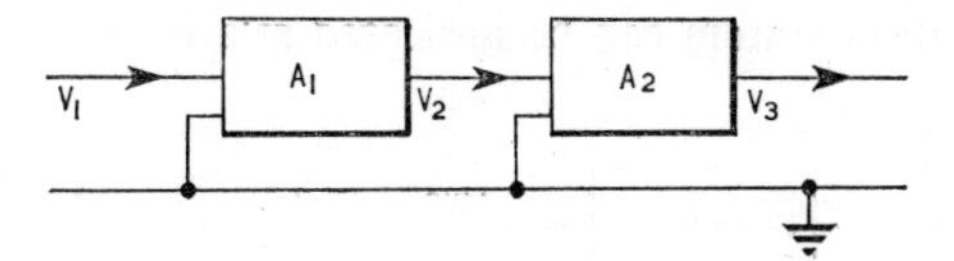

Fig. 6.1. Amplifiers in tandem

output of one stage provides the input to the next (*see* Fig. 6.1).

If the two amplifiers of Fig. 6.1 each have a gain of 30, the total gain could be calculated in the following way

$$V_3 = 30 \times V_2$$

and
$$V_2 = 30 \times V_1$$

$\therefore$
$$V_3 = 30 \times 30 \times V_1$$

$\therefore$
$$\frac{V_3}{V_1} = 900$$

In dB this is a gain of 20 log 900

$$= 20 \times 2{\cdot}9542$$
$$= 59 \text{ dB}$$

But converting the gain of *one* amplifier to dB we have

$$20 \log 30 = 20 \times 1{\cdot}4771$$
$$= 29{\cdot}5 \text{ dB.}$$

The total gain can therefore be found by *adding* the gains expressed in dB of the individual amplifiers, i.e.

$$\text{Total gain} = 29{\cdot}5 + 29{\cdot}5$$
$$= 59 \text{ dB}$$

In general let amplifying stages in tandem have gains of G_1, G_2, G_3, G_4, etc., then the total gain is

$$20 \log[G_1 \times G_2 \times G_3 \times G_4 \ldots .]$$

which, since the logarithm of a product is the sum of the logarithms of the terms in that product, may be written as

$$20[\log G_1 + \log G_2 + \log G_3 + \log G_4 + \ldots .] \text{ dB}$$
$$= [20 \log G_1 + 20 \log G_2 + 20 \log G_3 + 20 \log G_4 + \ldots .] \text{ dB}$$

which shows that the total gain is found by adding the decibel gains of the individual stages.

6.3. Energy Level Diagrams

Using the decibel notation, the rise and fall of signal level through a communication system can be indicated relative to its level at the

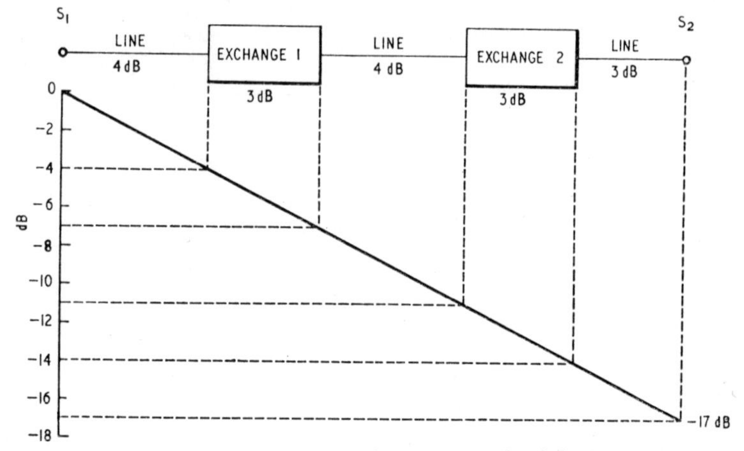

Fig. 6.2. dB notation used in an energy level diagram

origin of the signal. Fig. 6.2 shows the fall of signal level from a telephone subscriber S_1, to a second subscriber S_2, connected to the first via two telephone exchanges. The fall of energy level is marked on each section. Line losses naturally depend on the distances covered but, if the loss is likely to exceed about 4·5 dB, line amplifiers are needed to replace some of the loss. Losses due to exchange equipment are shown as 3 dB. The total fall from S_1 to S_2 is seen to be 17 dB.

Fig. 6.3 is an energy level diagram for a line which includes repeaters (amplifiers) approximately every 20 miles to compensate for line loss.

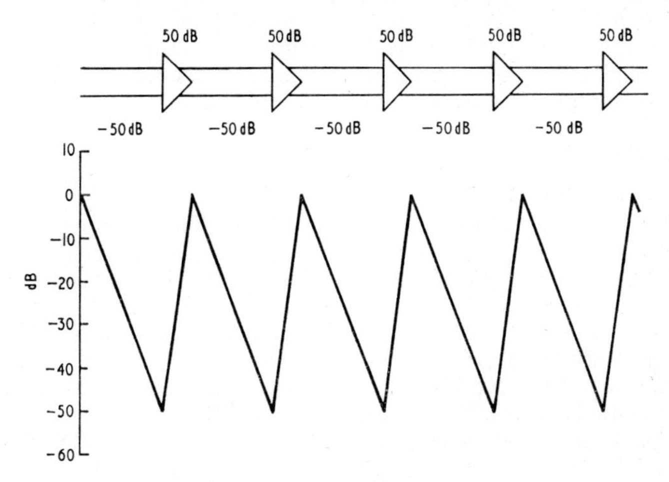

Fig. 6.3. Addition of gains and losses in a trunk line

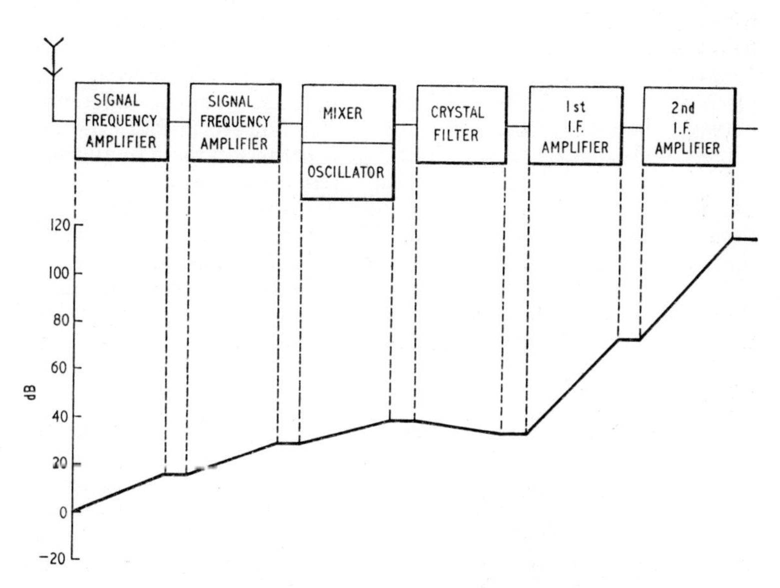

Fig. 6.4. Addition of receiver stage gains

In a similar way an energy level diagram might be drawn for two or more stages of a radio receiver.

Fig. 6.4 traces possible changes of signal level up to the detector stage of a receiver. From such a diagram it is clear at which stages the maximum gain is provided. Only the crystal filter incurs a loss of energy, since at this stage no amplification is provided.

6.4. Response Curves

6.4.1. CHOICE OF SCALES

It is often necessary and useful to show graphically how the gain of an amplifier or a receiver varies with frequency. The graph is most useful if it shows the way in which the ear interprets a change in response caused by a given change in input frequency. We have already seen that the decibel unit compares realistically different power amplitudes because loudness tends to vary as the logarithm of the sound wave intensity. The amplitude scale of the response curve should therefore be marked in decibels.

But the frequency scale deserves a logarithmic representation too. Fig. 6.5 shows two ways of dividing a scale which is five inches long into a frequency range from 0 to 10 kc/s. Scale A is a linear scale in which equal distances represent equal changes in frequency. Scale B is a logarithmic scale in which equal distances represent equal differences in the logarithms of the frequencies. It is scale B which most truly shows the effect on the ear of a given change in frequency. Sound frequencies of 100 c/s, 200 c/s, 400 c/s, 800 c/s, 1,600 c/s and so on appear to the ear to be uniformly spaced. The pitch seems to go up by uniform increments. Yet the differences between these frequencies are not similar but are in fact: 100 c/s, 200 c/s, 400 c/s, 800 c/s, etc. But the logarithms of these frequencies differ by a constant increment. Each frequency is twice the previous one so that each frequency has a logarithm which is 0·3021 greater than that of the preceding one. Therefore equal distances along scale B represent equal changes in pitch as appreciated by the ear, although this would not be so for scale A. Logarithmic scales should thus always be used for frequency in audio-response graphs, while the ordinate scale should be marked in dB.

6.4.2. RELATIVE RESPONSE

While it would be possible to plot a response graph for an amplifier using the voltage gain expressed in decibel notation as the ordinate scale, a relative response scale is often favoured. In a relative response curve the output at any given frequency is compared with the output at the frequency marked at the middle of the frequency scale. In most audio amplifiers the mid-frequency gain is the largest gain and the gains at other frequencies are expressed as so many dB below the mid-frequency output.

The data from which a response curve can be drawn is obtained by keeping the input voltage to an amplifier constant in amplitude

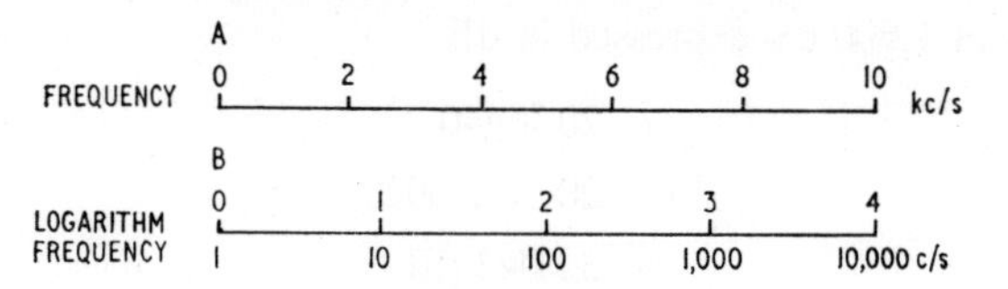

Fig. 6.5. Comparison of direct and logarithmic frequency scales

while changing its frequency in suitable increments over the required frequency range and measuring the output voltage at each new frequency of input.
Let

the input voltage be v_i,
the output voltage at the mid-frequency v_m,

and

the output voltage at some frequency f, v_f volts.

Then the gain at the mid-frequency is $20 \log\left(\frac{v_m}{v_i}\right)$ dB and the gain at f c/s is $20 \log\left(\frac{v_f}{v_i}\right)$ dB.
The relative response at f c/s is therefore,

$$20 \log\left(\frac{v_m}{v_i}\right) - 20 \log\left(\frac{v_f}{v_i}\right)$$

$$= 20 \left(\log\frac{v_m}{v_i} - \log\frac{v_f}{v_i}\right)$$

$$= 20 \log\left(\frac{v_m}{v_i} \Big/ \frac{v_f}{v_i}\right)$$

$$= 20 \log\left(\frac{v_m}{v_f}\right) \text{ dB}$$

Example 6.1

A transformer-coupled amplifier has a voltage gain of 15 at 50 c/s. Its response at 1,000 c/s gives a voltage gain of 40. What is the relative response at 50 c/s in dB to its response at 1,000 c/s?
The gain at 50 c/s expressed in dB

$$= 20 \log 15$$
$$= 20 \times 1{\cdot}1761$$
$$= 23{\cdot}522 \text{ dB}$$

The gain at 1,000 c/s expressed in dB

$$= 20 \log 40$$
$$= 20 \times 1{\cdot}6021$$
$$= 32{\cdot}042 \text{ dB}$$

The response at 50 c/s relative to the response at 1,000 c/s is thus

$$23{\cdot}522 - 32{\cdot}042$$
$$= -8{\cdot}52 \text{ dB.}$$

An alternative approach which is a little shorter is as follows.

$$\text{The relative response} = 20 \log\left(\frac{\text{output at 50 c/s}}{\text{output at 1,000 c/s}}\right)$$

(assuming the input voltage to be the same at both frequencies)

$$= 20 \log\left(\frac{15}{40}\right)$$
$$= -20 \log\left(\frac{40}{15}\right)$$
$$= -20 \times 0{\cdot}426$$
$$= -8{\cdot}52 \text{ dB}$$

An example of a gain-frequency graph and its corresponding relative response curve is now considered.

Table 6.1

Frequency (c/s)	*Gain*	*Frequency* (c/s)	*Gain*
10	36	600	85
20	50	700	85
30	64	800	85
40	72	900	85
50	77	1,000	83
60	80	2,000	80
70	81	3,000	76·5
80	82	4,000	75
90	83	5,000	72·5
100	84	6,000	70
200	85	7,000	68
300	85	8,000	65
400	85	9,000	62
500	85	10,000	60

Example 6.2

Table 6.1 shows the voltage gains obtained with a resistance-capacitance coupled amplifier for frequencies between 10 c/s and 10 kc/s. Plot the gain/frequency characteristic of the amplifier and its relative response curve.

Fig. 6.6 (a) can be plotted from the values given, using the gain as the ordinate value for $\frac{v_2}{v_1}$ where v_2 is the output voltage and v_1 is the input voltage of the amplifier. The frequency scale is logarithmic.

Before the relative response can be drawn, we must find the gain at each frequency in dB relative to the maximum gain which is seen

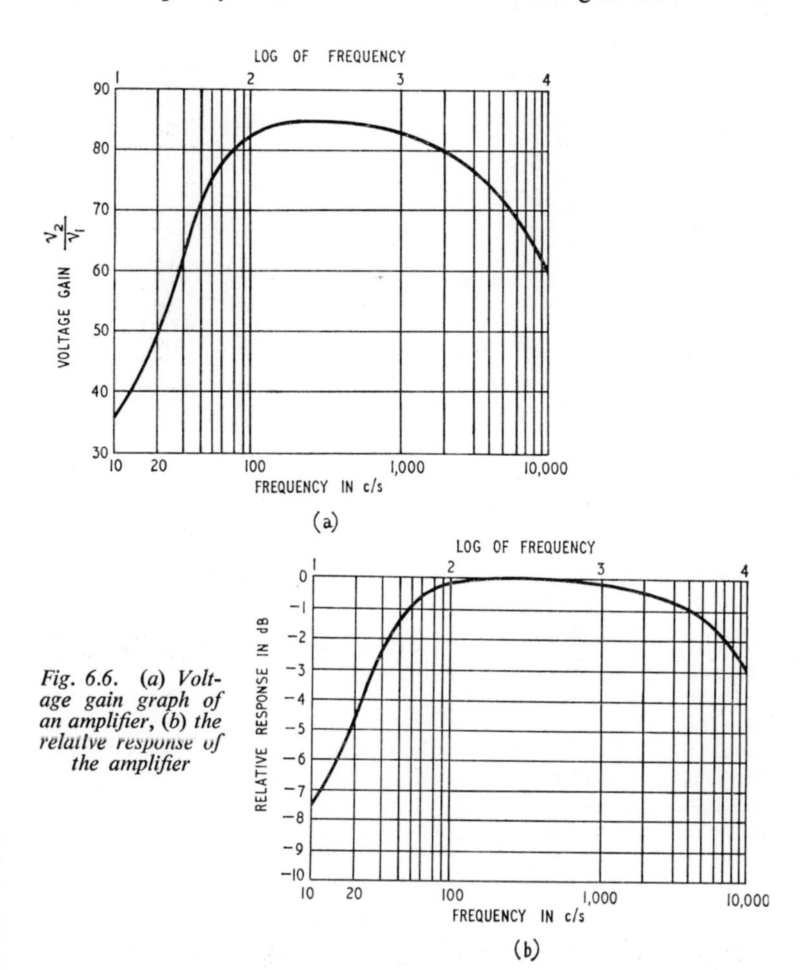

Fig. 6.6. (a) Voltage gain graph of an amplifier, (b) the relative response of the amplifier

to be 85, e.g. at 10 c/s the gain is 36. Relative to the maximum gain this is

$$20 \log\left(\frac{85}{36}\right) \text{ dB down, or } -7{\cdot}46 \text{ dB.}$$

Similarly for other frequencies the relative responses in dB below 85 are found and then plotted against the logarithm of the frequency as in Fig. 6.6 (b).

6.4.3. RELATIVE RESPONSE OF R.F. CIRCUITS

A graph can be drawn to show how the response of a tuned r.f. circuit varies as the frequency of the applied voltage is changed,

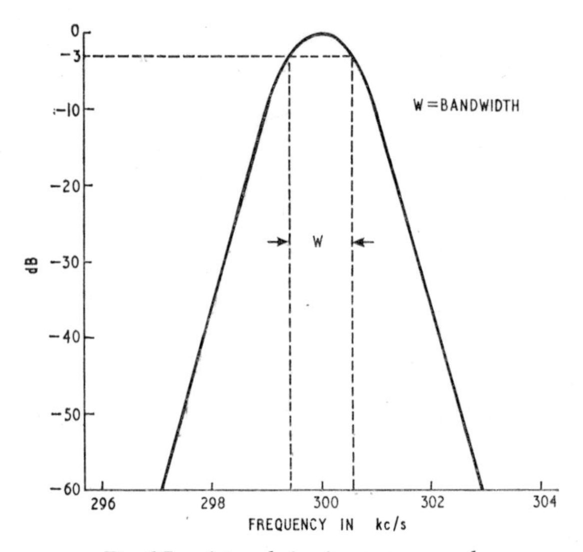

Fig. 6.7. A tuned circuit response graph

marking the ordinate scale in dB relative to the response at the resonance frequency. This is illustrated in Fig. 6.7.

If the circuit is a series or acceptor circuit, the data for the graph may be obtained by varying the frequency of the applied e.m.f. over the required frequency range while keeping the amplitude of the voltage constant and measuring the current amplitude for each frequency applied. If i_0 is the current measured at resonance and i_f is the current measured at some other frequency f c/s, then the relative response at f c/s is $-20 \log\left(\frac{i_0}{i_f}\right)$ dB.

Should the circuit be a parallel or rejector circuit then the supply current should be kept constant while the frequency is varied through the appropriate range and the voltage across the circuit is measured for each frequency applied. If v_0 is the voltage across the rejector circuit at resonance and v_f is the voltage for the same current at some other frequency f c/s, then the relative response at f c/s is $-20 \log\left(\frac{v_0}{v_f}\right)$ dB.

The effective bandwidth of the circuit is often taken as the difference in frequency between the two points on the response curve which are 3 dB below the response at resonance.

6.5. Other Applications

6.5.1. SIGNAL-TO-NOISE RATIO

In addition to the wanted signal arriving at the output of a communication system there is an unwanted voltage owing to a number of possible causes in the transmission system and which is classified as *noise voltage*. With aural reception this noise voltage may be heard as a faint background hiss or crackle, or under bad conditions as a loud and annoying noise which even tends to mask the wanted signal. A comparison of the signal power with the power due to random noise sources is the signal-to-noise ratio. If the r.m.s. signal voltage is v_s and the r.m.s. noise voltage is v_n, the signal noise ratio is given by

$$20 \log\left(\frac{v_s}{v_n}\right) \text{ dB}$$

For example the signal-to-noise ratio of the output of a receiver might be 20 dB. (*See also* Chapter 7.)

6.5.2. GAIN CONTROL

The range of control which a volume control of an amplifier or receiver has on the output may be expressed in dB. If for example the range is 100 dB, then

$$100 = 20 \log\left(\frac{\text{maximum gain}}{\text{minimum gain}}\right)$$

In this instance the ratio of maximum voltage gain to the minimum voltage gain is 10^5.

6.6. The Neper

The *neper* is a logarithmic unit employed for expressing the attenuation of currents or voltages along a transmission line or cable.

The rate of fall of voltage and current along a line is proportional to the line voltage and current at each point. Where the voltage is high the rate of decrease of voltage is also high. Where the voltage is smaller, the rate of fall of voltage along the line is smaller. The current value falls similarly with distance. Both voltage and current are said to fall exponentially.

The fall of line voltage can be represented by the graph of Fig. 6.8 whose equation is

$$V = V_1 e^{-Pl}$$

where V_1 is the line voltage at the beginning of the line,

l is the distance from the beginning of the line,

P is a constant called the *propagation constant* of the line, and

e is the constant 2·71828 which is the base of *Naperian* logarithms.

On re-arranging the above equation,

$$\frac{V_1}{V} = e^{Pl}$$

We have

$$\log_e\left(\frac{V_1}{V}\right) = Pl$$

since the logarithm of a value to a base e is the power to which e must be raised to equal that value. If only the difference in amplitude

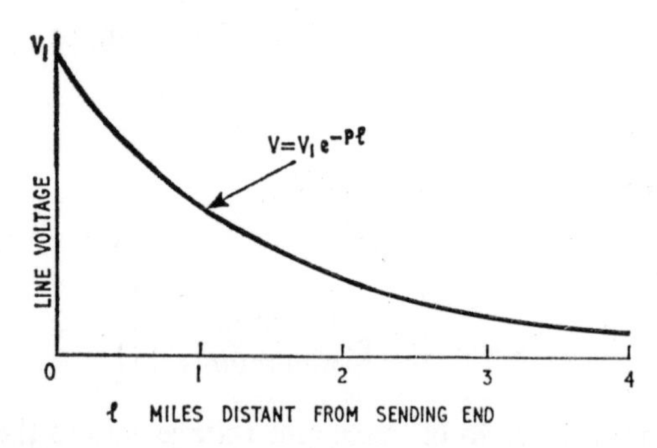

Fig. 6.8. The decay of signal voltage along a line due to loss

between V and V_1 is considered and the phase difference between them is neglected, P is the *attenuation constant* and is usually expressed in nepers per mile. For l miles the attenuation is Pl nepers which is seen to equal the *Naperian* or *Natural* logarithm of the ratio of the voltages at the beginning and the end of the section of the line l miles in length.

Similarly, if I_1 is the current at the beginning of a section of line and I_2 is the current at the end of l miles of line then the attenuation is

$$\log_e\left(\frac{I_1}{I_2}\right) \text{ nepers.}$$

The neper is primarily intended for a voltage or current comparison, but if we wish to express a power ratio in nepers we have

$$P_1 = \frac{V_1^2}{Z_0},\ P_2 = \frac{V_2^2}{Z_0}$$

where Z_0 is the line impedance

$$\therefore \qquad \frac{P_1}{P_2} = \frac{V_1^2}{V_2^2}$$

so that

$$\frac{V_1}{V_2} = \sqrt{\frac{P_1}{P_2}}$$

and

$$\text{attenuation} = \log_e\sqrt{\frac{P_1}{P_2}}$$

$$= \tfrac{1}{2}\log_e\left(\frac{P_1}{P_2}\right) \text{ nepers.}$$

Where P_1 and P_2 are the powers compared.

Line attenuation constants are assumed, however, to be expressed in nepers/mile unless they are stated to be in decibels/mile.

Nepers can be converted to decibels by multiplying by 8·686.

Let a power ratio be A dB, or B nepers, and let the voltages for the two powers be V_1 and V_2 across similar impedances.

Then
$$A = 20\log_{10}\left(\frac{V_1}{V_2}\right)$$

and
$$B = \log_e\left(\frac{V_1}{V_2}\right)$$

$$\therefore \qquad \frac{A}{20} = \log_{10}\left(\frac{V_1}{V_2}\right)$$

or

$$10^{A/20} = \frac{V_1}{V_2} \text{ and } e^B = \frac{V_1}{V_2}$$

$$\therefore \qquad 10^{A/20} = e^B$$

Taking logarithms of both sides to the base 10,

$$\frac{A}{20}\log_{10}10 = B\log_{10}e$$

$$\therefore \quad \frac{A}{B} = 20 \frac{\log_{10} e}{\log_{10} 10} = 20 \times 0{\cdot}4343$$

$$= 8{\cdot}686$$

$$\therefore \quad A \text{ dB} = 8{\cdot}686\, B \text{ nepers.}$$

Example 6.3

A cable used for the wire relay of television programmes has an attenuation of 1·6 dB/100 yd at 3·75 Mc/s. If the input voltage is 1 mV at this frequency find the voltage one mile from the sending end of the cable and state the attenuation per mile in nepers (1,760 yards = 1 mile).

$$\text{Attenuation in dB} = \frac{1{,}760}{100} \times 1{\cdot}6 \text{ dB}$$

$$= 17{\cdot}6 \times 1{\cdot}6$$

$$= 28{\cdot}16 \text{ dB}$$

Let V_2 = voltage at the end of 1 mile in millivolts.

Then

$$20 \log \frac{1}{V_2} = 28{\cdot}16 \text{ dB}$$

$$\log \frac{1}{V_2} = 1{\cdot}408$$

$$\therefore \quad \frac{1}{V_2} = 25{\cdot}59$$

$$\therefore \quad V_2 = \frac{1}{25{\cdot}59} \text{ mV}$$

$$= 0{\cdot}0392 \text{ mV}$$

$$\text{or } 39{\cdot}2\ \mu\text{V}$$

The attenuation in nepers is

$$\frac{28{\cdot}16}{8{\cdot}686} = 3{\cdot}242 \text{ nepers}$$

Questions

1. Explain what is meant by a decibel. The gain-frequency characteristics of two amplifiers are given in the following table:

Frequency (kc/s)	60	76	92	108
Gain of amplifier 1	310	330	340	390
Gain of amplifier 2	330	345	380	325

If the two amplifiers are connected in tandem separated by an attenuator having a loss of 15 dB, plot the overall gain-frequency characteristic of the combination, expressing the gain in dB. It may be assumed that the amplifiers and attenuator have the same input and output impedances. (*C & G*, 1959.)

2. In an audio-frequency response graph of an amplifier a logarithmic scale is used both for the gain and the frequency scales of the graph. Explain carefully why this is appropriate.

3. A network of resistors having equal input and output resistance value produces an attenuation of 25 dB. State (a) the ratio of input to output voltage, (b) the ratio of input power to output power.

4. The voltage gain of an audio amplifier at its mid-frequency is 150 but the gain falls to 106 at 50 c/s. State the relative response of the amplifier at 50 c/s. If the relative response at 15 kc/s is −6 dB, what is the voltage gain at this frequency?

5. Explain what is meant by the term " signal-to-noise ratio " in connection with a communication system and discuss its importance. A carrier is amplitude modulated by a sine-wave tone to a depth of 30 per cent and is applied to a receiver. The output signal-to-noise ratio is 20 dB. Assuming that the whole of the noise arises in the receiver input circuit, and that a linear detector is used, how is the signal-to-noise ratio affected: (i) when the transmission path loss is decreased by 3 dB, (ii) when the depth of modulation is increased to 60 per cent, (iii) when the receiver audio-frequency gain is increased by 6 dB?

6. An *L–C* acceptor circuit has a resistance of 25 ohms. What is the impedance of the circuit at frequencies for which the power expended is (a) half that of the power expended at resonance, and (b) a quarter of the power expended at resonance, the e.m.f. being the same for all frequencies considered. State the relative response of the circuit for each of the above frequencies in dB.

7. Under conditions of deep fading at night a signal from a distant medium-frequency broadcast station fluctuates between 4 mV/m and 200 μV/m. Express this signal strength change in dB.

8. A low-frequency amplifier has a gain of 56 dB. The input circuit is 600 ohms resistive impedance and the output is arranged for a load of 10 ohms. What will be the current in the load when an alternating voltage of 1 volt is applied at the input?

9. Show that an attenuation in nepers can be expressed in decibels by multiplying the number of nepers by 8·686.

10. A transmission line has an attenuation of 2 dB/mile and is 6 miles long. Express the total attenuation (i) in dB, (ii) in nepers, and state the ratio of the input voltage to the output voltage.

7

Noise in Communication Systems

Any spurious signal in communication systems is called noise. It is usually undesirable, though in certain circumstances it is sometimes deliberately generated.

The output due to noise tends to confuse or mask the desired output of the system. It causes audible noise in the output of a radio receiver, and white or black spots (depending on the system of modulation used) on the image on a television screen.

7.1. General Classification of Noise

Noise voltages often cover an extremely wide range of frequencies; in some instances the whole spectrum from zero to infinity, in others only a section of the range. Noise originating independently of the communication equipment is called *external noise.* If it is generated within the equipment it is termed *internal noise.*

7.2. External Noise

Phenomena in which electric sparks occur or where there is a rapid change of current, such as in lightning discharges, switching devices, lighting signs, rectifiers, car ignition systems, electric generators and motors are common causes of external noise.

External noise may reach the equipment by radiation or induction to the line, aerial system or equipment wiring or may be transmitted to the equipment by conduction along the mains supply leads.

7.2.1. REDUCTION OF EXTERNAL NOISE AT RECEIVING EQUIPMENT

Pick-up of external noise may be reduced or avoided in a number of ways. One of these is to place the receiver aerial an adequate distance from any known localised sources of noise and to screen the feeder connecting aerial and receiver. Similarly, for line systems the line should be spaced as far as necessary from interference sources such as electric traction systems.

When the frequency of operation of radio equipments is sufficiently high to enable the receiver aerial system to be made directive, the aerial can be aimed so as to minimise pick-up from noise sources.

Amplifying equipment, particularly the input and early stages (which, being followed by the most amplification, are most likely to cause a large noise signal in the output), may be screened.

The polarisation of radio systems may be chosen to avoid the worst effects of the forms of external interference most likely to be encountered. (One of the factors determining the choice of horizontal polarisation for early radar equipments was that most man-made interference likely to be encountered was vertically polarised.)

If a frequency-modulated system is employed instead of amplitude-modulation there is less noise in the output at the receiving end. (In the amplitude-modulated system already considered the amplitude of the transmitted carrier is varied in accordance with the information to be communicated: in frequency-modulation the amplitude of the carrier is maintained constant but the frequency is varied in accordance with the information to be communicated.)

As already noted in Chapter 3 a system of amplitude-modulation in which one sideband only is transmitted results in a marked reduction in noise at the output of the receiving equipment.

Should the bandwidth of amplifiers be narrowed, the quality of the wanted signal is reduced but at the same time the noise is diminished in proportion to the bandwidth reduction. Reduction of bandwidth to a certain value, which diminishes with increase in noise level, results in an improvement in the general acceptability of the output signal.

If change of frequency is practicable a change to a higher frequency of radio transmission may result in a lessening of the noise.

Filters, similar to those described in the next section, may be placed in the supply leads to prevent the entry of noise voltages from the supply.

7.2.2. SUPPRESSION OF INTERFERENCE AT THE SOURCE

Some forms of interference, for example that caused by lightning discharges, cannot be suppressed; their effects can only be minimised as much as possible by methods such as those outlined in the previous section. Effects of some forms of interference can, however, be reduced considerably by appropriate action at the source, as follows:

1. Direct radiation from sparking contacts may be suppressed by screening the offending equipment. Thus, vibrators used in power units are contained in earthed metal cans of good conductivity.
2. Radiation from leads to sparking contacts or to equipment in which sparking is occurring, may be reduced considerably by decoupling the leads as near to the source of the spark as possible. Decoupling filters of this kind usually consist of two capacitors (of about 0·1 μF) in series across the leads and centre-tapped to earth, with the addition, if necessary, of small inductors (of about 1 mH) in the leads. Such filters are themselves usually screened to prevent radiation.
3. Conduction of interference at lower frequencies along the supply lines from the offending equipment may be prevented or

minimised by decoupling. Chokes are placed in the supply leads to the equipment as close to it as possible and capacitors, centre-tapped to earth, are connected across the supply. The L and C component values are of the order of a few henrys and a few microfarads respectively.

7.3. Internal Noise

Noise which is generated within an equipment may originate in a number of different ways which are treated separately in the following section.

7.3.1. THERMAL NOISE

Noise of this type, also referred to as *Johnson noise*, or *circuit noise*, is the result of the random motion of electrons which occurs in any circuit even in the absence of an applied e.m.f. Over a period of time these random currents (for the motion of an electron is a current) cause no net potential difference across the circuit; instantaneously, however, they do and the result is the development of very small voltages having components at all frequencies in the spectrum from zero to infinity. (For obvious reasons noise covering such a spectrum is sometimes called *white noise*.) The magnitude of the noise voltage increases with the resistance of the circuit and with temperature. The noise voltage in the output of the equipment is proportional to the bandwidth. (Because, although the noise occurs at all frequencies, only those voltages accepted by the equipment, i.e. those within the pass-band, appear in the output.)

The r.m.s. value of the noise voltage is given by

$$V_{\text{r.m.s.}} = 2\sqrt{kRT(\Delta f)} \text{ volts}$$

where k is Boltzmanns constant ($1{\cdot}38 \times 10^{-23}$ joules/°K)

R is the resistance of the circuit or component concerned

T is the temperature of the circuit or component in °K

Δf is the pass-band of the equipment in c/s; as an approximation it is often adequate to take Δf as being the width between half-power points.

Example 7.1

Calculate the r.m.s. value of thermal-noise-voltage generated in a one-megohm resistor at 40° C in an amplifier of bandwidth 25 kc/s.

$$V_{\text{r.m.s.}} = 2\sqrt{1{\cdot}38 \times 10^{-23} \times 10^{6} \times 313 \times 25 \times 10^{3}} \text{ volts}$$

$$= 20{\cdot}8\ \mu\text{V}$$

As the above example shows, noise of this type is of low level and unlikely to be troublesome unless the signals to be handled are also of very low level. If they are it is necessary to reduce to a minimum all possible noise sources and the use of resistors of high ohmic value must be restricted.

7.3.2. NOISE IN CARBON RESISTORS

The noise voltage referred to in Section 7.3.1 occurs because of the random motion of electrons. In carbon resistors the presence of carbon granules results in a much higher noise level and the use of such resistors in the early stages of equipments in which the noise level must be kept very low should, therefore, be avoided.

7.3.3. VALVE AND TRANSISTOR NOISE

Because of the particle nature of electrons the flow of current across a valve from the cathode is not entirely smooth. A very small fluctuation takes place about the mean value of current and this random variation contains components of current at all frequencies. The resulting noise, *shot noise*, is thus similar in its effects to thermal noise.

Additionally, if the current is divided by the presence of grids which are positive to the cathode the noise increases (*partition noise*). Thus multi-electrode valves are noisier than triodes.

In transistors, likewise, the current is due to individual discrete charges (electrons and holes) so that a similar effect to that described above as shot noise occurs.

Also in transistors the rate at which current carriers (electrons and holes) are generated and recombine fluctuates in a random manner and results in noise.

Again, current carriers adjacent to junctions are carried across very rapidly and thus create current pulses which reveal themselves in the output as noise.

The level of noise introduced in transistors varies with frequency, source resistance, collector voltage and emitter current in a complicated manner.

The magnitude of noise voltage introduced by transistors and valves may be specified in a number of ways. One way is to quote the value of resistance which, placed at the input to the same valve or transistor (now assumed to be noiseless), yields the same noise level in the output. This value of resistance is known as the *equivalent noise resistance*.

In Section 7.5 the term *noise factor* is introduced and a good minimum value of this for transistors is quoted as 2 dB.

7.3.4. OTHER FORMS OF INTERNAL NOISE

Loose or poor contacts contribute a considerable quota of noise to the output of an equipment.

The electrodes of valves, particularly those early in an amplifier chain, which are not sufficiently rigid may generate a signal due to the fact that the spacing between them varies. This is known as *microphonic noise*. Transistors do not generate this form of noise.

Inadequate smoothing of a.c. power supplies (*Radio and Line Transmission, Volume* 1, Chapter 16), leakage between heater and cathode, stray fields from inductors and transformers all cause the introduction of another form of noise, hum.

7.4. Signal-to-noise Ratio

The relative magnitudes of the specified feature of the wanted signal and the appropriate feature of the noise response is often expressed as a ratio, the *signal-to-noise* ratio, usually expressed in dB, for specified conditions of sensitivity, bandwidth and input level.

Acceptable signal-to-noise ratios vary enormously according to the type of service. At one extreme, a radio-telegraphic link, under bad but still usable conditions, may have signal and noise in approximately equal proportions, i.e. a ratio of unity (0 dB); on the other hand for sound broadcasting reception under conditions of low noise the ratio may be 60 dB.

7.5. Noise Factor

Signal-to-noise ratio is often a useful criterion of the extent to which a signal or the output of an equipment is free from noise. The ratio, however, does not reveal directly the noise quality of an equipment: an estimate of this requires a comparison of the signal-to-noise ratio at the input with that at the output of the equipment, or, a knowledge of the noise at the input to the equipment and of the noise at the output over and above that due to the input noise.

The figure which gives this information is the *noise factor* (sometimes called the *noise figure*): this specifies the extent to which an equipment, or item such as a valve or transistor amplifier, causes a deterioration in the signal-to-noise ratio. The noise factor can be expressed as the ratio of the total mean-square noise output e.m.f. to that part of it which is due to the thermal noise of the source circuit. (For fuller information see *British Standard* 2065: 1954.) Thus, if an equipment or item were to add no noise to a signal the noise factor would be unity (0 dB). Any amount of noise which is added in the equipment results in a noise factor of more than 0 dB, the smaller the amount of added noise the nearer is the noise factor to 0 dB. As an illustration of orders of magnitude the noise factor of 2 dB quoted in Section 7.3.3 as a good minimum value for valves and transistors may be mentioned.

Alternatively the noise factor can be stated as the ratio of the signal-to-noise ratio at the input to that at the output of the equipment being considered.

7.6. Effects of the Influence of Noise on the Design and Operation of Communication Equipment

Some noise is inevitably present at the input to any amplifier. Thus, although there is virtually no limit to the gain which can be built into an amplifier there is an upper limit to the level of amplification which may usefully be applied. Consider, for instance, an amplifier the minimum noise level to the input of which is 5 μV and the required output from which is 10 V at a signal-to-noise ratio not inferior to 5 : 1. The minimum useful signal input is [(5 $\times$ 5) μV $=$ 25 μV] so that the maximum useful gain is $\frac{10\ \text{V}}{25\ \mu\text{V}}$, that is, 400,000. The provision of any additional gain is superfluous.

Noise introduced at the source or in early stages of an amplifying system produces in the output a greater noise voltage than it would if introduced at a later stage. Thus, equipments in which much gain is required (that is, those to which the input signal may be very small, such as certain radar and communication equipments operating at very high frequencies) are designed so that the noise introduced in the first stage, in particular, shall be reasonably small compared with the external noise at the input. For this the first, and perhaps the second, amplifier is chosen to be as noise-free as possible:

(1) by the use of high quality resistors of low noise level (and therefore not too high in value and of low-noise manufacture),

(2) by the use of low-noise valves (including, therefore, the avoidance of multi-electrode valves because of partition noise) of rigid construction to prevent the slightest microphony. When transistors are used they are of low-noise type,

(3) by employing maximum selectivity (i.e. minimum bandwidth), consistent with necessary requirements to provide the quality of signal needed in the output, in order to restrict the band in which noise components are amplified,

(4) by avoidance of the possibility of the introduction of hum (*see* Section 7.3.4 of this volume *and* Chapter 16 of *Radio and Line Transmission, Volume* 1).

Radio receivers for the reception of such signals are of the superheterodyne type. As much amplification as possible precedes the frequency changer (*see* Chapter 13) because multi-electrode frequency changers are noisy (partition effect) and any frequency changer increases the likelihood of noise. This is because noise voltages which themselves are outside the pass-band of succeeding amplifiers may beat between themselves to produce frequencies within the pass-band. Receivers designed for good performance at the higher frequencies (say 15 Mc/s upwards) usually have one or two stages of signal-frequency amplification before the frequency changer.

At lower radio frequencies the level of external noise (and often also that of the wanted signal) is probably higher than that at the higher frequencies. There is often, therefore, little to be gained by seeking very low-noise-gain in the early stages. Such equipments, therefore, commonly employ a frequency changer in the first stage.

Telephone lines are susceptible to the introduction of noise from nearby power cables and from electric traction systems. (An example has been quoted of noise voltages as great as 28 volts at 50 c/s, 9·5 volts at 150 c/s and 2·2 volts at 1,050 c/s induced in a railway telephone line from an adjacent overhead 50 c/s electric railway supply.) Clearly, telephone lines should be sited with care.

Agreed standards exist as to permissible noise levels in telephone circuits and the amount of noise which given items of equipment may introduce. (For instance, an internationally agreed standard for 2,500-kM length multi-channel telephone circuits is that noise should not exceed 10,000 picowatts in any audio channel for more than 1 per cent of a busy hour: of this not more than 2,500 picowatts may be introduced by terminal multiplexing equipment.) Lines and equipments must conform to the standards laid down for a particular circumstance.

Questions

1. In what ways may noise be generated within an amplifier?

2. Explain how certain noise sources may be suppressed.

3. In what ways can resistors contribute to the noise level in an amplifier? What precautions can be taken to reduce to a minimum noise due to resistors in high gain amplifiers?

4. A tuned circuit has $L = 100\ \mu$H, $C = 100$ pF. The loss resistance of the inductor is 10 ohms and that of the capacitor is negligible. Calculate the r.m.s. value of the noise voltage at the resonance frequency of the circuit if the temperature is 20° C. Assume that the pass-band is that between half-power points.

5. (a) Explain why there is a limit to the amount of amplification which may usefully be made available in any amplifier.

(b) Communication-type radio receivers are usually provided with at least two gain controls; one controlling the gain of the r.f. amplifiers and the other the level of signal fed to the a.f. stages. Explain why this is a better arrangement than that in which the only control is that which operates at the input to the first a.f. stage.

6. List some of the types of noise which may enter radio receiving equipment from outside sources. Explain how the effects of these types of noise may be minimised.

7. (a) In what ways may noise voltages be developed in valves and transistors?

(b) Distinguish between signal-to-noise ratio and noise factor.

8. (a) Define the following terms: circuit noise, white noise, shot noise, partition noise.

(b) Why is it that vacuum cleaners and hair driers are potential sources of noise?

9. What do you understand by the term random noise? Name one type of noise which is of a random nature and one which is not. Give reasons for your choice.

8

Audio Frequency Amplification

8.1. Class-A Amplifiers

A class-A amplifier is one in which current flows in the output circuit throughout each cycle of input signal. In this chapter we are concerned solely with amplifiers of this type. Later work will include a study of amplifiers in which current in the output circuit is cut off during part of each cycle.

8.2. Determination of the Amplification of a Single-stage Resistance-loaded Thermionic-valve Amplifier

The fundamental principles of operation of resistance-loaded amplifiers are outlined in *Radio and Line Transmission, Volume* 1.

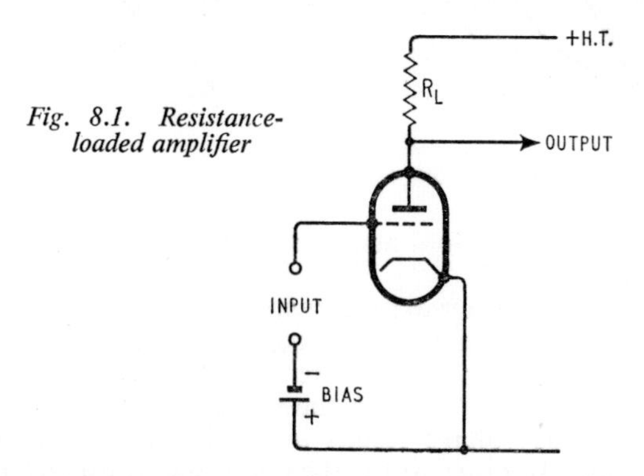

Fig. 8.1. Resistance-loaded amplifier

In brief: a varying voltage applied between the grid and cathode of a suitably-biased valve causes changes in anode current. These in turn result in corresponding voltage changes across the anode resistance load. These are of the same form as the input but of greater magnitude and inverted. The ratio of the output to the input voltage is the gain of the stage and is given by the expression

$$A = \frac{\mu R_L}{r_a + R_L} \qquad (1)$$

where A is the gain, μ the amplification factor and r_a the a.c.

resistance of the valve. R_L is the anode load. The basic circuit is in Fig. 8.1.

In *Radio and Line Transmission, Volume* 1 the circuit is analysed in terms of the mutual (or anode current-grid voltage) characteristic of the valve. It will now be considered with the aid of the output (or anode current-anode voltage) characteristic.

8.2.1. OUTPUT CHARACTERISTICS: TRIODE AND PENTODE

Typical curves for a triode are in Fig. 8.2. The curves are of similar shape and are approximately equally spaced and parallel.

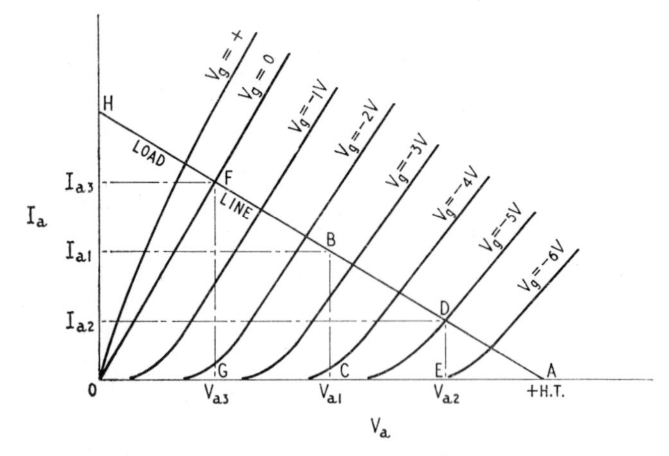

Fig. 8.2. Triode: anode current–anode voltage curves; load line

They are curved at the foot and reasonably straight over the rest of the operating range.

The corresponding curves for a pentode are in Fig. 8.3. These are of very different form and are less equally spaced, becoming increasingly crowded together at the more negative grid voltages. In the region to the right of the knee the slope is small, that is, a large increase in anode voltage results in a relatively small increase in anode current. In other words the anode a.c. resistance is large.

8.2.2. LOAD LINE

The characteristics shown in Figs. 8.2 and 8.3 are the static characteristics; static because it is assumed that a change in one potential or current will not cause changes elsewhere. In practice the changes are not independent. If the static curves are used to evaluate the action of a complete stage without making due allowance for the secondary effects the results will be inaccurate and optimistic. In *Radio and Line Transmission, Volume* 1, Chapter 10, the matter is dealt with by deriving from the mutual static curves the dynamic

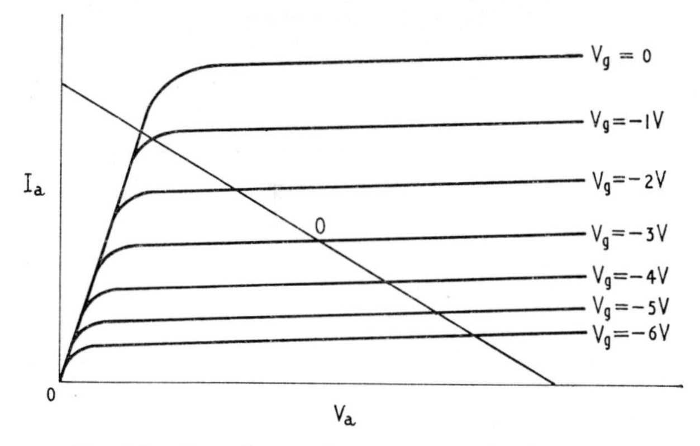

Fig. 8.3. Pentode: anode current–anode voltage curves

characteristic for the valve under the conditions of its particular use. Another method is to use the static output characteristics in conjunction with the appropriate load line.

In Fig. 8.2 *AH* is the load line for the assumed value of H.T. supply and load resistance. It terminates in the anode voltage axis in a point *A* (corresponding to the H.T. supply voltage) and is drawn at the slope appropriate to the load R_L (so that, for instance, $\frac{CA \text{ (volts)}}{BC \text{ (amps)}} = R_L$).

Before proceeding further it may be advisable to refresh our memory of the use of load lines by working the following example.

Example 8.1

Sketch load lines for the values of R_L and H.T. supply given in Table 8.1.

Table 8.1

No.	R_L (Kilohms)	*H.T.+* (Volts)
1	10	200
2	10	150
3	20	200
4	100	150
5	zero	200

The lines are shown in Fig. 8.4. Note:

(a) (1) and (2) are parallel because they represent the same value of load resistance.

(b) (1), (3) and (5) terminate in the same point on the anode voltage axis because the H.T. voltage is the same.

(c) (4) is nearly horizontal because it represents a very large resistance.

(d) (5) is vertical because it represents zero resistance: thus the

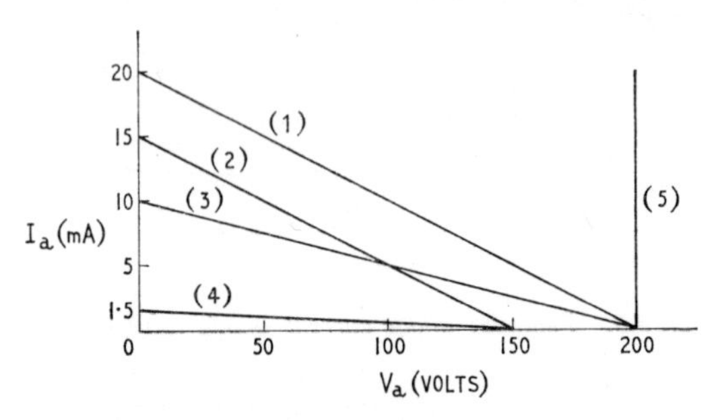

Fig. 8.4. Illustrating Example 8.1

anode voltage is equal to that of the supply whatever the value of the anode current.

Example 8.2

Under the conditions illustrated in Fig. 8.2 determine (a) the anode voltage, (b) the anode current and (c) the voltage across the anode load resistor when the grid voltage is (i) zero, (ii) $-2\frac{1}{2}$ volts.

(i) The operating point is *F*, therefore,

(a) anode voltage $= V_{a3}$

(b) anode current $= I_{a3}$

(c) voltage across anode load $=$ (H.T.+) $- V_{a3} = GA$

(ii) The operating point is *B*, therefore,

(a) anode voltage $= V_{a1}$

(b) anode current $= I_{a1}$

(c) voltage across anode load $=$ (H.T.+) $- V_{a1} = CA$

8.2.3. USE OF LOAD LINE TO DETERMINE AMPLIFICATION OF RESISTANCE-LOADED THERMIONIC VALVE AMPLIFIER

It has been shown in Chapter 10 of *Radio and Line Transmission, Volume* 1, that the gain of a stage is influenced by the extent to which the anode load is modified by other parts of the circuit—coupling capacitor, grid resistor, input capacitance to the next stage, stray capacitances, etc. In what follows all such effects will be ignored. In Section 8.3.5 the modification necessary to allow for the effect of shunting resistance will be mentioned.

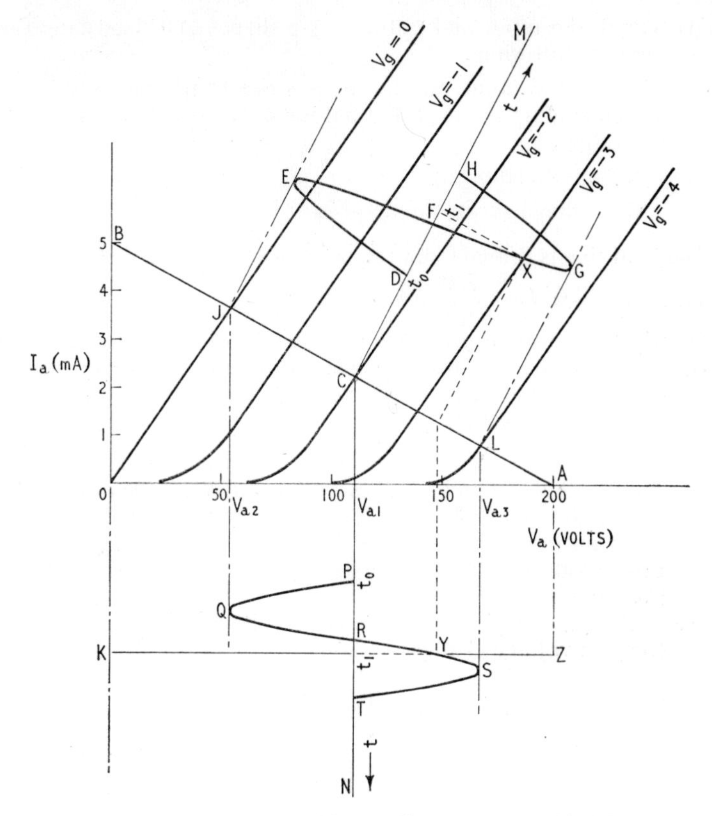

Fig. 8.5. Use of load line. Illustrating Example 8.3

Let the static output characteristic and the load line be as shown in Fig. 8.2; assume a grid input of 2·5 volts peak. To avoid grid current a negative bias of 2·5 volts must be applied to the grid so that the grid voltage swings between zero and −5 volts; that is between the points F and D. The corresponding anode voltage swing is between V_{a3} and V_{a2}. The voltage gain of the stage is the ratio of anode voltage swing to that at the grid, i.e.

$$\text{gain} = \frac{V_{a2} - V_{a3}}{5}$$

Example 8.3

A triode having the characteristics shown in Fig. 8.5 is used with an anode load resistance of 40 kΩ and H.T. supply voltage of 200 V. An input sine wave of 4 volts peak-to-peak is to be applied.

(i) Sketch the input and output wave shapes and determine the approximate gain,

(ii) Choose any point on the input wave shape and show the corresponding point on the output wave; specify at this point the value of

(a) anode voltage,

(b) voltage across anode resistor.

The load line is drawn from the point A ($V_a = 200$ V, $I_a = 0$) to point B ($V_a = 0$, $I_a = \frac{200}{40}$ mA $= 5$ mA).

Clearly the negative grid bias must be 2 volts because less will permit the grid to run positive while more will take the operation too far into the bend of the characteristics and result in undue distortion (*see* Section 8.3.1).

(i) The operating point is C at the junction of the load line and $V_g = -2$ V. The time axes CM and CN can be drawn and the datum time of $t = t_0$ located where convenient on each. After an input sine wave DEFGH has been inserted suitable projections EJ, JQ, etc., enable the output wave shape to be sketched in ($PQRST$).

The gain is given by the anode voltage change divided by the grid voltage change

$$\text{gain} = \frac{V_{a3} - V_{a2}}{4}$$

$$\simeq \frac{170 - 55}{4}$$

$$\simeq 29$$

(ii) Choose any point X corresponding to a time t_1. Measure the distance appropriate to this time period along the input axis DM and set off the equivalent distance along the output axis PN. Project from X via the load line to obtain the required point Y. Here the anode voltage is KY, about 147 volts, and the voltage across the anode resistor is YZ, about 53 volts.

8.3. Choice of Anode Load and Grid Bias

A number of factors must be considered in the choice of load and operating point:

(i) For the resistance-loaded amplifiers which are now being considered one end of the load line terminates in the anode voltage axis at the value of the H.T. supply. This value is limited by a number of considerations such as type of valve (the maximum voltage and current ratings must not be exceeded), economic factors and so on.

(ii) To prevent grid current, operation must not extend to a grid voltage more positive than zero. In some amplifiers, however, some grid current is allowed and this, in passing through a grid resistor of high value, sets up the required grid bias voltage (*see* Section 8.6, Fig. 8.19).

(iii) The load resistance must have as high a value as is permissible in order to obtain maximum gain (Equation 1).

(iv) The load resistance must be small enough to ensure that the grid voltage excursions take place across the straight and parallel parts of the output characteristics; otherwise equal increments of grid voltage do not result in equal changes in anode voltage and the output voltage is distorted.

8.3.1. DISTORTION

A full consideration of the factors bearing on the optimum condition of load line is a complicated and lengthy undertaking. Here

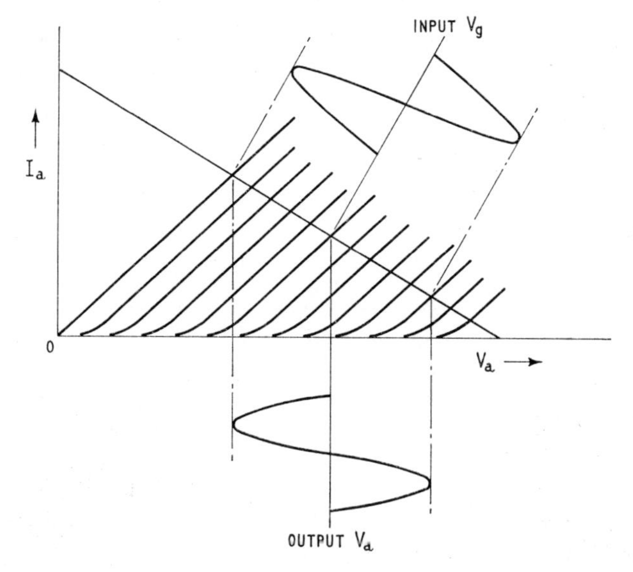

Fig. 8.6. Distortionless amplification

it is intended to mention only a few generalisations which may be helpful.

If the load line operates across a series of curves which are equally spaced there is no distortion; this is the ideal to be aimed at (Fig. 8.6). If the load line lies across a series of curves which become

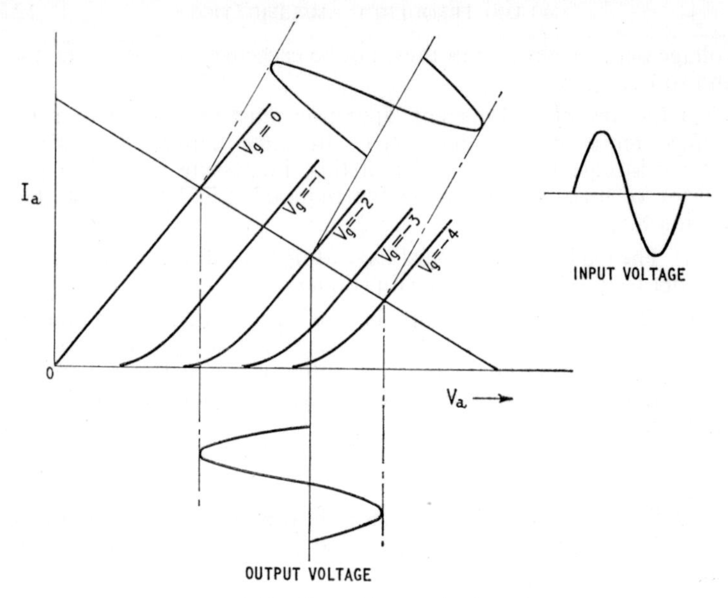

Fig. 8.7. Second harmonic distortion

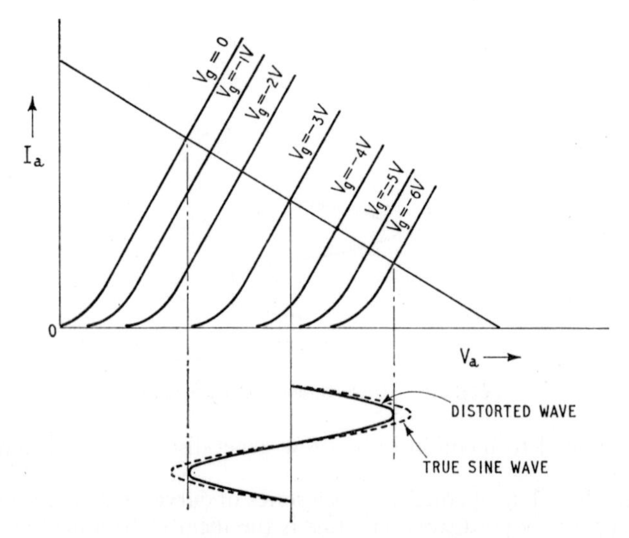

Fig. 8.8. Third harmonic distortion

progressively closer for equal input voltage increments a symmetrical input results in the asymmetric type of output in which one half cycle is flatter than the other. Fig. 8.7 shows a symmetrical input of 2 volts peak value applied at $V_g = -2$ V. Note that although symmetrical, the input curve appears distorted in the diagram. This is because of the crowding of the grid lines in the characteristic curves. On the other hand the anode voltage scale is linear, thus the fact that the output voltage curve appears distorted betokens the fact that distortion really exists here. Earlier work, as for instance that of Chapter 3 of *Radio and Line Transmission, Volume* 1, shows that such distortion is due to the introduction of even harmonics, notably second.

When, on the other hand, the load line operates across characteristics which become closer together (or farther apart) to either side of the mid-point, the output, although symmetrical, is nevertheless distorted, the wave being somewhat flattened (or peaked) on both half-cycles. This is illustrated in Fig. 8.8: an input signal of 3 volts peak applied at $V_g = -3$ volts results in the anode waveform shown by the full line. The true sine wave which would have resulted had the grid lines been equally spaced is shown dotted. Earlier work shows that such distortion is due to odd harmonics, particularly the third.

8.3.2. TRIODE RC-COUPLED VOLTAGE AMPLIFIER

Consider Fig. 8.9 and suppose that the H.T. supply voltage is represented by the point *A*. The correct bias is about $-1{\cdot}5$ volts. The load line *AZ* is for a high value of load resistance and gives maximum gain. The whole of the operation, however, takes place in the bottom bend region and consideration of, say, an input wave of 3 volts peak-to-peak shows that the anode output voltage is considerably flattened on the positive half-cycles. There is, therefore, much second harmonic distortion.

A fourfold reduction in the anode load leads to load line *AE*: here is a manifest improvement in output waveform but the amount of second harmonic is still appreciable.

A further load resistance reduction to that represented by *AF* seems to be about right. The gain is notably less than that resulting from the first anode load but is still quite good and the distortion has been decreased sufficiently to be regarded as satisfactorily small.

A further reduction in load resistance to that represented by *AG* causes a considerable reduction in gain with little compensating reduction of distortion.

If the H.T. voltage could be increased, say up to the value represented by the point *B*, then a value of load resistance higher than that represented by *AF* could be employed. Load line *BH*, parallel to *AE* and therefore corresponding to the same resistance, gives no more distortion than *AF* and an appreciably larger output. However, the H.T. voltage must not be increased to so high a value that the permissible dissipation of the valve is exceeded.

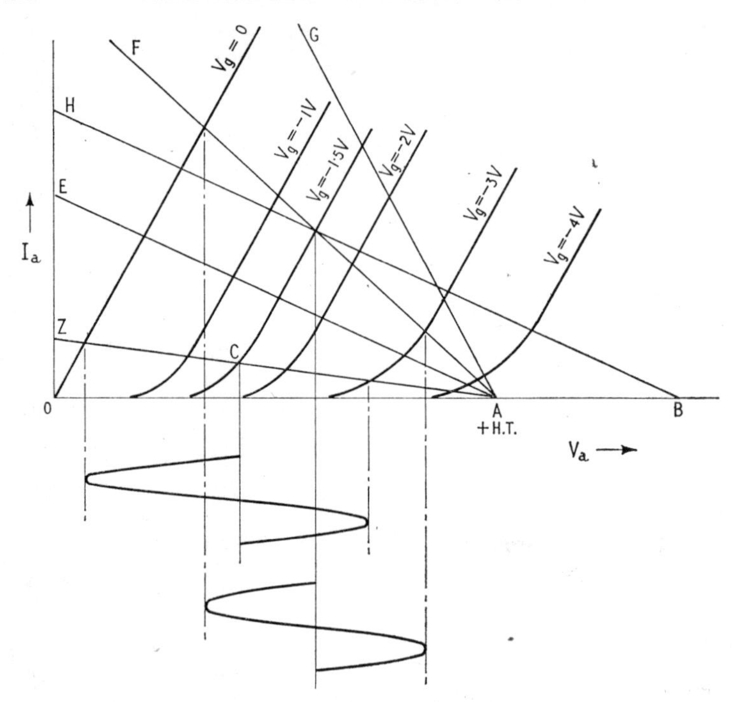

Fig. 8.9. Triode RC-coupled amplifier; choice of load line

8.3.3. PENTODE RC-COUPLED VOLTAGE AMPLIFIER

A load line with a slope suitable for a pentode is drawn in Fig. 8.3. Pentode output curves for equal increments of grid voltage are not so well spaced as those for triodes and greater distortion is likely to result. From what has been said in Section 8.3.1 it is plain that with the load line in the position shown the distortion is mostly third harmonic. Increasing the load resistance results in flattening the negative-going half-cycles of the anode voltage waveform; decreasing the resistance flattens the positive-going half-cycles. Each change, therefore, results in the introduction of second harmonic distortion in addition to the existing third. Third harmonic is usually less tolerable than second so it is common to use a relatively low value of load resistance which gives reduced third harmonic at the cost of some increase of second. With this value of load resistance the output is much less than the maximum possible but is still greater than would be obtained with a triode.

8.3.4. TRANSISTOR RC-COUPLED AMPLIFIER

Fig. 8.10 shows a typical common-emitter transistor *RC*-coupled stage. R_1 and R_2 set the base potential while R determines the

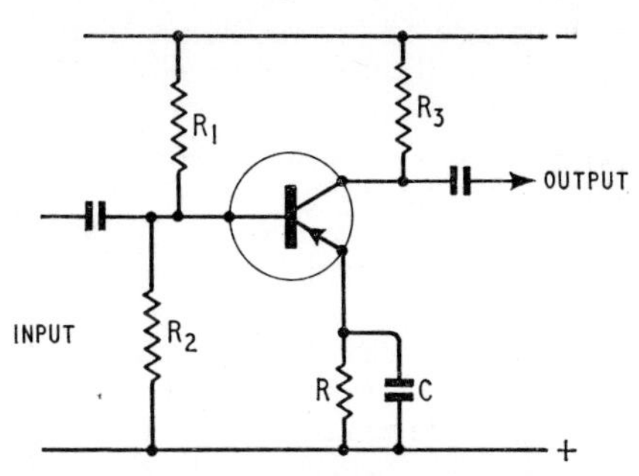

Fig. 8.10. Transistor RC-coupled amplifier

emitter current and provides stabilisation. *C*, if sufficiently large—up to 100 μF for audio-frequency amplifiers—renders the bias entirely smooth and prevents any negative feedback.

The collector current–collector voltage characteristics for various base currents are of the form shown in Fig. 8.11. A load line can be drawn across the characteristics by employing the same general principles as adopted for thermionic valve circuits. The line is

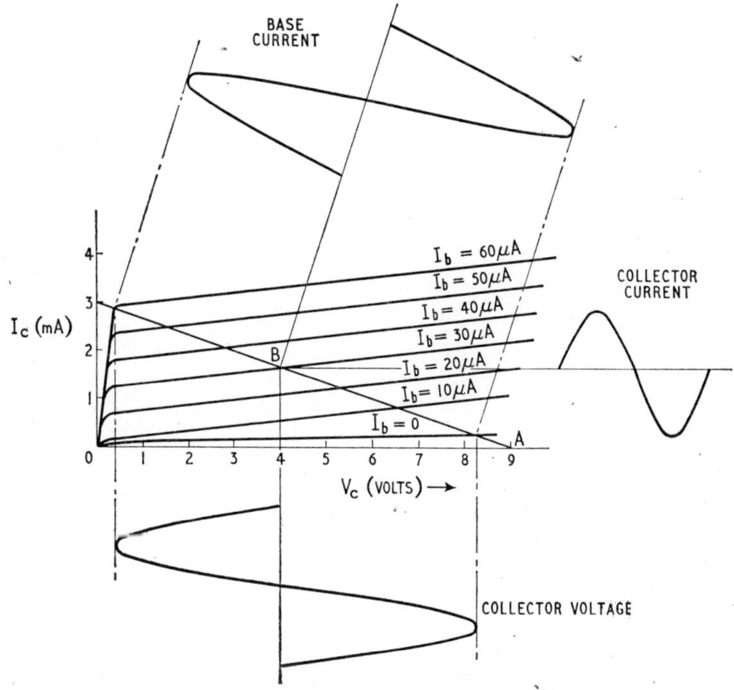

Fig. 8.11. Transistor: collector current–collector voltage characteristic. Load line

given a slope appropriate to the resistance in the collector circuit and it terminates in the collector voltage axis at a point representing the supply voltage: *A* in the figure.

For pnp transistors the collector voltage is negative but it is customary to draw the figures as though it were positive. The operating point *B* indicates the standing values of base current (30 μA in Fig. 8.11), collector voltage (about 4 volts) and collector current (about $1\frac{1}{2}$ mA). An input signal causes changes in the base current and these, together with the resultant collector current and collector voltages, can be graphed as shown. The swings of collector current and voltage can be taken down almost to zero before any appreciable distortion is introduced; thus the characteristics can be far more effectively employed than those of a triode valve. The higher the collector resistance the greater is the gain but the smaller the permissible current swing before bottoming (i.e. the excursions of the swings into the bend of the characteristic) is encountered.

In calculating the gain of a transistor stage it must be borne in mind that the following stage is probably of low input impedance. If current gain is being considered the effect of the following stage can probably be disregarded; if the calculation refers to voltage gain then account must be taken of the following low impedance.

The general theory shows that the current gain α' of a common emitter amplifier is given by

$$\alpha' = \frac{\alpha}{1 - \alpha}$$

provided that the collector load is not unduly large. α is the current amplification factor and since it is close to unity the value of the expression for α' varies considerably with small changes in α. α' is commonly about 50.

The voltage gain depends on the value of the resistance through which the input and output currents flow and is given by the relationship

$$\text{Voltage gain} = \text{Current gain} \times \frac{\text{Load resistance}}{\text{Input resistance}}$$

as may readily be shown as follows:

$$\text{Input current, } i_{in} = \frac{\text{Input voltage, } v_{in}}{\text{Input resistance, } r_{in}}$$

$$\therefore \quad v_{in} = i_{in} r_{in}$$

$$\text{Output current, } i_{out} = \frac{\text{Output voltage, } v_{out}}{\text{Load resistance, } R_L}$$

$$\therefore \quad v_{out} = i_{out} R_L$$

$$\text{Voltage gain} = \frac{i_{out}R_L}{i_{in}r_{in}}$$

$$= \text{Current gain} \times \frac{R_L}{r_{in}}$$

$$= \text{Current gain} \times \frac{\text{Load resistance}}{\text{Input resistance}}$$

If there is a following transistor of like characteristics $R_L = r_{in}$ and thus the voltage gain equals the current gain. If there is no following transistor the second half of the expression for voltage gain may well be considerably greater than unity so that the voltage gain may be several times the current gain. If the following transistor has a lower input resistance than the first, the voltage gain is less than the current gain.

Example 8.4

A stage has a total input resistance of 2,000 ohms, a current gain of 40 and a load resistance of 4,000 ohms.

Calculate (a) the voltage gain and (b) the power gain.

$$\text{Voltage gain} = 40 \times \frac{4{,}000}{2{,}000}$$

$$= 80$$

$$\text{Power gain} = \text{voltage gain} \times \text{current gain}$$

$$= 40 \times 80$$

$$= 3{,}200$$

Example 8.5

The stage of the previous example feeds a following stage of input resistance 900 ohms. Calculate the voltage gain and the power gain from the input of one stage to the input of the next.

The current gain is still 40.

The load resistance, 4,000 ohms shunted by 900 ohms, is about 730 ohms so that the voltage gain is

$$40 \times \frac{730}{2{,}000}$$

$$= 14{\cdot}6$$

Therefore the power gain is

$$40 \times 14{\cdot}6$$

$$= 584$$

The marked reduction in voltage and power gain which results from connection to the next stage can be minimised by suitable transformer matching. This (a) enables the collector circuit impedance to be matched to the input resistance of the transistor circuit and (b) provides a step-down in resistance in the ratio n with a transformer step-down ratio of $\sqrt{n} : 1$ (i.e. with a voltage step-down of only $\sqrt{n}$).

Example 8.6

A common-emitter amplifier stage has an output resistance of 36,000 ohms: when it is provided with a load resistance of the same value the voltage gain is 600. When the stage feeds directly a following stage of input resistance 900 ohms the voltage gain is only 30. What would be the gain if a suitable transformer were interposed in the output circuit to match the output circuit to the 900 ohm load?

The output resistance of the stage is 36,000 ohms. For optimum results the transformer primary should show this resistance across its primary terminals. Since the resistance across the secondary is 900 ohms the transformer ratio should be $\sqrt{\dfrac{36{,}000}{900}} = 6{\cdot}3 : 1$. The voltage gain to the transformer primary winding will be 600; to the secondary winding it will be $600/6{\cdot}3 = 95$.

Note that with the transformer the gain is 95 compared with 30 without it. In practice, however, taking due consideration of all the factors, it is often found better to use a resistance as the collector load rather than introduce a transformer.

8.3.5. MODIFICATION TO THE LOAD LINE IN THE PRESENCE OF A SIGNAL

The load lines so far used have taken account only of the load resistance R_L. Modification may be necessary when a signal is

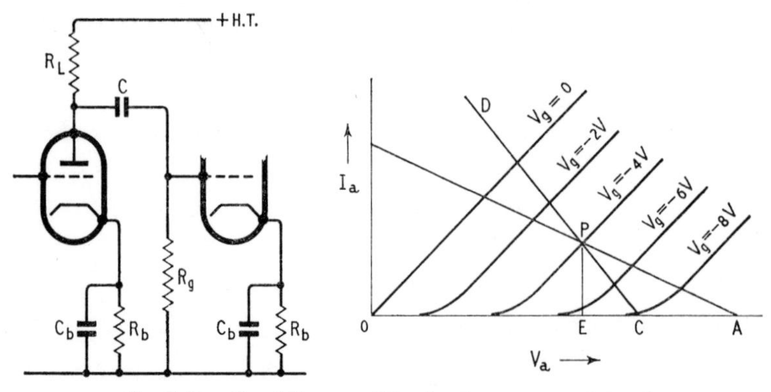

Fig. 8.12. Load line: modification in presence of signal

received. Consider Fig. 8.12. If the H.T. supply is *A*, the grid bias −4 volts and the load line as shown, the operating point will be at *P*. The anode voltage will, therefore, be *OE*. The advent of a signal, which is transferred to the next valve via the coupling capacitor *C*, puts the following grid resistor R_g in parallel with R_L so that the effective load is now that of R_L and R_g in parallel. Operation, about the point *P*, now takes place along the load line (*CD*) representing the new value of load. If R_g is very large compared with R_L the difference may be negligible, if it is not then the tendency to increased distortion, because of bottom bend, must be allowed for—by reducing the grid bias and the input or by increasing the H.T. supply voltage if feasible.

The functioning of R_b and C_b will be explained in Section 8.6; they provide the grid bias.

8.4. Loads Other Than Resistive

8.4.1. INDUCTIVE LOAD

The anode load for an audio-frequency amplifier need not be resistive, although it usually is. An inductor may be employed. The advantage of the use of a choke (as an inductor of this type is often called) is that a high value of reactance down to the lowest frequency which it is desired to amplify may readily be obtained while the d.c. resistance of the choke is small. Thus the average anode voltage is virtually that of the H.T. supply instead of being materially reduced below this value by the voltage drop in the anode resistance.

To find the operating point for an inductive load the static condition must first be determined by noting the point of intersection, *C*, between the appropriate grid bias curve (say $V_g = -3$ V) and the load line *AB* corresponding to the d.c. resistance of the choke (Fig. 8.13). This resistance will be so small that the load line is almost vertical. The load line *DE*, representing the impedance of the load is then drawn through *C*. When a signal is received the relationship between the grid voltage and the anode current and voltage is given by this line and not by the static line *AB*.

Note that the overall grid swing which can be accommodated along *DE* without running into bottom bend curvature of characteristics is about 6 volts. Along *AF*, the load line for a resistive load used in the same circuit with the same anode voltage (*OA*), the maximum grid swing is only about 4 volts.

Two additional points should be mentioned. The impedance of the load varies with frequency and is not constant as has been assumed: the gain also varies and the choke must therefore be selected so that its reactance is adequately high at the lowest frequency to be handled. Because the voltage and current are not in phase in an inductive circuit the load line is not, in fact, a straight line as drawn in Fig. 8.13 but is an ellipse. The general principles which have been evolved, however, hold good.

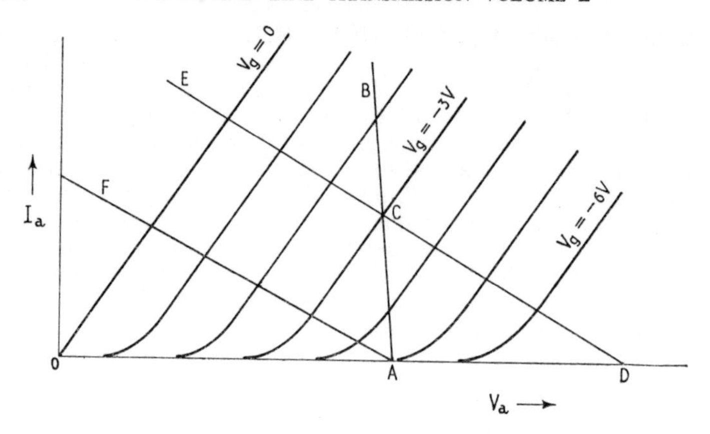

Fig. 8.13. Load line: inductive load

Despite its manifest apparent advantages the choke method of loading is seldom employed because chokes are heavy, expensive and bulky and these drawbacks are sufficiently serious to offset the advantages of their use. Moreover, modern pentodes produce adequate gain from a simple resistor.

8.4.2. TRANSFORMER LOAD

A transformer may be substituted for the choke of the last section. As a result some step-up of voltage may be obtained, but often not as much as might be expected (*see* Section 5.3.1).

For good amplification down to the lowest frequencies the transformer primary reactance must be large compared with r_a at the lowest frequencies. This necessitates the use of a large primary inductance. The step-up ratio is given by the ratio of the secondary to primary turns. For a large step-up, therefore, the number of secondary turns must be exceedingly large. The greater the number of secondary turns, however, the greater the self-capacitance of the secondary and the greater the fall-off in amplification at the higher frequencies (*see* Section 10.6.2 of *Radio and Line Transmission, Volume* 1). Hence there is an upper limit to the number of secondary turns which may be used and hence to the step-up ratio of the transformer. A ratio of about 3 : 1 is about as much as is normally practicable in the type of application to which this whole discussion is restricted—that of step-up transformers with unloaded secondaries.

In addition to the fall-off in output at the lower frequencies because of the inadequacy of the primary inductance, and the fall-off at higher frequencies because of the self-capacitance of the secondary winding, the circuit is liable to give a peaked response at some frequency at which it is resonant. When a uniform response over a relatively wide band of frequencies is required, say one of several kc/s, transformers are not used for the role discussed. For speech, where a bandwidth of about 2 kc/s may sometimes have to suffice, they are occasionally employed and for morse reception, where a narrow bandwidth, centred on about 1 kc/s, is a positive advantage —in that it provides an additional aid to selectivity by discriminating against an unwanted adjacent signal—transformers are sometimes used with a high step-up ratio, say 10 : 1.

Like chokes, transformers are heavy, bulky and expensive and so are very seldom used for the purpose discussed in this section except for specialised work such as indicated in the previous paragraph. They have the minor advantage that there is no d.c. connection

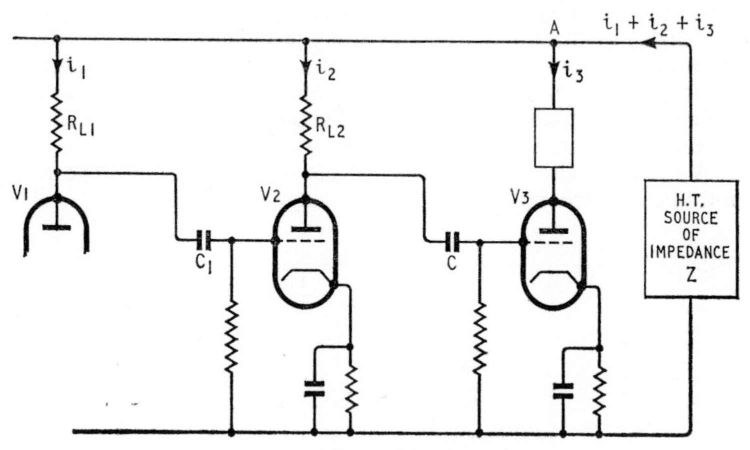

Fig. 8.14. The need for decoupling

between primary and secondary so that the coupling capacitor and grid resistor can be dispensed with.

8.5. Simple Decoupling Arrangements

For many purposes it is satisfactory to regard the source of H.T. supply as being of negligible impedance. In fact, of course, the impedance is not zero and can cause trouble if suitable precautions are not taken.

The H.T. supply is usually common to all stages of an equipment as illustrated in connection with three thermionic valve stages in Fig. 8.14. Here the H.T. supply is shown as having an impedance Z and as supplying the fluctuating anode currents i_1, i_2 and i_3 to the valves $V1$, $V2$ and $V3$.

The voltages appearing at A in the common H.T. line have an alternating component, $(i_1 + i_2 + i_3)Z$, added to the steady H.T. voltage. Since i_3 is usually much greater than the other currents this spurious voltage can be written as approximately equal to i_3Z.

This alternating component is transferred through Z to the anode of $V1$ and thence, along with the signal i_1R_{L1}, via the coupling capacitor C_1 to the grid of $V2$ and on to $V3$.

Since i_3Z can easily be as large as i_1R_{L1} it is clear that the effect cannot be ignored. Depending on circumstances, the spurious voltage i_3Z can bear any phase relationship to the signal i_1R_{L1}. When they are in phase they can cause instability, when out of phase very low gain is possible. In the amplifier illustrated and assuming 180° phase change in $V2$ and $V3$ the voltage fed back through R_{L1} is in phase with the signal: that through R_{L2} is in anti-phase.

The result, usually of instability, can sometimes be cured by the simple expedient of reducing the impedance of the power supply by increasing the capacitance of the final smoothing capacitor (or, when the trouble is due to the rising impedance of a failing battery, by putting a capacitor across the battery terminals). This expedient, however, is only likely to be successful in a few instances because no great reduction in impedance can normally be achieved; if the output capacitance is doubled, for instance, the output impedance is reduced by no more than half.

A better method of reducing effects to supply impedance is to introduce a decoupling circuit R_1C_1 as shown in Fig. 8.15. The

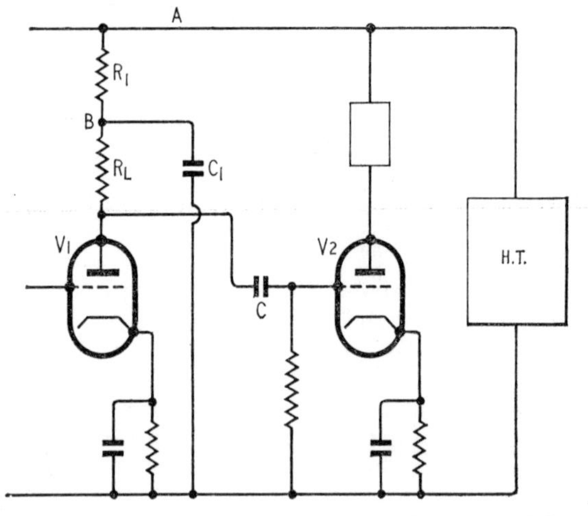

Fig. 8.15. Decoupling: thermionic valve a.f. amplifier

operation is as follows: the values of C_1 and R_1 are so chosen that the reactance of C_1 is much less than the resistance of R_1 even at the lowest frequency which it is desired to amplify. Thus any fluctuations in the H.T. supply which appear at A will be divided across R_1 and C_1 in the ratio of their impedances: if the resistance of R_1 is, say, at least twenty times the reactance of C_1 the fraction of the H.T. fluctuations which appears across C_1, i.e. at B, can never exceed one-twentieth of that existing at A. Very little of the H.T. variation, therefore, is superposed on the signal at the anode of $V1$ for transfer to the grid of $V2$.

For a.f. circuits R_1 is likely to be about 10 kΩ and C_1 about 8 μF. At a frequency of 100 c/s the reactance of C_1 is $1/(2\pi f C_1)$

$$= \frac{10^6}{16 \times \pi \times 100} \text{ ohms}$$

$$\simeq \frac{10^4}{50} \text{ ohms}$$

That is, one-fiftieth of the resistance of R_1. At higher frequencies the ratio is greater.

The presence of R_1 does, of course, reduce the anode supply voltage to $V1$. This, however, is a small price to pay for the benefits accruing.

Example 8.7

A decoupling circuit uses a resistance of 10 kΩ and a capacitance of 4 μF in a circuit similar to that of Fig. 8.15. Draw the graph of the approximate percentage of any H.T. line variations between 50 c/s and 1,600 c/s which appear at B.

When $R_1 \gg X_1$ the total impedance of the decoupling circuit is approximately equal to R_1. The fraction of the H.T. line variations which appears at B will be X_1/R_1. X_1 is the reactance of C_1. As a percentage this is 100 X_1/R_1.

$$X_1 = \frac{1}{2\pi f C_1}$$

$$\simeq \frac{40{,}000}{f} \text{ ohms}$$

$$\therefore \qquad 100\frac{X_1}{R_1} \simeq \frac{400}{f}$$

Substitution of values within the required range results in the values given in Table 8.2. These figures are plotted in Fig. 8.16.

Table 8.2

Frequency (c/s)	50	100	200	400	800	1,600
$100\frac{X_1}{R_1}$	8	4	2	1	0·5	0·25

From the foregoing it is seen that any difficulties which arise are likely to occur only at the lowest frequencies. It is not the presence

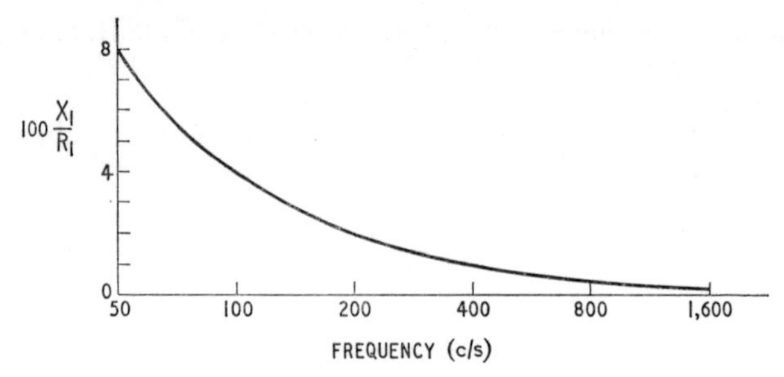

Fig. 8.16. Efficiency of decoupling: illustrating Example 8.7

at *B* of the fluctuating voltage from the H.T. line which causes instability, it is the fact that the fluctuations are passed on to the grid of *V*2 where they are amplified, which causes the trouble. In the majority of amplifiers, as in Fig. 8.15, the coupling to *V*2 is through a capacitor, *C*. Hence a measure of correction is automatically applied in that the lower the frequency, the greater the reactance of *C* and the less the transfer to *V*2. When instability is experienced the capacitance of *C* may deliberately be decreased to inhibit the trouble. This expedient, of course, reduces the frequency response of the amplifier.

Similar arrangements can be applied to transistor equipments to combat feedback due to increasing battery resistance. The circuit of Fig. 8.17 shows a suitable arrangement in which C_1 and R_1 provide the decoupling. In these relatively low-voltage low-impedance circuits values of resistance are much lower and of capacitance much higher than in thermionic valve circuits. In the audio-frequency circuit shown R_1 would be about 1 kΩ and C_1 100 μF.

8.6. Simple Bias Arrangements

The obvious way in which to provide grid bias is by means of a battery for small equipments and by means of a generator for more powerful types of apparatus. The latter method is adopted in appropriate circumstances: the former very seldom; instead, a scheme similar to that shown in Fig. 8.18 is more generally used in battery valve sets. The H.T. circuit current traverses *R* in the H.T. negative lead and the voltage developed is taken to the grid through the grid resistor, R_g, which prevents the grid bias supply from short-circuiting the input. In radio receivers and small amplifiers R_g usually has a value between 0·1 MΩ and 2 MΩ and the effect of its presence on the gain will be as discussed in Section 8.3.5. *C*, of some 16 μF smooths the voltage across *R* and also preserves a low H.T. circuit impedance even when the battery is running down and thus prevents feedback troubles (*see* Section 8.5).

In some amplifiers a much larger grid resistance is employed, about 10 MΩ, the valve deriving its bias from the voltage set up

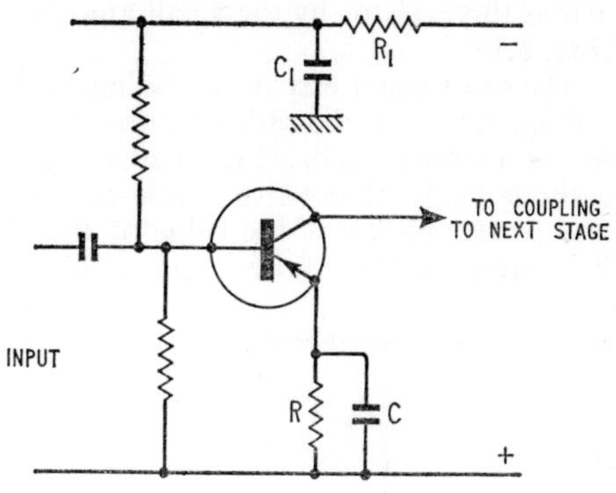

Fig. 8.17. Decoupling: transistor a.f. amplifier

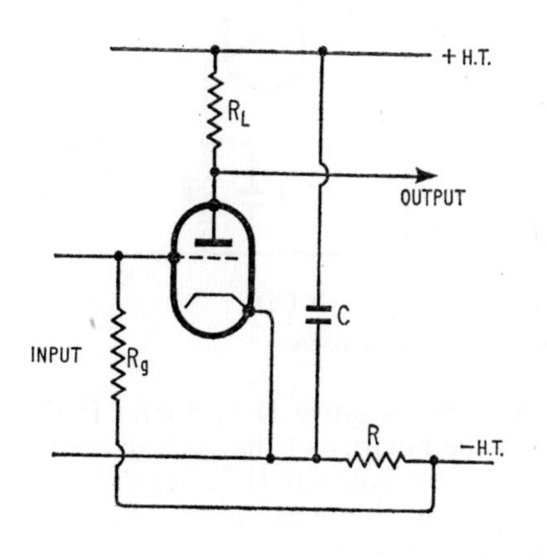

Fig. 8.18. Grid bias arrangement : battery receiver

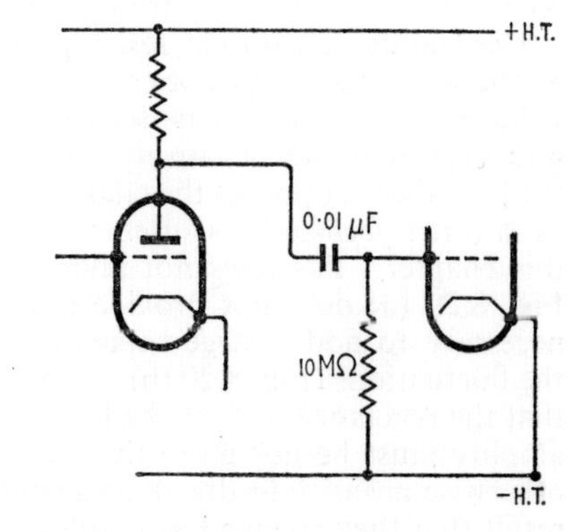

Fig. 8.19. Grid current biasing

across this resistor by the small amount of grid current which flows (Fig. 8.19).

The most usual way of providing automatic grid bias is to use the voltage set up by the cathode current of the valve across a resistance in the cathode lead. Fig. 8.20 (a) shows a resistance, *R*, in the cathode lead. The valve anode current sets up a voltage across *R* and correct choice of the value of *R* ensures that the voltage across it is equal to the required bias. This voltage is positive at the top

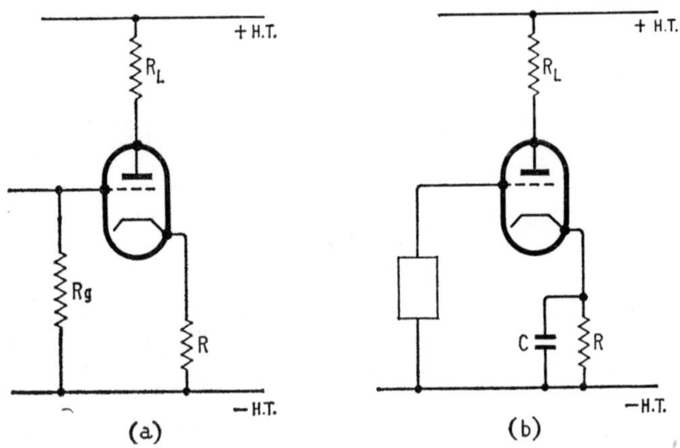

Fig. 8.20. Development of self-bias

(cathode) and negative at the bottom (negative H.T. line). If the bottom of *R* is joined to the grid the polarity of the voltage is that required for grid bias. The connection from the H.T. negative line to the grid often exists through the input circuit as shown schematically in Fig. 8.20 (b). If this circuit does not have d.c. continuity connection can be made through a grid resistor as in Fig. 8.20 (a).

Used alone a cathode resistor does not provide a steady bias: as the input varies so does the anode current and with it the bias voltage. It can readily be seen that if the input rises positively the grid bias increases negatively and vice versa. Thus the change in the bias always opposes the change in the signal. This effect, known as *negative feedback*, is often very useful and is discussed later in this chapter; this does not alter the fact that the arrangement of Fig. 8.20 (a) does not provide a steady bias as it stands and it is necessary to add a large capacitor, *C*, in shunt with *R* to smooth the fluctuations (Fig. 8.20 (b)). As an approximation it can be said that the reactance of *C* at the lowest frequency which it is desired to amplify must be not more than about one-tenth of the value of *R* otherwise negative feedback occurs at the lower frequencies with the result that they receive less amplification than higher frequencies.

The value of *R* is commonly of the order of 1,000 ohms, depending on the valve and circuit. The stipulation made above, therefore, requires that the reactance of C shall be not more than 100 ohms at the lowest frequency to be amplified, say 50 c/s. These figures give a value for C of about 30 μF. This is a large value but the voltage across the capacitor is small so that a low voltage electrolytic type is quite satisfactory. Capacitors of 50 μF and much more are in common use.

Bias voltages for transistors can be derived in a similar way. RC in Fig. 8.17 provides the emitter bias and stabilisation voltage. R is usually a few hundred ohms so that, at audio frequencies, C needs to be about 100 μF.

Example 8.8

Use the curves and load line of Fig. 8.5 and assume an H.T. supply of 200 volts to determine approximately the value of the cathode bias resistor needed for operation with as large an input signal as possible.

The largest possible input is about four volts peak-to-peak. This requires a grid-bias voltage of 2 volts. The operating point will be at C so that the average anode current is about 2 mA. The required value of bias resistor is thus

$$R = V/I = 1{,}000 \text{ ohms}$$

Note that the bias is, to some extent, self-adjusting; should the anode current rise for any reason, perhaps because of a rise in anode voltage, the bias automatically increases and thus restricts the rise in current.

Methods of providing bias in oscillator circuits are dealt with in Chapter 11.

8.7. Power Amplification

So far only problems of voltage amplification have been dealt with and it has been assumed that the power requirement was negligible.

A power amplifier must supply an appreciable current swing as well as a voltage swing. Such an output is needed to drive a loudspeaker or to provide an input to a valve which is drawing grid current. Reference to Fig. 8.9 shows that while a load of large value provides the maximum, albeit distorted, voltage output, such a load absorbs little power because of the small current variations. A low-resistance load, on the other hand, gives large current variations but very little voltage change; i.e. once again little power is generated in such a load. An intermediate value of load resistance gives significant voltage and current swings and hence appreciable output power.

In confirmation, Fig. 8.21 shows how the power output varies with the ratio of the load resistance to the valve or transistor a.c. resistance when the input voltage is fixed. Maximum power is

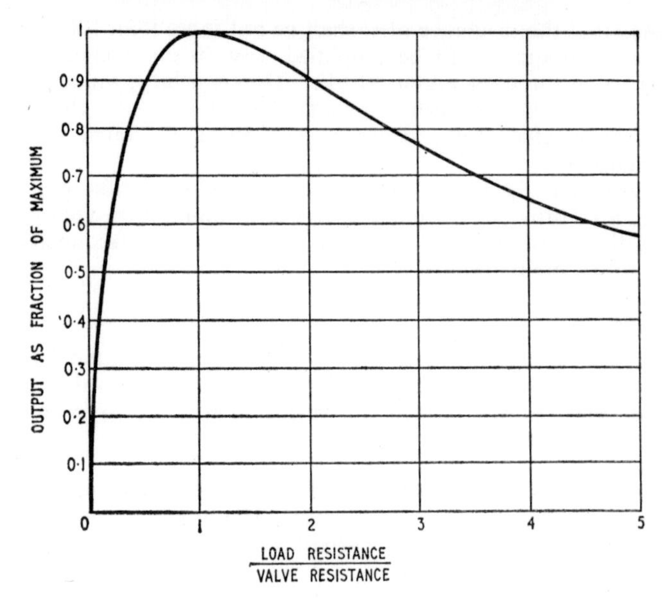

Fig. 8.21. Variation of power output with load

obtained when the load resistance is equal to the a.c. resistance but the maximum of the curve is very flat and the power does not fall below 90 per cent of this value for an increase or decrease in the load resistance value in the ratio 2 : 1. (A general law for resistive loads and generators is that maximum power is supplied to the load when the load resistance is equal to that of the generator. This is known as the maximum power transfer theorem.)

The factor which usually dominates the choice of load resistance is that of obtaining maximum power output without overheating the valve or transistor, the input voltage being adjusted appropriately. For this condition the load resistance, for a triode, generally requires to be of the order of twice the valve a.c. resistance. For pentodes the optimum load resistance is about one-tenth of that of the valve a.c. resistance.

The power output of the stage is given by the product of the r.m.s. current squared and the effective load resistance. The input power, to the anode or collector, is the product of the supply voltage to that electrode and the current taken from the supply. The efficiency is the ratio of the power output to the power input: if the overall efficiency is needed the power input must include not merely that to the anode or collector but also all other inputs (as, for instance, to heaters).

The load of a power amplifier is usually coupled via a choke or transformer (for instance, Figs. 8.22 (b), 8.23 and 8.31). Under these conditions the mean anode voltage is only slightly less than that of the supply and the voltage swings above and below this value. (This has already been noted in Section 8.4.1 in connection with the load line for the inductive load shown in Fig. 8.13.)

Example 8.9

The sinusoidal anode current swing in a certain power amplifier ranges from 3 to 19 mA. The corresponding anode voltage swing is between 250 and 100 volts. Calculate (a) the power output and (b) the anode efficiency if the voltage delivered by the power supply is 175 volts.

(a) The peak anode current change is

$$\frac{19-3}{2}\text{ mA} = 8\text{ mA}$$

The peak anode voltage change is

$$\frac{250-100}{2}\text{ V} = 75\text{ V}$$

Therefore, the power output is

$$I_{\text{r.m.s.}} \times V_{\text{r.m.s.}} = \frac{8}{1{,}000\sqrt{2}} \times \frac{75}{\sqrt{2}}\text{ W}$$
$$= 0{\cdot}3\text{ W or }300\text{ mW}$$

(b) The anode current drawn from the supply is

$$\frac{19+3}{2}\text{ mA} = 11\text{ mA}$$

The anode supply voltage is 175 volts. Therefore, the input power to the anode is

$$I \times V = \frac{11}{1{,}000} \times 175\text{ W} = 1{\cdot}93\text{ W}$$

and the efficiency is

$$\frac{0{\cdot}3}{1{\cdot}93} = 0{\cdot}158\text{ or }15{\cdot}8\text{ per cent.}$$

8.7.1. MATCHING

In practice the resistance of the valve or transistor load is usually too far removed from the optimum value for adequate power to be developed. If a transformer is interposed as shown in Fig. 8.22 (b) then the apparent resistance in the output circuit becomes

$$R_e = R \times \left(\frac{T_1}{T_2}\right)^2$$

where T_1 and T_2 are the number of turns on the primary and secondary windings of the transformer respectively. By suitable choice

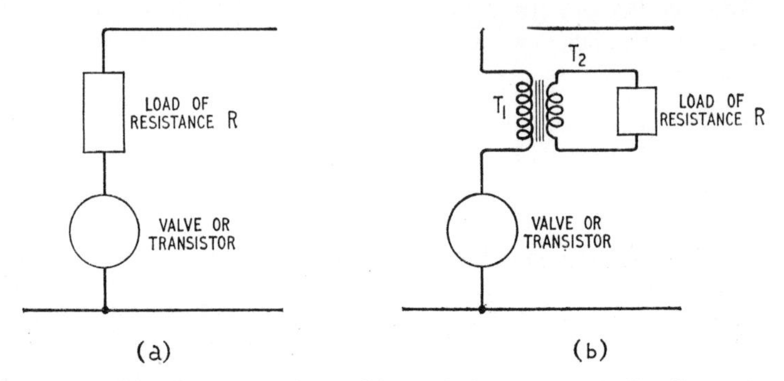

Fig. 8.22. Transformer matching: (a) Load of resistance R, (b) effective load resistance $= \left(\frac{T_1}{T_2}\right)^2 R$

of T_1 and T_2 the effective resistance in the output circuit can be made to have a value near to the optimum so that the power output is close to the maximum possible. This is called *matching*.

Example 8.10

A 3-ohm loudspeaker speech coil is to be matched to a valve of $r_a = 3{,}000$ ohms. Calculate a suitable turns ratio.

Suppose that the optimum resistance is 6,000 ohms.

$$\frac{T_1^2}{T_2^2} \times 3 = 6{,}000$$

$$\therefore \quad \frac{T_1}{T_2} = \sqrt{2{,}000}$$

$$= 10\sqrt{20}$$

$$= 45$$

The required turns ratio is thus about 45 : 1.

8.8. Methods of Increasing Stage Output

When a single valve will not provide adequate output a number of alternatives are possible to obtain increased output: a larger valve may be used, two or more valves may be used in parallel (*see* Section 8.8.1), valves or transistors may be employed in push–pull (*see* Section 8.8.2), or in some combination of parallel push–pull.

It is not always practicable or desirable to use a single valve for a given position and advantage can sometimes be derived from the use of two or more smaller ones. For instance, and the list is not exhaustive:

(a) there may not be available a single valve having appropriate characteristics,
(b) the problem of providing spares is simplified if a single valve type is used as far as possible throughout an equipment, using the given type singly in the early stages and in parallel or push–pull in later ones,
(c) if one valve of a number in parallel fails the service can continue on reduced power without interruption,
(d) the use of valves in push–pull confers certain advantages which will be discussed later (*see* Section 8.8.2).

8.8.1. PARALLEL WORKING

For parallel working, valves are used with corresponding electrodes interconnected (grid to grid, anode to anode, etc.). The voltages required are the same as for single valve operation but allowance must be made for the fact that currents are increased in the ratio of the number of valves paralleled. The effective g_m is increased and r_a is decreased in the same ratio.

Valves connected in parallel must have characteristics which are as nearly as possible identical: the greater the number of valves so connected the less the permissible tolerance.

Because of inter-electrode and wiring capacitances and lead inductance, valves connected in parallel are very prone to oscillate at some (usually high) frequency which is often remote from the designed frequency of operation. Such unwanted oscillation, *parasitic oscillation* as it is termed, must be avoided because in addition to the possibility of creating interference it will probably lead to the overloading of the stage in which it occurs. To prevent such oscillation it is usual to fit suppressor resistors as close as possible to the anode and grid pins of the valves. The resistors have a value of about 25 ohms and are large enough to inhibit the start of parasitic oscillations, yet small enough to give no more than slight loss at the operating frequency. Sometimes they are shunted by small inductors so that at the high frequency at which parasitic oscillation is likely to occur the reactance of the inductors is large and causes the parasitic oscillation current to pass through the resistors. The resistors, in fact, then prevent oscillation from starting, while at the lower operating frequency the, now, low reactance of the inductors permits the passage of the signal currents which thus do not need to traverse the resistors and no loss is entailed.

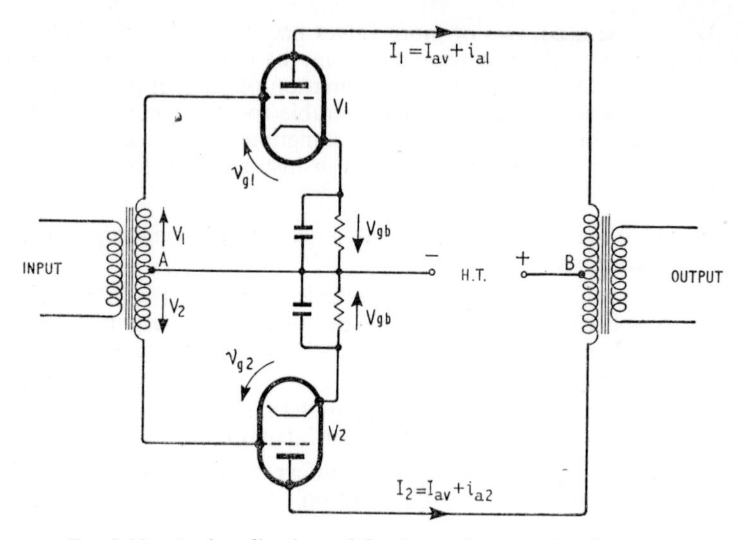

Fig. 8.23. Push–pull a.f. amplifier (v_{g1} and v_{g2} are the alternating components of grid–cathode voltage)

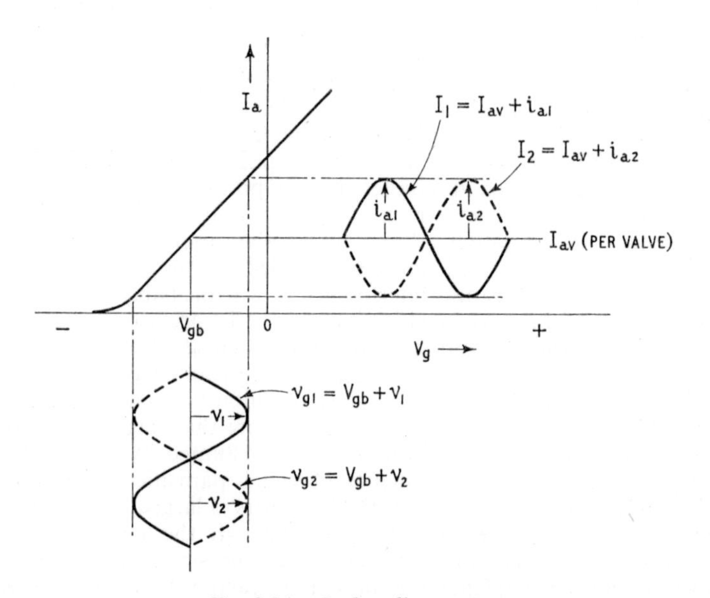

Fig. 8.24. Push–pull operation

8.8.2. PUSH–PULL WORKING

A push–pull circuit is one in which the total available grid-cathode input is divided into halves, respectively in anti-phase, each being applied to one of an identical pair of circuits. The outputs are so combined as to be additive.

The push–pull connection may be used for amplifiers or oscillators and at audio or radio frequencies. A typical audio-frequency circuit is given in Fig. 8.23. The input and output transformer windings are centre-tapped at *A* and *B* respectively and the circuit is symmetri-

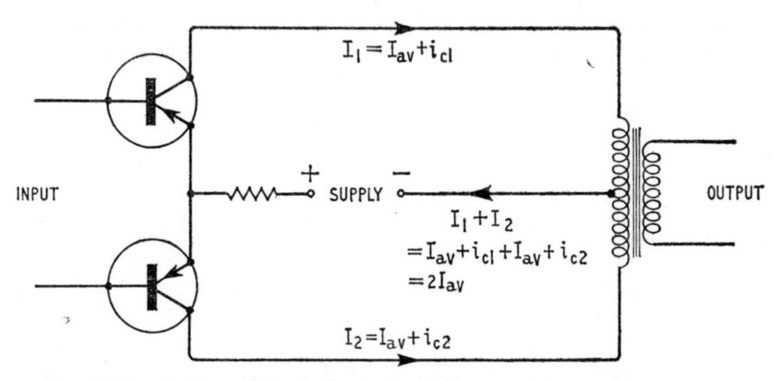

Fig. 8.25. Push–pull operation. Cancellation of d.c. components in transformer primary winding and of signal currents in the supply

cal about the centre. From the figure observe that the grid voltages, being those at opposite ends of a transformer winding, are in anti-phase: it follows that the anode currents are also in anti-phase but since they enter the output transformer from opposite ends the effect is one of reinforcement. Fig. 8.24 shows the dynamic mutual characteristic for either of the push–pull valves. It is assumed that the two valves and their respective circuits are absolutely identical and symmetrically arranged. In practice they will not be and the statements which follow require modification to take this discrepancy into account. Class-*A* working is assumed but later work will show that major advantages of push–pull working accrue when operation takes place with the bias, of valves or transistors, approximating to the cut-off value.

v_{g1} represents the input to *V*1 and v_{g2}, the equal and oppositely phased input to *V*2. i_{a1} and i_{a2} are the respective anode currents. These, as already stated, are phased so as to produce an additive effect in the transformer. Other interesting and useful effects result from the form of the circuit, these include:

1. The average anode currents to each valve, or collector currents to each transistor, travel in opposite directions in the two halves of the output transformer primary winding so that their effects cancel so far as direct current magnetisation is concerned.

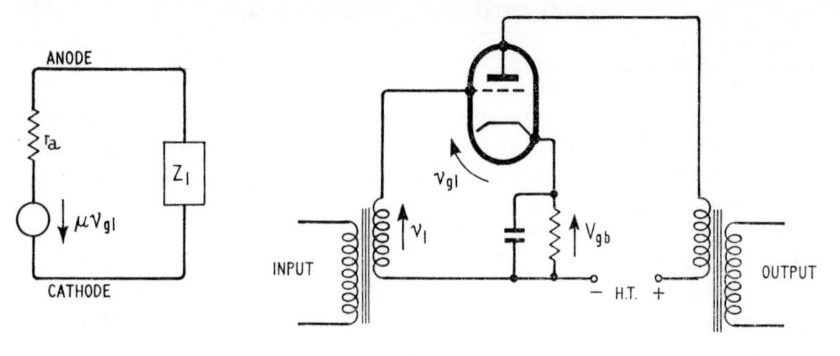

Fig. 8.26. One half of push-pull circuit and constant voltage equivalent (v_{g1} is the alternating component of the grid–cathode voltage)

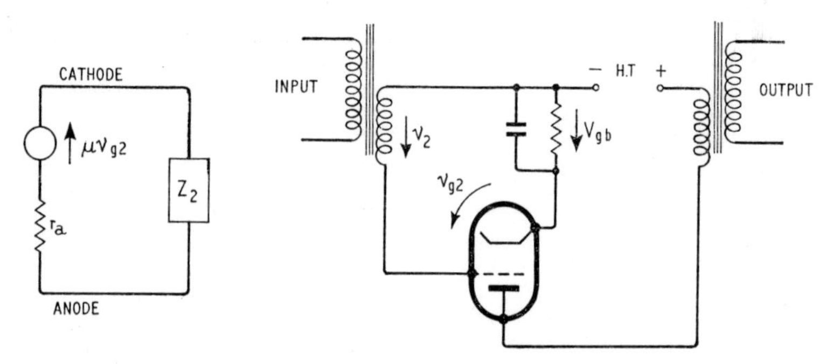

Fig. 8.27. Other half of push–pull circuit and constant voltage equivalent (v_{g2} is the alternating component of grid–cathode voltage)

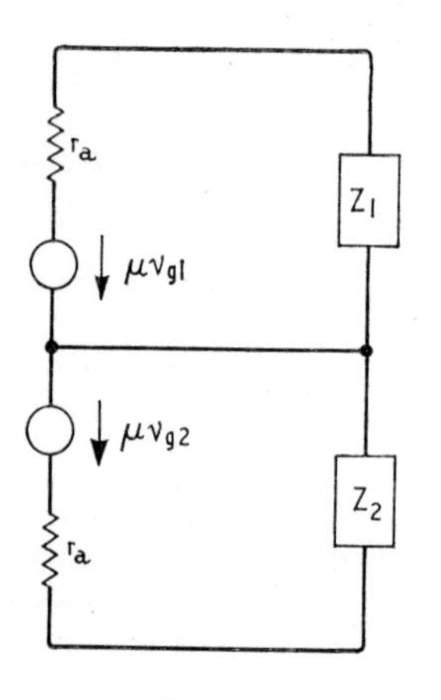

Fig. 8.28. Push–pull circuit: constant voltage equivalent taking account of relative phases of μv_{g1} and μv_{g2}

Magnetisation is thus due solely to the alternating component: with a given core, therefore, permeability is higher than for a single-ended amplifier and a given value of inductance can be obtained with a smaller and, therefore, cheaper and lighter core. This is illustrated in Fig. 8.25 for a transistor circuit.

2. In the common supply or H.T. line i_{c1} (or i_{a1}) and i_{c2} (or i_{a2}) are in the same direction; the variations cancel leaving a steady current $2I_{av}$ (Figs. 8.24 and 8.25). Thus, with Class-*A* operation, there can be no voltage drop at signal frequency across the supply impedance and none of the troubles outlined in Section 8.5. Hence decoupling is simplified. Also, any departures of the H.T. supply from absolute constancy (mains hum, etc.) produce opposing effects in the two valves and do not affect the output. H.T. supply smoothing can therefore be less elaborate than would otherwise be necessary. In transistor circuits the emitter bias smoothing might be omitted for the same reason.

(3) The fundamental-frequency components in the two halves of the circuit reinforce one another in the output transformer. Clearly the third and other odd harmonics do likewise since they retain the same relative phase as the fundamental which itself can be regarded as the first, and therefore an odd, harmonic. The second and other even harmonics, however, have the opposite phase relationship and currents at these frequencies in the two halves of the output transformer primary produce opposing effects and, therefore, no output. Thus the push–pull connection results in the elimination of even harmonics from the output. Consequently, for a given permissible amount of distortion, more power can be taken from valves in push–pull than from valves used singly or in parallel.

4. Although it is outside the scope of this book it is worth mentioning here that when bias is applied to near cut off (to give Class-*B* working) considerable economies and increase in efficiency of working are obtained. This idea is arrived at from a consideration of the standing current in Fig. 8.24 as drawn and as it would be if the valves were biased to near cut-off, the saving in H.T. supply current, particularly when the signal is small or zero, is plain to see.

To conform with the general principles already derived the equivalent circuit for one half of a push–pull stage may be drawn as in Fig. 8.26: a thermionic valve circuit is used but the same principles can be applied to a consideration of transistor circuits. Fig. 8.27 shows the corresponding circuit for the other half of the push–pull stage assuming the phase of the input to be the same, i.e. μv_g acting from cathode to anode in the external circuit. However, the two inputs are not independent; they are in anti-phase. To allow for this, the phase of one of them must be reversed. This is done in Fig. 8.28 which shows the two previous figures combined into one with one of the inputs reversed in phase. It is seen that the two e.m.f.s (μv_g) act in the same direction round the circuit and that the total a.c. resistance is $2r_a$.

8.9. Circuits and Features of Low-power Audio-frequency Amplifiers

To illustrate the various aspects of amplifier operation which have been dealt with, a number of complete circuits will now be discussed. When valves are included they will be drawn as if each were included in a separate envelope; it will be realised that a single valve may often contain two or more sections.

Heaters are not shown, external leads to heaters from a.c. sources are run as twisted pairs to reduce the magnetic field which otherwise might cause hum in the output. The heater supply (*XX* in Fig. 8.29, for instance) is often centre-tapped and earthed to fix the potential and prevent electrostatic induction of hum.

Fig. 8.29 shows a two-stage voltage amplifier designed to accept a signal of small value (a few millivolts) and to amplify it to about one volt at which level it is suitable for feeding to a line or for further amplification in another amplifier, such as that of Fig. 8.31. R_1, of about half a megohm, is the grid resistor to *V*1 and transfers the grid-bias voltage from the bottom of R_2 to the grid. R_2C_1 provides self-bias for *V*1 (*see* section 8.6).

$$R_2 = \frac{\text{Grid bias required}}{\text{Average cathode current}}$$

$$C_1 \simeq 50\ \mu\text{F}$$

The input lead to *V*1 grid, and probably that from *V*1 anode to *V*2 grid, is usually screened to prevent electrostatic pick-up of hum. *V*1 itself is a low-noise pentode, of extremely rigid construction and conservatively rated. R_4 is the anode load of large value, 0·1 MΩ or more, to enable maximum amplification of the small signal to be obtained (*see* Section 8.3). Both R_4 and R_5 are high-stability resistors to give low noise. R_3C_2, about 20 kΩ and 8 μF respectively, constitutes the anode-decoupling circuit (*see* Section 8.5). R_5 drops the supply voltage to the value required for the screen, usually about equal to the quiescent anode potential. C_3 is sufficiently large, about 0·1 μF, to prevent the screen potential varying at audio frequency which would cause negative feedback and reduced gain.

C_4 transfers the signal to the grid of *V*2 and isolates the latter from H.T. positive. C_4 has a capacitance, about 0·01 μF, large enough to offer a low reactance to the lowest frequency to be amplified and to satisfy other conditions (*see* Chapter 10, *Radio and Line Transmission, Volume* 1). In a low signal-level circuit of this sort where R_4 is very large and the anode voltage of *V*1, therefore, very

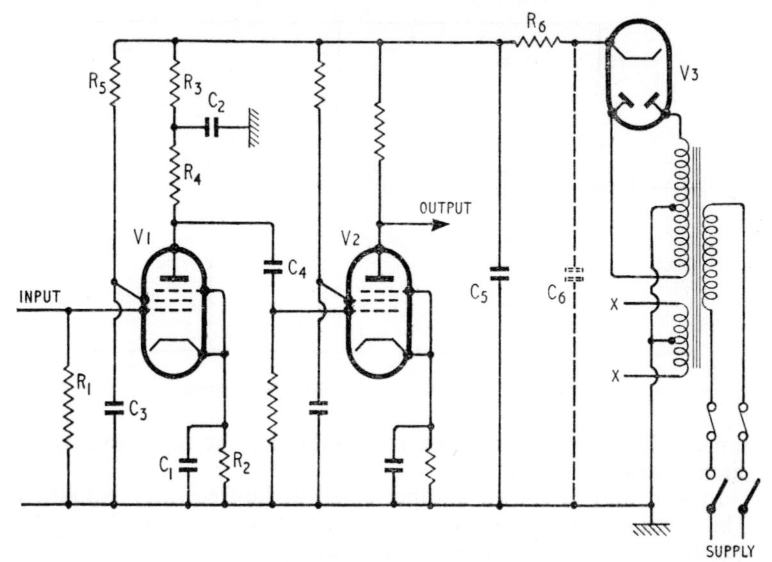

Fig. 8.29. Two-stage valve pre-amplifier

small, it is easy to dispense with C_4 altogether. All that is necessary is to arrange to raise the cathode voltage of *V*2 slightly to allow for the fact that the grid is now at the same potential as *V*1 anode.

Components in the second stage fulfil similar functions to those in the first.

The power supply is derived from a standard full-wave rectifier the output of which is smoothed by R_6, of about 10 kΩ, and C_5, about 16 μF. The current drain of the amplifier is very small so that *RC*-smoothing is quite satisfactory. A reservoir capacitor, C_6, shown dotted, is normally included; it is omitted here because of the very small current.

Fig. 8.30 shows a similar transistor circuit. Resistance networks of the form R_1, R_2 and R_3 fix the base and emitter potentials at suitable values. The component values depend on conditions but would be of the order of 50 kΩ, 10 kΩ and 1 kΩ respectively. The by-pass capacitors C_2 and C_3 and the coupling capacitors C_1 and C_4 have higher values than in the higher impedance valve equipments, say 100 μF and 10 μF respectively.

The load resistor R_4 is probably about 3 kΩ while R_6 and R_7 are of the same order as R_2 and R_3. R_5 will be smaller than R_1, say about 20 kΩ.

As drawn the input impedance is low, not more than a few kilohms. For use with an input source of high impedance a high impedance potentiometer, say about 1 MΩ, may be connected across the input terminals and a quarter-megohm resistance put in series with the source. Used in this way the input coupling capacitor may have the

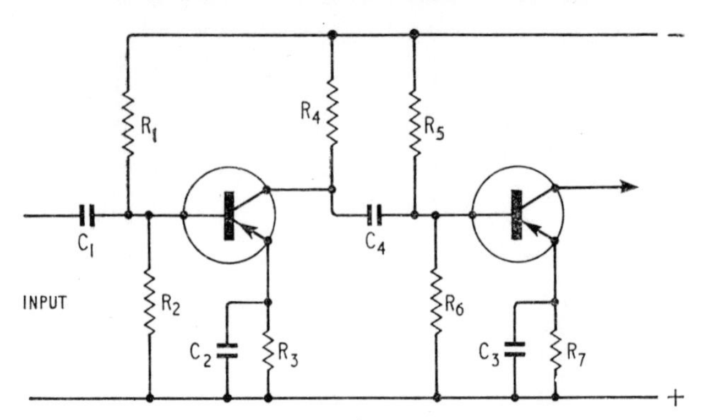

Fig. 8.30. Two-stage transistor pre-amplifier

same order of value as would be suitable for the input to a valve circuit—about 0·1 μF.

In Fig. 8.31 is shown another small audio-frequency amplifier. This gives an output of two or three watts with an input of about a quarter of a volt. It could, for instance, receive its input from the circuit of Fig. 8.29, or constitute the last two stages of a radio re-receiver. *V*1 is a triode and is provided with a potentiometer in its input circuit. This can be used as a volume control, it also serves to transfer the bias voltage to the grid. The arrangement C_1R_2 is a tone control in which design is sacrificed to simplicity, the performance varying with the setting of R_1 and with the source resistance. The basis of the action is as follows. With R_2 set to zero, C_1, of about 0·001 μF, by-passes the upper frequencies which, therefore, do not appear in the output. Increasing the amount of resistance decreases the shunting effect of C_1 and permits more of the high frequencies (" top ") to pass through the amplifier.

Most of the components function as described for the circuit of Fig. 8.29. There is no decoupling capacitor to the screen of *V*2 because this is taken direct to H.T. positive and is therefore effectively earthed to audio-frequencies by the final capacitor of the power supply (not shown in the figure). The output tetrode feeds the loudspeaker through a matching transformer.

A 10-watt amplifier with push–pull output is shown in Fig. 8.32. The pentode valve amplifier *V*1 is similar to stages already described; decoupling is provided by R_1C_1. *V*2 is a phase splitter; its function is to provide the equal and opposite voltages needed for the inputs to the push-pull valves *V*3 and *V*4 (*see* Section 8.8.2). R_2 and R_3 are of large and equal value so that the input to *V*2 causes the voltages across R_2 and R_3 to change by equal amounts in opposite directions; these changes are transmitted to *V*3 and *V*4 via coupling capacitors C_2 and C_3. The push–pull valves provide the power output into the output transformer which matches to the loudspeaker speech coil. Separate self-bias components are shown for the two push–pull

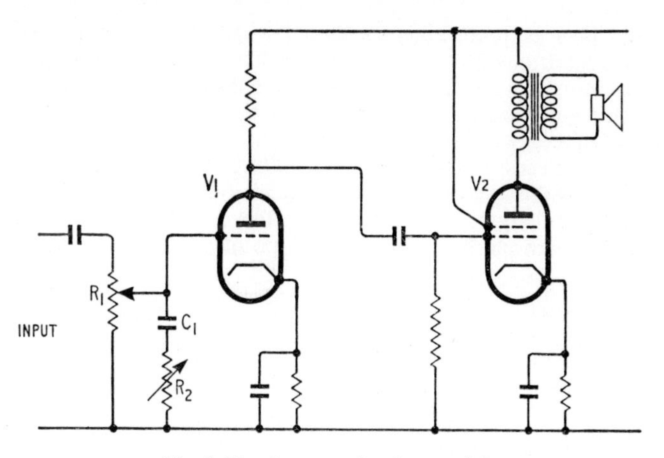

Fig. 8.31. 2-watt a.f. valve amplifier

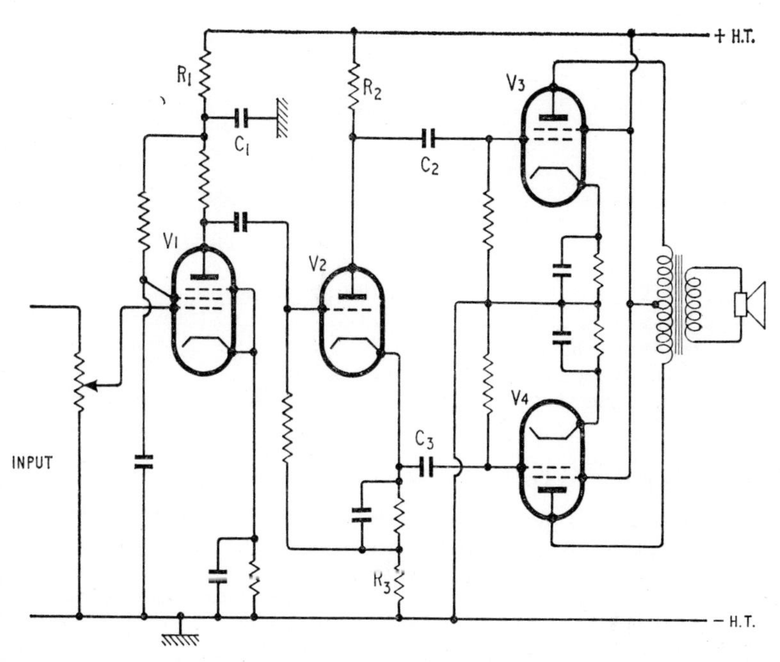

Fig. 8.32. 10-watt amplifier: push–pull output

valves. This is preferable to the arrangement of common bias resistor sometimes used because it means that the circuits are self-compensating to a considerable extent and if the valves age differently the circuit still works satisfactorily. With common bias the circuit is not satisfactory unless the valves are almost identical.

It is often necessary to send signals along a length of transmission line. The length involved may be measured in miles or feet but, whatever the length, the system is often matched to the line by means of a cathode follower. This type of circuit is the subject of

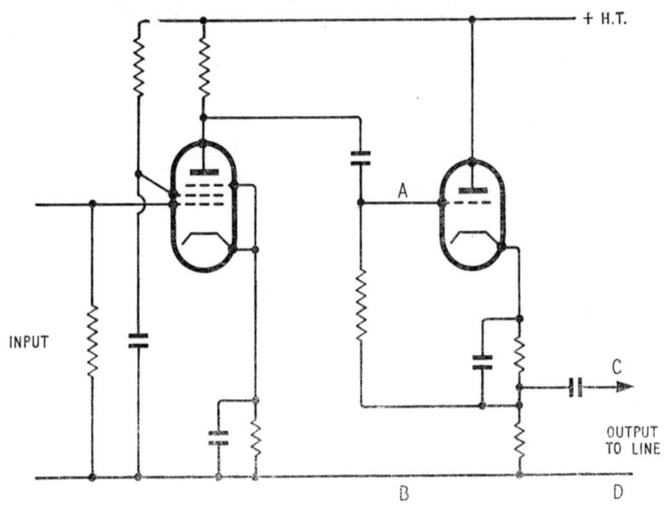

Fig. 8.33. Cathode follower: feed to line

later work but a brief outline of the properties of such circuits is given in Section 8.10.10 of this volume. Among transistor circuits the common-collector type of connection has similar properties and is sometimes used for the same purpose.

Fig. 8.33 shows a pentode amplifier stage followed by a cathode follower.

8.10. Negative Feedback

By feedback is meant the transfer of part or all of a signal from some stage of an amplifier to an earlier stage. When the voltage which is fed back is in the same phase as the original the feedback is said to be *positive*. If it is in opposite phase the feedback is *negative*. If the voltage fed back is proportional to the voltage in the output circuit it is called *voltage feedback*, if it is proportional to the current it is *current feedback*.

Positive feedback gives a tendency to instability. A possible source of unwanted positive feedback was discussed in Section 8.5; an example of desirable positive feedback is that introduced in oscillator circuits in Chapter 11, and in regenerative detectors in Chapter 12.

8.10.1. EFFECT ON GAIN

Fig. 8.34 shows an amplifier of gain A to which an input v_{in} is applied. Of the output some fraction β is fed back to the input in antiphase to v_{in}. The difference represents the net input v. The output will be Av and the voltage fed back $A\beta v$.
Thus:

$$v = v_{in} - A\beta v \text{ or, } v = \frac{v_{in}}{1 + A\beta}$$

$$\text{Output } Av = \frac{Av_{in}}{1 + A\beta}$$

$$\text{Net gain} = \frac{\text{Output}}{\text{Input}}$$

$$= \frac{A}{1 + A\beta} \qquad (2)$$

The application of negative feedback, therefore, reduces the gain in the ratio $1 + A\beta$.

This reduction need not necessarily mean a reduction in output. Provided the input is stepped up in the ratio $(1 + A\beta)$ when negative

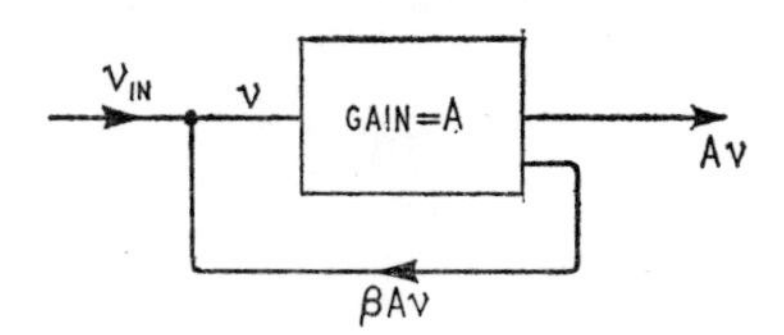

Fig. 8.34. Negative feedback

feedback is applied the output will be unchanged. This, of course, may require the use of an additional amplifier.

Example 8.11

A stage of gain 20 has 20 per cent negative feedback applied. By how much does the introduction of the feedback reduce the gain? If 1 volt input just provides the requisite output before the application of negative feedback, to what level must the input be raised when the feedback is applied?

$$\text{Gain without feedback} = 20$$

$$\text{Gain with feedback} = \frac{20}{1 + 20 \times \frac{1}{5}} = 4$$

Therefore the gain is reduced in the ratio 5 : 1. For a given output the input with negative feedback must be five times greater than without. That is, 5 volts input will be needed.

Note that although the application of negative feedback requires an increase in input for the same output the valve or transistor is still handling the original input swing and can therefore operate over the same load line from the same operating point and with the same bias voltage.

8.10.2. EFFECT ON STABILITY OF GAIN

Equation 2 above reduces to $1/\beta$ if the product $A\beta$ is several times greater than unity. That is, as long as this assumption holds good the gain is independent of the stage gain A. This means that it is independent of any of the factors which affect the stage gain, e.g. the amplification factor of the valve or transistor, the supply voltage.

Example 8.12

A circuit of gain 50 has 20 per cent feedback applied. The valve is changed so that the nominal gain becomes 25. Calculate the change in overall gain.

(i) $\beta = \frac{1}{5}$; $A = 50$

Therefore gain with feedback

$$= \frac{50}{1 + 10} = 4{\cdot}6$$

(ii) $\beta = \frac{1}{5}$; $A = 25$

Therefore gain with feedback

$$= \frac{25}{1 + 5} = 4{\cdot}2$$

Thus a change in nominal gain in the ratio 2 : 1 results in a change in overall gain of less than 10 per cent.

8.10.3. EFFECT ON DISTORTIONS INTRODUCED IN THE AMPLIFIER

Suppose that an amplifier to which negative feedback is applied has an output which includes d volts distortion produced within the amplifier. A distortion voltage βd is fed back to the input; after amplification this appears in the output as βdA volts. Let the corresponding distortion voltage produced in the amplifier, i.e. the one which would appear in the output in the absence of negative feedback, be D volts.

It follows that

$$D - \beta dA = d, \text{ or, } d = \frac{D}{1 + A\beta}$$

Thus the internally-produced distortion is reduced in the same ratio as the gain. This may be deduced without mathematics as follows. Any signal going through the amplifier is reduced by negative feedback in the ratio $\frac{1}{1 + A\beta}$. This reduction applies to signal and to internally-generated distortion. The signal, together with any distortion associated with it in the input, can be restored to its original level by increasing the input. This does not, however, increase the internally-produced distortion which remains at its reduced level.

The advantages are clear. A given voltage amplifier can be over-run somewhat to obtain greater output without the addition of

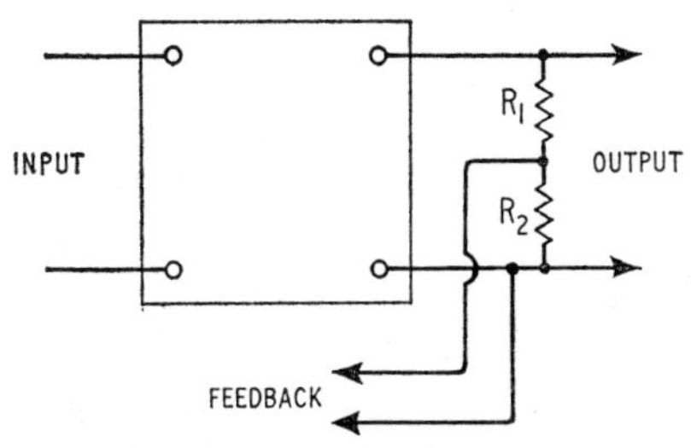

Fig. 35. Voltage feedback

extra stages; without negative feedback this would result in excessive distortion. Where no more than extremely small distortions are permissible the application of negative feedback can bring about the desired result. Hum introduced at high level in an amplifier can be greatly reduced by the judicious application of negative feedback from this stage: this means that smoothing, an expensive process involving bulky and heavy components, need not be carried out as effectively as would be necessary in the absence of negative feedback.

It is fair to add, of course, that some distortions may possibly be introduced in providing the additional amplification to make good the actual loss of gain in the stages over which the negative feedback is applied.

8.10.4. SELECTIVE FEEDBACK

It has so far been assumed that a given percentage of feedback has been applied irrespective of frequency. It is sometimes desired to confine feedback to a certain range of frequencies, as for instance, to minimise the effect of an interfering signal, to reduce the output at a particularly-undesirable frequency of distortion, or to produce a non-level response for some particular application, e.g. to equalise the output of a pick-up. To bring this about the feedback can be applied in a frequency-selective manner. Fig. 8.35 shows a resistive potential divider across the output of an amplifier. Here the fractional feedback is

$$\frac{R_2}{R_1 + R_2}$$

and is independent of frequency. In Fig. 8.36 the resistors are replaced by impedances one of which, at least, is assumed to be reactive: the fractional feedback,

$$\frac{Z_2}{Z_1 + Z_2}$$

clearly depends on frequency. Thus to diminish the feedback for a given narrow band of frequencies and so to give a greater output

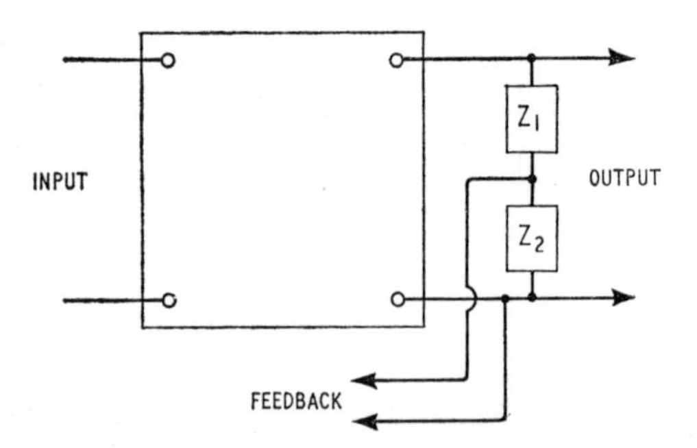

Fig. 8.36. Selective voltage feedback

at these frequencies, Z_2 could be an acceptor circuit tuned to the mid-frequency. To provide attenuation which increases with frequency Z_2 could be inductive and Z_1 resistive, and so on.

8.10.5. PHASE SHIFT IN FEEDBACK SYSTEMS

Feedback systems are potentially dangerous in that although negative feedback is required the circuit which has been set up is just as suitable for conveying positive feedback. In fact negative feedback systems are prone to instability, usually at a frequency outside the range for which the equipment is designed to work. Since the frequencies of instability are usually outside the normal pass-band of the circuit the tendency to instability can often be removed by introducing sharp cut-off filters into the circuit.

The matter of instability in negative feedback amplifiers is beyond the scope of this volume but it should be noted as one needing special attention.

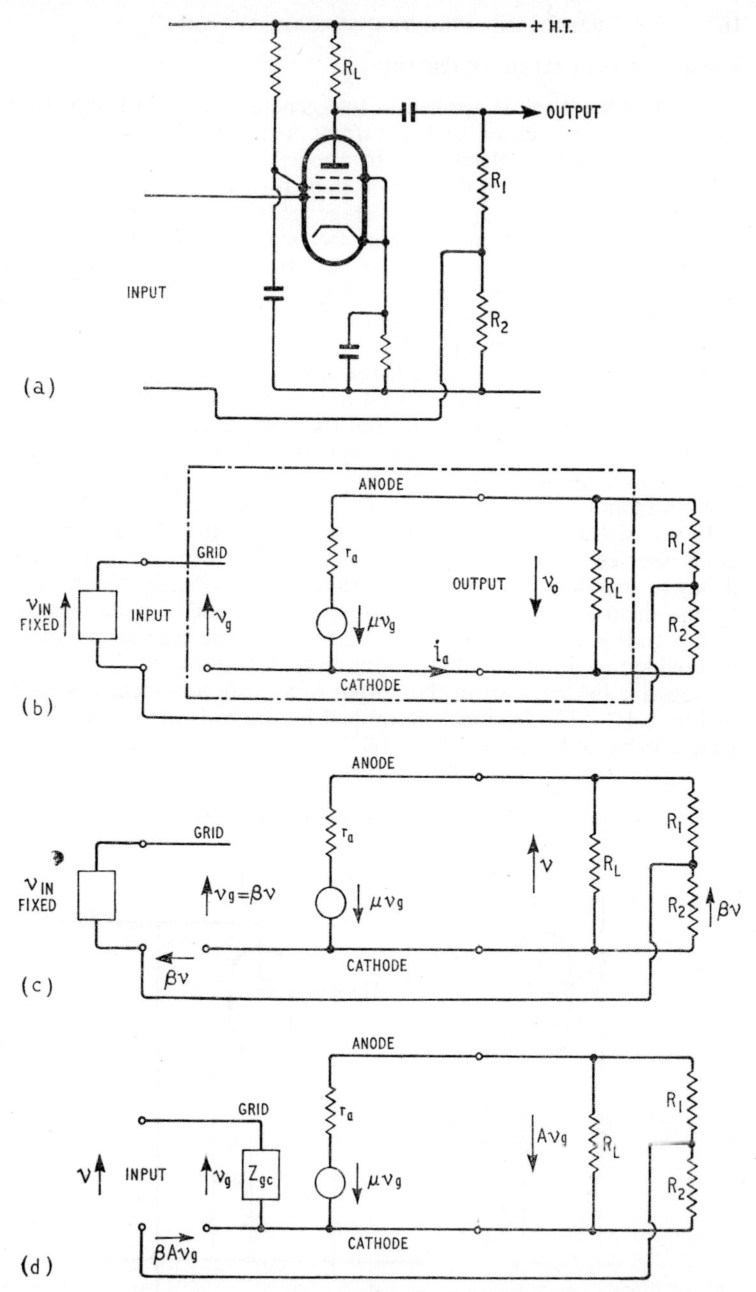

Fig. 8.37. (a) Voltage feedback applied to one valve stage, (b) equivalent circuit illustrating Example 8.13, (c) equivalent circuit illustrating Example 8.13, (d) equivalent circuit illustrating Example 8.13

8.10.6. TYPES OF FEEDBACK CIRCUIT

Negative feedback is applied in two general ways: in proportion either to the voltage or to the current in the final stage of those stages over which feedback is to be applied.

The circuits of Figs. 8.35 and 8.36 show forms of voltage feedback. Fig. 8.37 is of a circuit in which voltage feedback is applied over one stage and Fig. 8.38 of one in which two stages are involved. Note that the connections are made so that the sense of the feedback shall be correct. Over one stage the feedback is from the anode of the valve to its grid while over two stages it must be from the anode of *V*2 to the cathode of *V*1.

Fig. 8.39 shows a current feedback circuit; here the voltage which is fed back depends on the current in the output circuit. The load current is also the current in R_2 and the voltage across this resistor, which is thus proportional to the load current, is applied to the grid and cathode in series with the input voltage. The grid bias is derived from the self-bias circuit, R_1C, in the usual way.

In transistor circuits, in which, in general, it is inadvisable to apply the feedback over more than two stages because of the tendency to instability, a common method of applying the feedback is by the inclusion of a small resistor (not by-passed) in the emitter lead. In Fig. 8.40, which is of an output stage, the negative feedback is provided by *R*, of about one ohm, in the emitter lead.

Negative feedback from the output of a push–pull stage is shown in Fig. 8.41. The feedback is applied by the resistor *R* which may have a value of the order of 100,000 ohms.

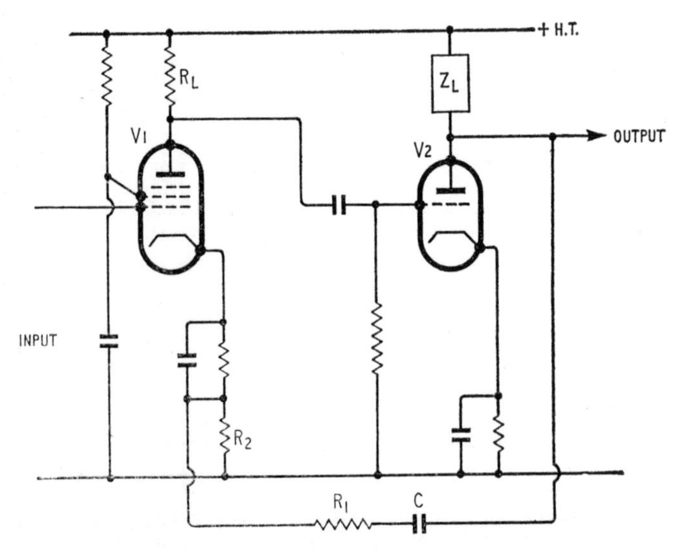

Fig. 8.38. Voltage feedback over two valve stages

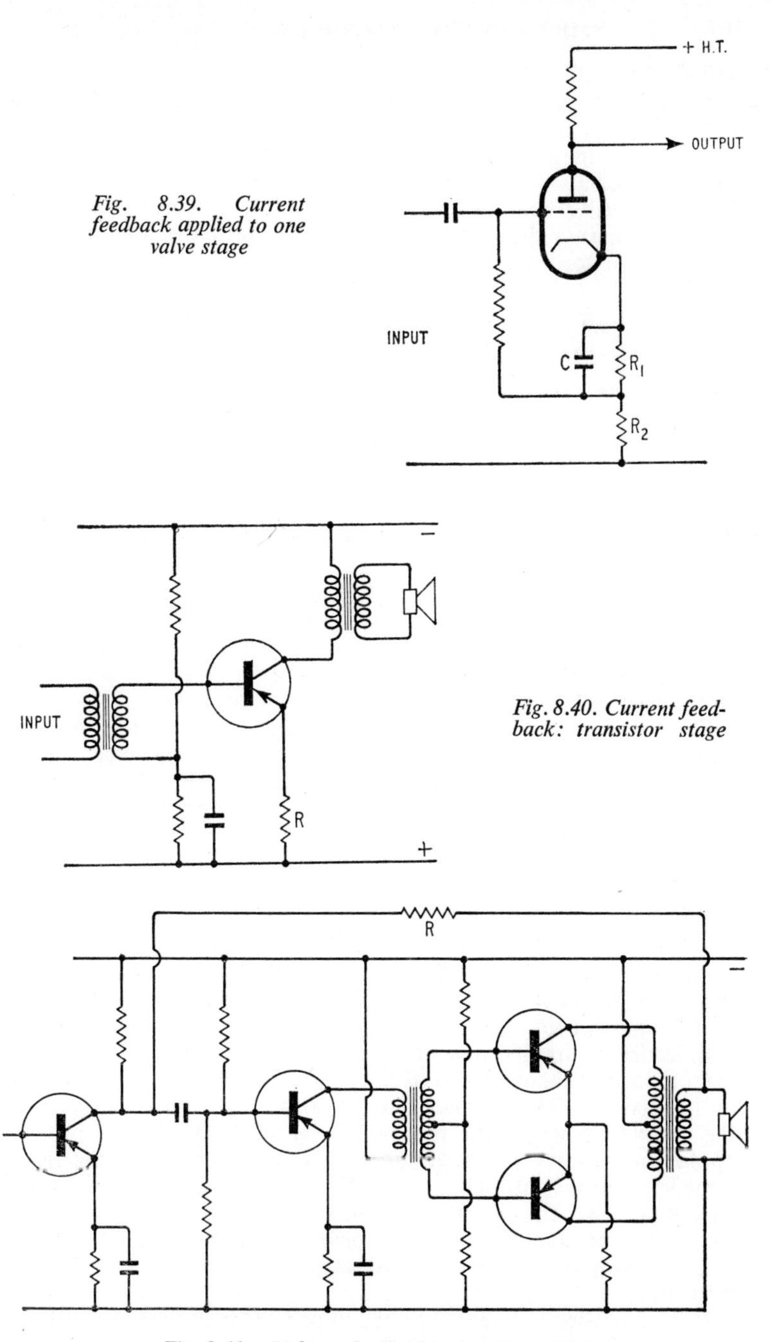

Fig. 8.39. Current feedback applied to one valve stage

Fig. 8.40. Current feedback: transistor stage

Fig. 8.41. Voltage feedback: transistor stages

8.10.7. EFFECT ON OUTPUT IMPEDANCE

The output impedance of a stage is that seen looking into the stage from its output terminals. Thus the output impedance is given by the magnitude of a change in voltage applied across the output terminals divided by the resulting current change. For a thermionic valve without feedback it is simply r_a, because a change in voltage across the output terminals has no effect on μv_g (Fig. 8.42). With feedback the situation is different because a change

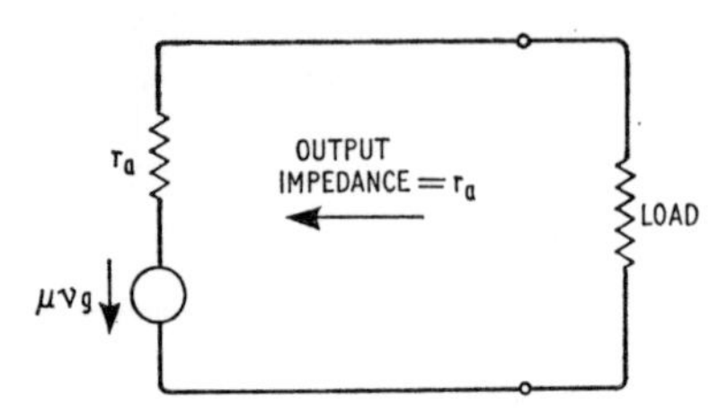

Fig. 8.42. Output impedance

in the output voltage results in a change in the input voltage as well. The change in output current, therefore, may be more or less than that which would result if r_a were the only controlling factor. That is, the output resistance may be less or more than r_a depending on the feedback.

For valve and transistor circuits, negative voltage feedback decreases and current feedback increases the output impedance. Thus voltage feedback tends to maintain the output constant irrespective of the value of the load. Although the output impedance is changed by the application of feedback the optimum load for a stage to which feedback is applied is not altered from that needed in its absence.

8.10.8. EFFECT ON INPUT IMPEDANCE

The input impedance of a stage is that seen looking into the stage from its source of signal. It is thus given by the ratio of the change in voltage across the input terminals to the resultant current change in the same circuit.

In feedback circuits some of the change in the output which results from the input change is fed back to the input and thus modifies the initial change. The ultimate effect of this modification depends on the manner in which the feedback is applied. If the feedback is applied in series with the input the effect is to increase the input impedance in the ratio $(1 + A\beta)$. Such an increase is often a great advantage, particularly in transistor circuits in which the basically low input impedance is often a considerable drawback. The effect also operates strongly in the cathode follower circuit (*see* Section 8.10.10).

If the feedback voltage is applied in parallel with the input voltage the input resistance is decreased.

8.10.9. SUMMARY OF EFFECT ON OUTPUT AND INPUT IMPEDANCE

The foregoing two sections may be summarised as follows. The effect on the output impedance is determined by the type of feedback; voltage feedback decreases and current feedback increases the output impedance. The effect on input impedance depends on whether the feedback is in series or parallel with the input; series application increases and parallel application decreases the input impedance.

Example 8.13

With the aid of an equivalent circuit of a single-stage amplifier show that the application of series-connected negative voltage feedback results in (a) a decrease in the output resistance and (b) an increase in the input resistance of the stage.

(a) Let the stage be that of Fig. 8.37 (a). The equivalent circuit is in 8.37 (b). In view of the importance of the correct assigning of directions to voltages and currents in this type of example the circuit is first considered without negative feedback; this part of the circuit is shown within the dotted line in the figure. To recapitulate the basic principles: an increase in v_g in the direction shown, i.e. at the grid with respect to the cathode, results in an increase in anode current from cathode to anode round the external circuit (this is the conventional direction of flow—the electrons go the other way). μv_g, therefore, acts in the direction shown and the output voltage increases in the direction of the arrow v_0 (i.e. the cathode potential rises with respect to that at the anode, or, the anode voltage falls with respect to that of the cathode).

Returning to the complete equivalent circuit, re-drawn in Fig. 8.37 (c) and taking account of the feedback, let the input voltage v_{in} be fixed and the output voltage changed so as to make the anode voltage (v volts) more positive. The voltage fed back as negative feedback is that across R_2; its magnitude is βv volts and it is positive to the grid. Since v_{in} is fixed it can be omitted from consideration and the equivalent circuit, therefore, shows only βv volts acting in the grid circuit. The change in grid voltage is thus $v_g = \beta v$. The valve generates a change μv_g $(= \mu\beta v)$ in the direction of the arrow μv_g. Thus the applied change in voltage v across the output terminals brings into being a further voltage $\mu\beta v$ in the output circuit. From the figure it can be seen that the two voltages assist one another (both are acting anti-clockwise round the circuit) and the resultant e.m.f. is, therefore,

$$v + \mu\beta v = v(1 + \mu\beta).$$

The circuit resistance is r_a (the resistors R_L, R_1 and R_2 have been accounted for because the voltage v which shunts them is to be used in the calculation).

The current change, therefore, is

$$i = \frac{v(1 + \mu\beta)}{r_a}$$

The output resistance is the applied change in voltage across the output (v) divided by the resulting current change; i.e. it is $\frac{v}{i}$.

$$\text{Output resistance} = \frac{v}{i}$$

$$= \frac{v}{\frac{v(1 + \mu\beta)}{r_a}}$$

$$= \frac{r_a}{1 + \mu\beta}$$

In the absence of negative feedback the output resistance is r_a. Thus the application of negative feedback reduces the output resistance in the ratio $(1 + \mu\beta)$.

(b) Again use the circuit of Fig. 8.37 (a).

The input impedance in the absence of feedback is the grid-cathode impedance Z_{gc}.

Draw the equivalent circuit Fig. 8.37 (d) including the grid-cathode impedance.

Change the input voltage by v volts, positive to the grid, say, i.e. in the direction shown in Fig. 8.37 (d); let the resultant grid-cathode voltage change be v_g. As a result the valve generates μv_g which acts in the output circuit and, if the stage gain is A, yields an output voltage Av_g in the direction shown (i.e. positive to the cathode). The feedback voltage is βAv_g and is thus negative to the grid; i.e. it opposes the applied voltage v. Thus:

$$v_g = v - \beta Av_g$$

$$v_g + \beta Av_g = v$$

$$v_g = \frac{v}{1 + A\beta}$$

In the absence of feedback, $v_g = v$, and the input current (through the input impedance Z_{gc}) is v/Z_{gc}. The introduction of negative feedback reduces the effective grid-cathode voltage from v to

$$\frac{v}{1 + A\beta}$$

and the input current from $\frac{v}{Z_{gc}}$ to

$$\frac{1}{1 + A\beta} \times \frac{v}{Z_{gc}}$$

Thus, for a given input voltage v the introduction of negative feedback reduces the input current in the ratio $(1 + A\beta)$, i.e. it causes an increase in the effective input impedance in the ratio $(1 + A\beta)$.

8.10.10. CATHODE FOLLOWER

An extreme example of negative feedback is the cathode follower. In this circuit the whole of the output voltage is fed back as negative feedback. A simple circuit is given in Fig. 8.43 (compare the

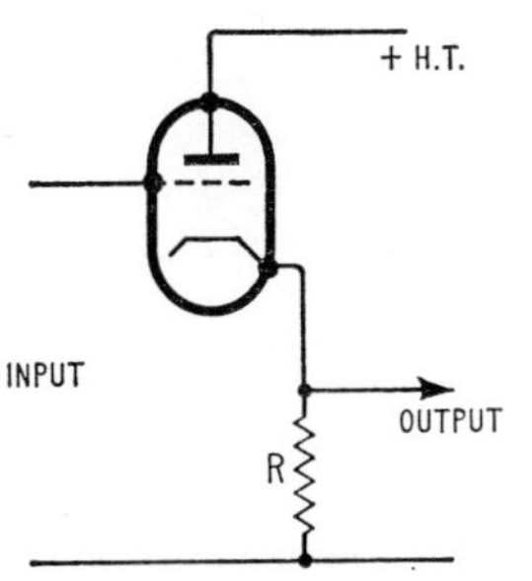

Fig. 8.43. Cathode follower

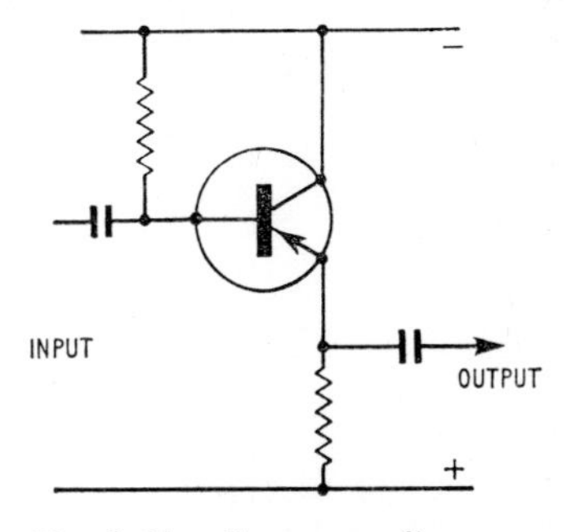

Fig. 8.44. Common collector or emitter follower

common-collector, or, emitter-follower, circuit of Fig. 8.44 which acts similarly).

The cathode resistance is R, the anode being taken direct to H.T.+. The entire output voltage is fed back in series with the input so that $\beta = 1$. It follows:

1. From Equation 2: the stage gain $= A/(A + 1)$, i.e. it is slightly less than 1 and is largely independent of the load.
2. The input resistance is high and the input capacitance is low. Compare the values with those for a triode normally connected (*see Radio and Line Transmission, Volume* 1, Section 10.6). This means that the loading of the previous stage is very light and the risk of distortion due to current drain or input capacitance is reduced.
3. Distortion is negligible and large inputs can often be handled without fear of overloading.
4. The output impedance is low and approximates to the value $1/g_m$. The stage can, therefore, be used as a matching device

to feed circuits of low impedance. Even if the capacitance of the output load is considerable and the frequency of operation high the output of the cathode follower will be maintained.

In its role of matching device the cathode follower is similar to a transformer but is superior because of the fact that its output voltage is practically equal to its input and for the reasons listed above.

A practical example of a cathode follower stage is shown in Fig. 8.33.

8.10.11. EMITTER FOLLOWER

The operation of the emitter follower circuit may be exemplified by the circuit known as the AND circuit which is used in switching

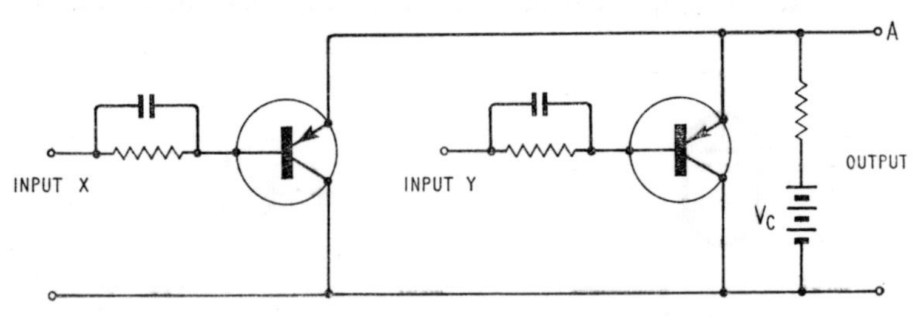

Fig. 8.45. AND circuit: illustrating emitter follower

systems. The AND circuit is shown schematically in Fig. 8.45. Briefly, if both input voltages are held steady at a value somewhat below V_c both transistors conduct and the potential difference between emitter and base is very small. Thus, the output at *A* is nearly the same as the input value. (It is, in fact, slightly above it.)

If both the inputs are raised both the emitter voltages rise simultaneously, that is, the output at *A* follows the input voltage.

If only one of the input voltages is raised, the emitter voltage of that transistor (which otherwise would faithfully follow the input), cannot follow because it is tied to the emitter of the other transistor which does not move in potential because its input has not changed.

Thus, while a single emitter follower provides a collector output voltage which follows the base voltage (within the range which permits the transistor to conduct), the arrangement of Fig. 8.45 provides a means of registering (by the output at *A*) when both the inputs at *X* and *Y* rise simultaneously.

Questions

1. An audio-frequency amplifier uses a triode valve with an anode slope resistance of 10 kΩ and an amplification factor of 50. What value of anode load resistor will be required to give a voltage amplification of 40 times?

If the direct current taken by the valve is 5 mA what should be the power rating of the anode load resistor? (*C & G*, 1958.)

2. Give the circuit diagram of an audio-frequency amplifier capable of an output of about 10 W. The circuit of the H.T. and L.T. power supply unit may be omitted.

Explain how (a) the level and (b) the tone quality (frequency response) of the audio output could be controlled. (*C & G*, 1951.)

3. Give examples of (a) coupling and (b) decoupling circuits in an audio-frequency amplifier. State the purpose of each of these circuits and explain how they function. (*C & G*, 1952.)

4. Give the circuit of a two stage resistance-capacitance coupled audio-frequency amplifier using triode valves. If the slope resistance (anode resistance) and amplification factor of the valves are 20 kΩ and 100 respectively and the resistors in the anode circuits are each of 10 kΩ, what is the overall gain of the two stage amplifier? (*C & G*, 1949.)

5. Sketch the circuit of a two-stage transformer-coupled amplifier for audio-frequency operation, incorporating cathode resistors to provide grid bias.

Assuming the anode current in the output stage is 15 mA and the value of the associated cathode resistor is 400 ohms, what is the resulting grid bias voltage on this stage? (*C & G*, 1947.)

6. A valve used as an audio-frequency Class-A amplifier draws 30 mA at an anode voltage of 200 volts; the anode voltage varies between 40 and 360 volts and the corresponding anode current change is 50 to 10 mA.

Calculate:

(a) the efficiency of the stage,

(b) the output transformer turns ratio if the valve is matched for maximum transfer of power to a 2,000-ohm load, when the grid is driven by a sinusoidal voltage. (*C & G*, 1956.)

7. Sketch a typical family of output curves for a pentode r.f. voltage amplifier and explain, using load lines, the considerations affecting the choice of effective load resistance for optimum performance.

Use the above curves and the optimum load line to show how to determine the output wave shape for an assumed sinusoidal input.

8. Compare the methods, advantages and disadvantages of parallel and push–pull working as a means of amplification.

9. The gain of an a.f. amplifier normally falls off at high and low frequencies and is at a maximum over some middle-frequency range. Explain how the variations in gain can materially be reduced by the application of negative feedback.

10. Why is it that the application of negative feedback may result in a change in the input and output resistance of an amplifier stage?

Give two examples where such a change in impedance is advantageous.

9

Tuned Circuits

9.1. Response Curves

It has been shown in Chapter 10 of *Radio and Line Transmission, Volume* 1, that a circuit in which an inductor and a capacitor are connected in series or in parallel can be used as a means of selecting one narrow band of frequencies from others somewhat displaced of frequency. Such a circuit is called a tuned circuit. Graphs of

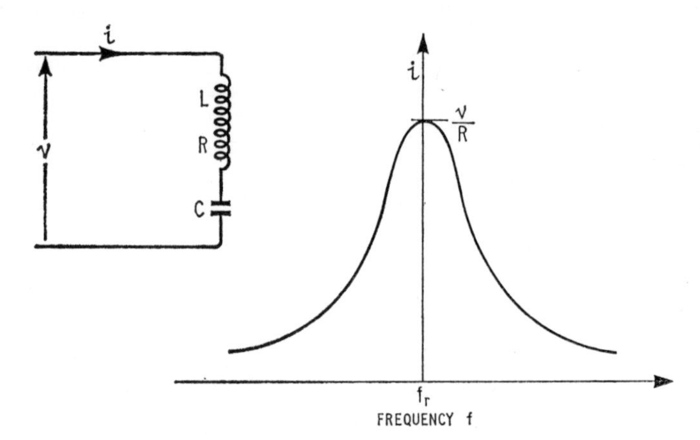

Fig. 9.1. Series resonance: response curve

current versus frequency for the series circuit and of impedance versus frequency for the parallel circuit have the same general shape (Figs. 9.1 and 9.2).

These curves are *resonance curves* or *response curves*. The condition of resonance for the parallel circuit is sometimes called anti-resonance and the response curve for such a circuit may be termed a curve of anti-resonance.

When the circuit parameters (i.e. L and C) are adjusted so that the circuit is in resonance with an applied signal the circuit is said to be tuned to the frequency of that signal.

For the series circuit the resonance frequency, f_r, is that at which the reactance of the inductor is equal in magnitude to that of the capacitor: at this frequency the current has its maximum value $\left(i_{\max} = \frac{e}{R}\right)$ and is in phase with the applied voltage.

$$2\pi f_r L = \frac{1}{2\pi f_r C}$$

$$f_r = \frac{1}{2\pi\sqrt{LC}} \text{ c/s} \tag{1}$$

(L in henrys; C in farads)

For the parallel circuit, unlike the series, the resonance frequency can be defined in several different ways: for instance, as the frequency at which the circuit impedance is a maximum; as the frequency at which the current is in phase with the applied voltage, or as the

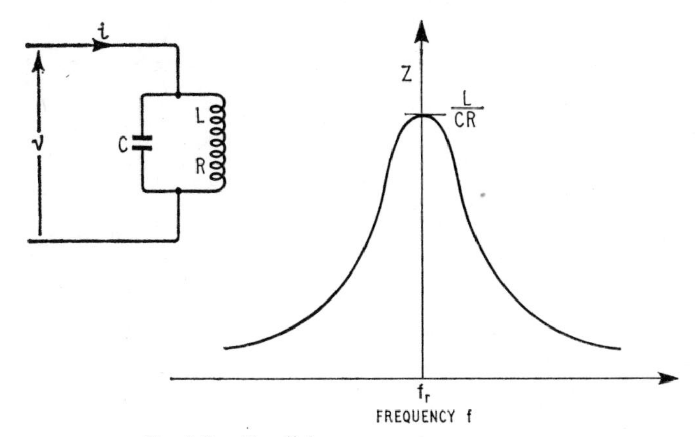

Fig. 9.2. Parallel resonance: response curve

frequency of resonance of the same components connected in series. If the circuit is of low loss (i.e. of high Q, *see* Section 9.3) the divergence between these various values is very small and it will here be assumed that resonance occurs at

$$f_r = \frac{1}{2\pi\sqrt{LC}} \text{ c/s}$$

and that at this frequency all the above definitions of resonance are satisfied.

9.1.1. EFFECT OF RESISTANCE

The frequency, f_r, about which the response curves are approximately symmetrical is given by the above expression. The shape of the curves depends on the circuit resistance: the smaller this is the steeper the gradient of the curves and the greater the discrimination between one frequency and another, i.e. the more selective the circuit is said to be. The effect is illustrated in Fig. 9.3. Here a series circuit is assumed: for the same components connected in a parallel circuit the curve is of the same shape but the ordinates are of impedance instead of current.

The current at resonance is given by $i = \frac{e}{R}$ where R is the total series loss resistance, normally simply that of the inductor and

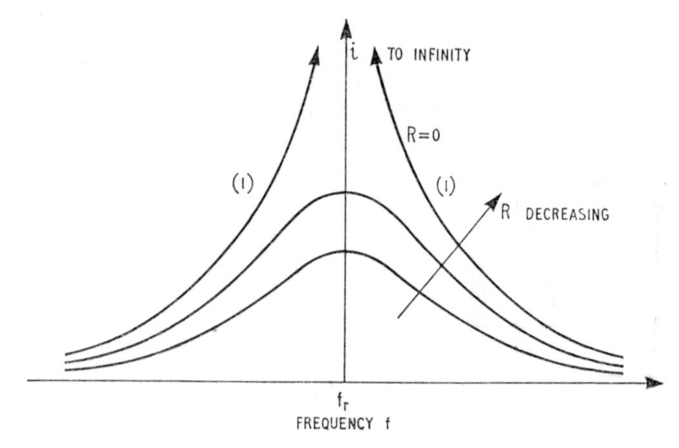

Fig. 9.3. Response curves: effect of resistance

capacitor. Of these the capacitor loss resistance is usually so much the smaller that it can be ignored and R is then the loss of the inductor. It must be remembered that at high frequencies the resistance is the r.f. resistance, which varies with frequency and is generally several times greater than the d.c. resistance.

If the resistance could be reduced to zero the current at resonance would be infinitely great (curve 1 in Fig. 9.3).

9.2. Series Circuit Impedance-frequency Characteristic

Fig. 9.1 shows the general shape of the current-frequency curve for a series circuit of low loss. The corresponding impedance-frequency curve is the inverse of this and is shown in Fig. 9.4. At zero frequency (i.e. d.c.) the capacitor reactance is infinite (because a capacitor cannot pass d.c.) and the inductive reactance is zero. As the frequency rises the capacitive reactance falls and the inductive reactance rises until the frequency reaches that of resonance. At this point the capacitive and inductive reactances cancel and $Z = R$. Further rise in frequency causes a steady increase in impedance because of the growth of the inductive reactance.

9.3. Concept of Q

9.3.1. SERIES RESONANT CIRCUIT

At resonance in a series circuit the voltages across the inductor and the capacitor are in anti-phase (because the reactances are of equal magnitude and the current common). Thus, their sum is zero

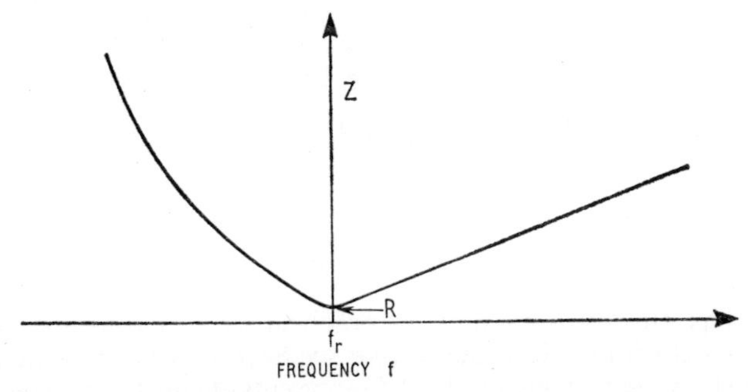

Fig. 9.4. Series resonance: variation of impedance with frequency

and the entire applied voltage is used in neutralising the voltage drop in the resistance.

Consider an e.m.f. applied to a series circuit at resonance. Let the reactance of the inductor, which at this frequency is equal to that of the capacitor, be X and the total circuit resistance R. Since the current is the same throughout a series circuit it is clear that the

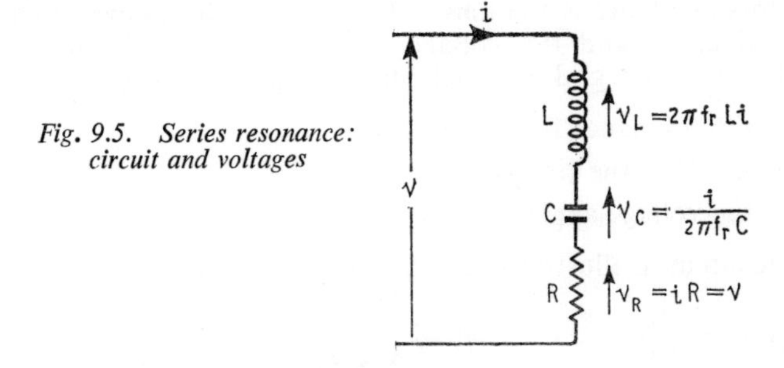

Fig. 9.5. Series resonance: circuit and voltages

ratio of the voltage across the inductor (or the capacitor) to that across the resistor is in the ratio X/R: that is, the inductor (or the capacitor) voltage is X/R times greater than the applied voltage. The ratio X/R is commonly of the order of 100 in low-loss circuits and is termed the magnification or *Q-factor* of the circuit. The Q-factor can be defined as the magnification factor of a circuit and indicates the number of times that the applied e.m.f. is magnified across the reactance of the inductor or capacitor in a series circuit at resonance.

Other things being unchanged, the lower the value of R the greater the Q value. Thus, the greater the Q value the lower the circuit loss so that Q can be regarded as a measure of the goodness of the circuit.

Fig. 9.5 shows the series circuit and the voltages referred to above.

$$Q = \frac{v_L}{v}\left(= \frac{v_C}{v}\right)$$

$$= \frac{2\pi f_r Li}{iR}$$

$$= \frac{2\pi f_r L}{R}$$

$$= \frac{1}{2\pi f_r CR} \qquad (2)$$

As has already been said, in practical circuits resistance is not usually deliberately introduced (*but see* Section 9.11) and the quantity R shown separately in Fig. 9.5 represents chiefly the r.f. resistance of the inductor. This increases with frequency so that Q does not necessarily increase with frequency. In fact it tends to remain fairly constant for a given inductor. (*See also* Sections 10.2.1 and 10.2.2.)

Example 9.1

An e.m.f. of 1 mV is delivered from a receiver aerial circuit to a series circuit consisting of $L = 200\ \mu$H and $C = 200$ pF. The effective loss resistance is 10 ohms. This circuit is in resonance with the applied e.m.f. and the voltage developed across the inductor is applied between grid and cathode of a valve. Calculate:

(a) the resonance frequency,

(b) the Q of the circuit,

(c) the voltage applied to the valve.

The circuit is illustrated in Fig. 9.6.

(a)

$$f_r = \frac{1}{2\pi\sqrt{LC}}$$

$$= \frac{10^7}{4\pi} \text{ c/s}$$

$$= 796 \text{ kc/s}$$

(b)

$$Q = \frac{2\pi f_r L}{R}$$

$$= 100$$

(c) $$v_L = Qv$$
$$= 100 \text{ mV}$$

and this is the voltage applied to the valve.

9.3.2. Q VALUE OF INDUCTOR AND CAPACITOR

The Q value of an inductor or a capacitor at a given frequency is the ratio of its reactance to its resistance.

$$Q_L = \frac{2\pi f_r L}{R_L}$$

$$Q_C = \frac{1}{2\pi f_r C R_C}$$

In a low-loss circuit of L and C in which most of the loss is in the inductor, the Q of the entire circuit is substantially the same as that of the inductor.

9.4. Parallel Circuit

9.4.1. IMPEDANCE-FREQUENCY CHARACTERISTIC

The impedance-frequency characteristic for a low-loss parallel circuit is of the form illustrated in Fig. 9.2. The sharpness of the

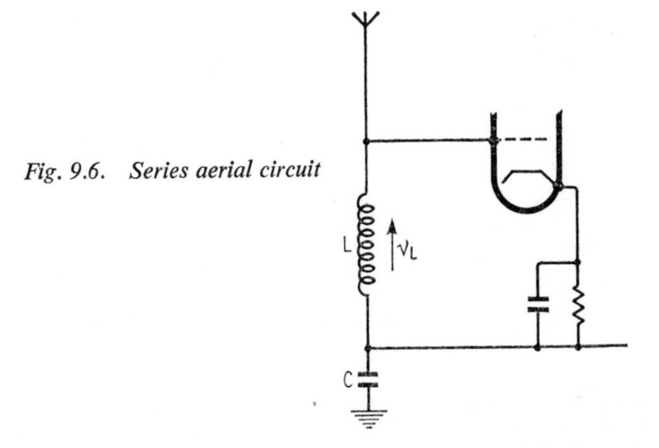

Fig. 9.6. Series aerial circuit

curve varies inversely with R (i.e. it varies with Q) in the same manner as does the current-frequency curve for the series circuit. The series circuit curves of Fig. 9.3, therefore, are equally valid for the parallel circuit if Z is substituted for i as the ordinate. In the parallel circuit the impedance would rise to infinity if the circuit losses could be reduced to zero.

9.4.2. DYNAMIC RESISTANCE

Fig. 9.7 shows a parallel LC circuit. A simplified way of obtaining the circuit impedance, Z, at resonance is as follows.

$$Z = \frac{Z_L Z_C}{Z_L + Z_C} = \frac{Z_L Z_C}{Z_s}$$

where Z_s is the series impedance round the circuit $Z_L Z_C$.
At resonance $Z_s = R = (R_L + R_C)$.

$$\therefore \qquad Z_{\text{res}} = \frac{Z_L Z_C}{R}$$

Since the circuit is assumed to be of low loss Z_L is approximately equal to X_L and Z_C likewise to X_C.

$$\therefore \qquad Z_{\text{res}} = \frac{X_L X_C}{R} = \frac{L}{CR}$$

(since, at resonance $X_L = 2\pi f_r L = X_C = \dfrac{1}{2\pi f_r C}$)

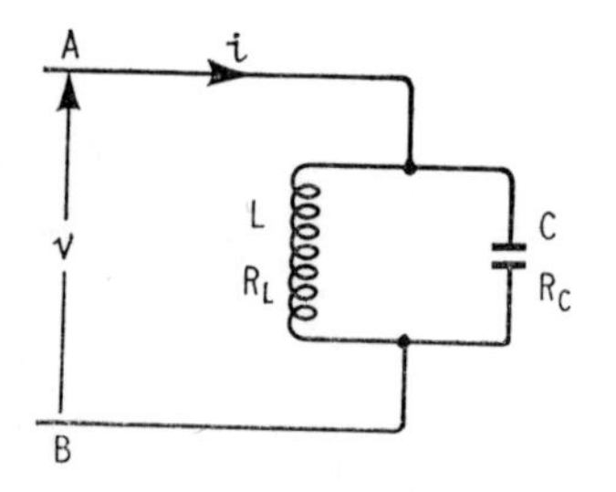

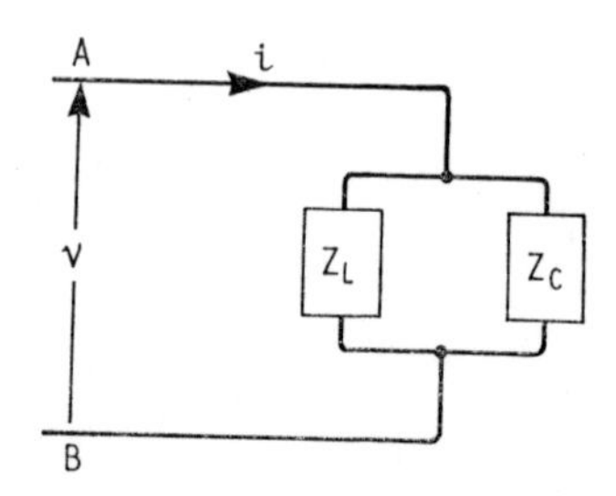

Fig. 9.7. Parallel circuit

Although, for simplicity, derived by means of approximations this is in fact an exact result. It shows that the impedance at resonance is independent of f; it is resistive. This resistance is termed the dynamic resistance, R_e, and it is interesting to note that it is Q times the reactance of either branch, for (using $\omega_r = 2\pi f_r$):

at resonance $\qquad \omega_r^2 LC = 1$, i.e. $C = \dfrac{1}{\omega_r^2 L}$

$$\therefore \quad R_e = \frac{L}{CR}$$

$$= \frac{\omega_r^2 L^2}{R}$$

$$= Q\omega_r L$$

$$= Q\left(\frac{1}{\omega_r C}\right)$$

The two resistances R and R_e must not be confused. R is the series resistance round the LC circuit and is small in a low-loss circuit. (If L and C were connected in series R would be the circuit impedance at resonance.) R_e is the resistance which is measured across the parallel circuit at resonance, i.e. across AB in Fig. 9.7. It has a high value in a low-loss circuit and is inversely proportional to R.

Example 9.2

A parallel LC circuit is resonant at 700 kc/s. Calculate the dynamic resistance if $L = 200$ μH and the series resistance round the circuit is 10 ohms. What would be the dynamic resistance if R could be reduced to 8 ohms?

$$R_e = \omega_r^2 L^2$$

$$= \frac{(2\pi \times 700 \times 10^3)^2 \times 200^2}{10 \times 10^{12}} \text{ ohms}$$

$$= \frac{4\pi^2 \times 49 \times 10^{10} \times 4 \times 10^4}{10 \times 10^{12}} \text{ ohms}$$

$$= 4\pi^2 \times 49 \times 40 \text{ ohms}$$

$$= 78{,}000 \text{ ohms}$$

If R were reduced to 8 ohms R_e would be increased in the ratio 10 : 8, that is to almost 100,000 ohms.

9.5. Response Curves Related to the Q-factor

Figs. 9.1 and 9.2 show, respectively, the current-frequency and impedance-frequency response curves for series and parallel resonant circuits. For circuits of a given Q value the curves are the same: that is, if a given L and C are connected in series a certain current-frequency curve results. If the same L and C are connected in parallel the resulting impedance-frequency response curve, if drawn to a suitable scale, is the same. This fact considerably simplifies the discussion because it is often unnecessary to consider whether a circuit is of the series or of the parallel type.

A good impression of the shape and sharpness of a response curve may readily be obtained if the Q value is known, as follows. Let the value of the curve at its apex (i.e. at resonance) be unity. If the frequency is moved up or down from resonance by an amount

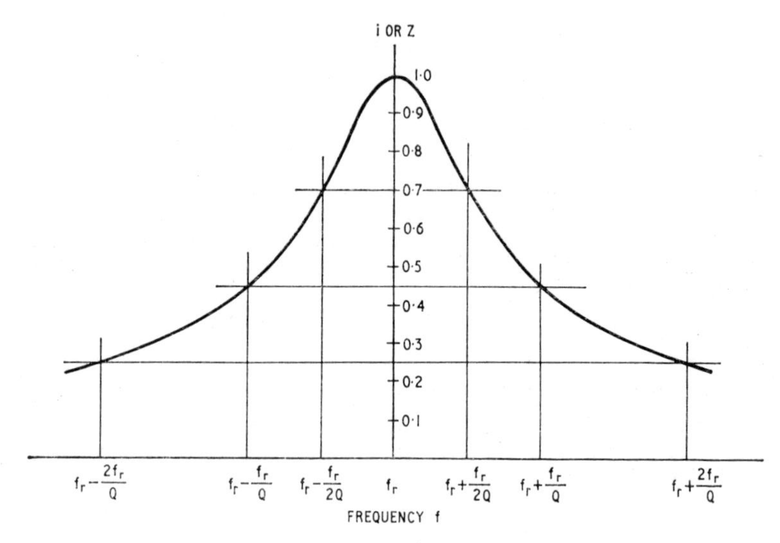

Fig. 9.8. Universal resonance curve

$_r/2Q$ then the ordinates of the curve at these points have the value $0{\cdot}707\left(=\frac{1}{\sqrt{2}}\right)$. If the frequency is shifted a further $f_r/2Q$, i.e. a total of f_r/Q from resonance, the value of the ordinates is reduced to about 0·447. A further change of frequency to make the total departure from resonance equal to $2f_r/Q$ reduces the ordinates to about 0·24. These points are plotted in the universal resonance curve of Fig. 9.8.

Values at any point may be calculated from the expression

$$N^2 = 1 + 4S^2 \tag{3}$$

Where N is the ratio of the response at resonance to that at the frequency considered, and

S is a measure of the frequency shift away from resonance, δf, in terms of Q and f_r:

$$S = \frac{Q\delta f}{f_r} \tag{4}$$

Thus, to take a value already quoted, if the frequency is shifted by $2f_r/Q$ from resonance

$$S = \frac{Q\delta f}{f_r}$$

$$= 2$$

$$\therefore \quad N^2 = 1 + 16$$

$$= 17$$

$$\therefore \quad N = 4{\cdot}1$$

that is, the response at resonance is about four times the response at the frequency removed from resonance by $2f_r/Q$.

Equation 3 is exact only if the loss resistance of the circuit is the same at all the frequencies being considered. This is generally a reasonably accurate assumption.

Example 9.3

A low-loss parallel circuit resonates at a frequency of 10 Mc/s. The inductor has a Q value of 100 and the dynamic resistance of the circuit is 100,000 ohms. Sketch the resonance curve.

Seven points on the curve can readily be put down:

(1) At resonance the circuit impedance is 100,000 ohms.

(2) and (3) At frequencies removed from resonance by $f_r/2Q$ = 10/200 Mc/s, i.e. by 50 kc/s, the impedance is reduced to 0·707 times 100,000 ohms, i.e. to 71 kilohms.

(4) and (5) At frequencies removed from resonance by f_r/Q = 10/100 Mc/s, i.e. by 100 kc/s, the impedance is reduced to about 45 per cent of its value at resonance, i.e. to 45 kilohms.

(6) and (7) At frequencies removed from resonance by $2f_r/Q$ = 10/50 Mc/s, i.e. by 200 kc/s, the impedance is reduced to about 24 per cent of its value at resonance, i.e. to 24 kilohms.

The points form the required curve, (a) in Fig. 9.9.

Example 9.4

Repeat the work of the previous example for an identical circuit but assume that the resistance is doubled.

Since R is doubled:

(a) the Q value is halved, i.e. $Q = 50$,

(b) the dynamic resistance R_e is halved, i.e. $R_e = 50{,}000$ ohms.

The apex of the response curve is at 50 kilohms.

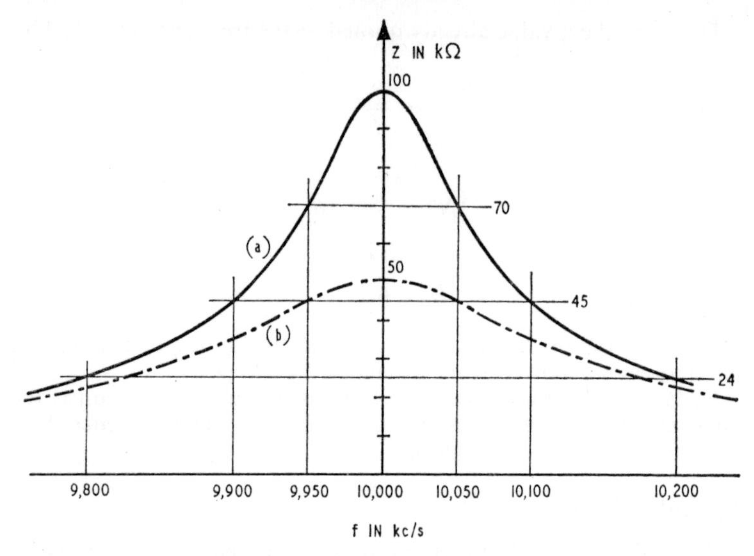

Fig. 9.9. Illustrating Example 9.3

To obtain the percentage reductions in impedance used in obtaining points (2) to (7) in the previous example it is necessary to apply twice the corresponding frequency shifts (because of the halving of Q).

The figures obtained in this way are plotted in curve (b) in Fig. 9.9.

Example 9.5

Calculate the impedance of a parallel circuit of resonance frequency 1,200 kc/s, $Q = 100$ and $R_e = 100$ kilohms at points (a) 3 kc/s, (b) 6 kc/s, (c) 9 kc/s and (d) 12 kc/s above and below resonance.

Equations 3 and 4 can be used

(a) $S = \frac{1}{4}$ and $N = 1{\cdot}12$

$$\therefore \qquad Z = \frac{100}{1{\cdot}12} \text{ kilohms}$$
$$= 89 \text{ kilohms}$$

(b) $S = \frac{1}{2}$ and $N = \sqrt{2}$

$$\therefore \qquad Z = \frac{100}{\sqrt{2}} \text{ kilohms}$$
$$= 71 \text{ kilohms}$$

This is recognised as the value of the impedance at a frequency removed from resonance by $f_r/2Q$.

(c) $S = \frac{3}{4}$ and $N = 1{\cdot}8$

$$\therefore \qquad Z = \frac{100}{1{\cdot}8} \text{ kilohms}$$
$$= 56 \text{ kilohms}$$

(d) $S = 1$ and $N = \sqrt{5}$

$$\therefore \qquad Z = \frac{100}{\sqrt{5}} \text{ kilohms}$$
$$= 45 \text{ kilohms}$$

(This is the impedance obtained at a frequency removed from resonance by f_r/Q.)

A consideration of the impedance-frequency curves for parallel circuits with various Q values, as for instance the two curves calculated and plotted in Fig. 9.9 and those shown in Fig. 9.10, shows that at frequencies remote from resonance Z is largely independent

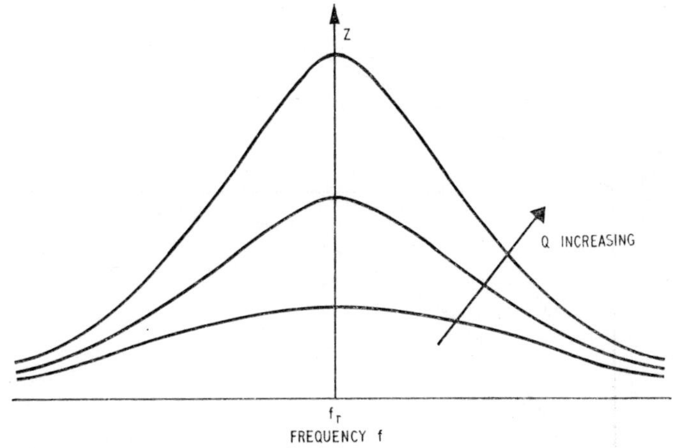

Fig. 9.10. Parallel resonance: effect of Q-value

of Q (or R), that close to resonance Z depends largely on R and that at resonance Z depends entirely on R.

9.5.1. HALF-POWER POINTS

Note that in the series circuit at a frequency $f_r/2Q$ c/s above or below resonance the current is reduced to 70·7 per cent of its value at resonance. At this point the circuit power is reduced to one half of its value at resonance $\left[\text{because power} = i_{\text{r.m.s.}}{}^2R = \left(\frac{i_{\text{max}}}{\sqrt{2}}\right)^2 R = \frac{i_{\text{max}}{}^2R}{2}\right]$. Hence the points corresponding to these frequencies are referred to as the *half-power* points.

In the parallel circuit, at the same frequencies, the impedance is reduced to 0·707 of the value at resonance and the reactive component of the impedance is reduced to one half of the impedance at resonance as is also the resistive component.

9.6. Selectivity

One of the functions of a tuned circuit is to enable selection to be made of one narrow band of frequencies from among others. The ability of a circuit to perform this task is measured by its selectivity. It is clear from what has already been said that the selectivity increases with the Q value.

Extreme selectivity is not always wanted. Very often what is needed is a response curve which remains level over some relatively

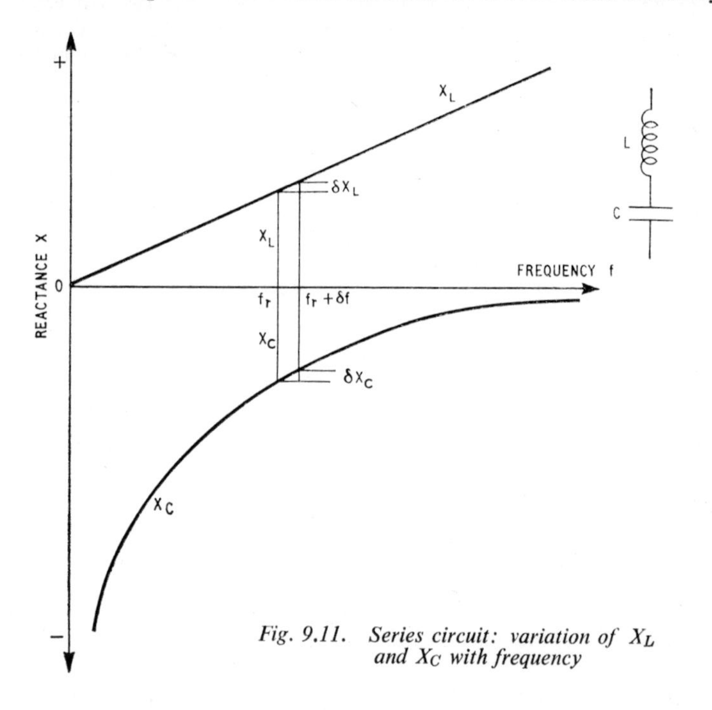

Fig. 9.11. Series circuit: variation of X_L and X_C with frequency

wide band of frequencies before falling off steeply at the sides (*see* Section 9.3.3 in *Radio and Line Transmission, Volume* 1).

9.7. Acceptor and Rejector Circuits

At resonance a low-loss series circuit offers a very low impedance. For this reason it is often called an acceptor circuit. A parallel circuit, on the other hand, offers a high impedance at resonance and is termed a rejector circuit.

Some uses of these circuits are mentioned in Section 9.4 of *Radio and Line Transmission, Volume* 1: this brief mention has been made here for completeness.

9.8. Rate of Change of Reactance in the Vicinity of Resonance

At resonance the reactance of the circuit inductor is equal to that of the capacitor. It is also true, and is proved in the next section, that at resonance the rate of change of X_L with respect to frequency is equal to that of X_C, thus (Fig. 9.11) $\delta X_L/\delta f = \delta X_C/\delta f$ at resonance.

At a frequency slightly above resonance, say $f_r + \delta f$, X_L will have increased by a small amount, δX_L, and X_C will have decreased numerically by almost exactly the same amount. The net change in reactance, therefore, is simply twice this amount and is inductive. If the frequency is shifted again by the same amount the net change in reactance is doubled.

If the frequency is reduced by the same amount the net reactance change is the same numerically but of opposite sign.

It must not be forgotten that this simplification only holds good in the immediate vicinity of resonance: it is obvious from Fig. 9.11 that the basis of the assumption is that the X_C curve is a straight line over the region considered and that it is parallel to the X_L curve (which is a straight line). The greater the departure from the resonance frequency the less true does the assumption become; it is, however, valid for most of the calculations which have to be made of the impedance of a circuit within its normal pass band.

The big advantage of the simplification is that calculations are required only of $2\pi fL$; the more complicated $\frac{1}{2\pi fC}$ need not enter into the working (*see* Example 9.6).

The same principle holds for the parallel circuit but here it is necessary to deal with susceptance instead of reactance. Susceptance is the reciprocal of reactance and is outside our present scope.

Example 9.6

A low-loss series circuit is resonant at 2 Mc/s and the inductor is of 100 μH. Calculate the net reactance at the following frequencies: (a) 1,995 kc/s, (b) 2,000 kc/s, (c) 2,005 kc/s and (d) 2,010 kc/s.

Taking (b) first: this is the resonance frequency so that $X = 0$.

(a) 1,995 kc/s is 5 kc/s removed from resonance and the change in X_L from that at resonance is

$$2\pi \times 5{,}000 \times \frac{100}{10^6} \text{ ohms} = \pi \text{ ohms}$$

Since the frequency has been reduced this represents a decrease in inductive reactance.

The change (an increase) in X_C is almost the same and the total change is, therefore, 2π ohms. The net reactance at this frequency is thus 2π ohms, capacitive.

(c) 2,005 kc/s is also 5 kc/s away from resonance. The net reactance at this frequency is, therefore, again 2π ohms but is now inductive because the frequency is above resonance.

(d) 2,010 kc/s is 10 kc/s above resonance. The net reactance is, therefore, 4π ohms inductive.

Note the simplicity of the calculations compared with those based on the computation of

$$2\pi fL \sim \frac{1}{2\pi fC}$$

9.8.1. TO PROVE THAT AT RESONANCE THE RATE OF CHANGE OF INDUCTIVE REACTANCE WITH RESPECT TO FREQUENCY IS EQUAL TO THAT OF THE CAPACITIVE REACTANCE

$$X_L = 2\pi fL$$

therefore, the slope of the curve,

$$\frac{dX_L}{df} = 2\pi L$$

also

$$X_C = -\frac{1}{2\pi fC}$$

therefore, the slope of the curve,

$$\frac{dX_C}{df} = \frac{1}{2\pi f^2 C}$$

At resonance

$$2\pi f_r L = \frac{1}{2\pi f_r C}$$

i.e.

$$2\pi L = \frac{1}{2\pi f_r^2 C}$$

i.e. at resonance,

$$\frac{dX_L}{df} = \frac{dX_C}{df}$$

That is, both curves have the same slope at resonance.

9.9. The Parallel Circuit under the Influence of Applied and Induced E.M.F.

Care must be taken to distinguish between the ways in which an e.m.f. can be applied to a parallel circuit. These are by direct

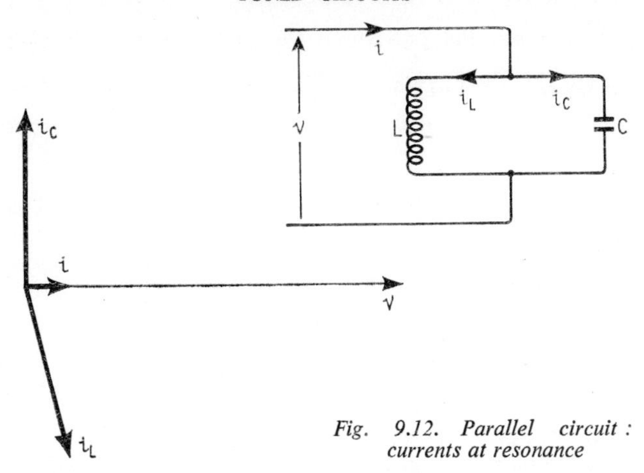

Fig. 9.12. Parallel circuit : currents at resonance

connection across the circuit and by injection in series with the circuit. The different effects are described in the next two sections.

9.9.1. E.M.F. APPLIED ACROSS PARALLEL CIRCUIT

When an e.m.f. is applied across a parallel circuit (Fig. 9.12) it sees an impedance which varies with frequency in the manner already described for parallel circuits and illustrated in Figs. 9.2 and 9.10. It has already been seen (in Section 9.4.2) that the impedance of the

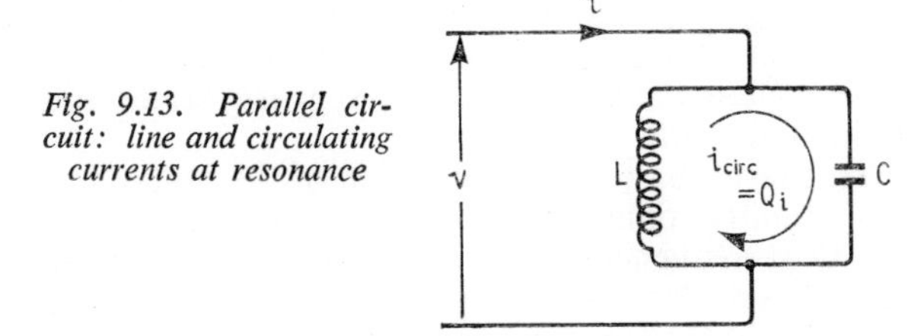

Fig. 9.13. Parallel circuit: line and circulating currents at resonance

parallel circuit at resonance is Q times the reactance of either branch of the circuit. Since (obviously) the same voltage is applied across (a) the inductor, (b) the capacitor and (c) the circuit as a whole it follows that the currents i_L and i_C in each branch are Q times larger than the net current i. There is no anomaly in this because the line current, i, is the vector sum of i_L and i_C: these, respectively, lag and lead the applied voltage by nearly 90° and are, therefore, almost in anti-phase. Thus their sum is small (Fig. 9.12). Since i_L and i_C in the directions shown in Fig. 9.12 are almost in anti-phase it follows that at any instant they are both acting in the same direction round the LC circuit and so constitute a circulating current round LC. Thus, when an e.m.f., v, is applied across the terminals of a parallel LC circuit the line current, i, is given by $i = \frac{v}{Z}$ where Z is the impedance of the parallel circuit. At frequencies near to resonance i is small but at the same time a current, i_{circ}, circulates round the LC-circuit and this current is Q times i (Fig. 9.13).

LC is sometimes called the tank circuit and i_{circ} the tank circuit current. In equipments in which the power is considerable, as for example in radio transmitters, the fact that the circulating current is several times larger than the line current must be borne in mind in deciding conductor and component sizes on the basis of current-carrying capacity.

9.9.2. VOLTAGE INDUCED INTO PARALLEL CIRCUIT BY MUTUAL COUPLING

When a voltage is transferred into a parallel circuit by mutual coupling as shown in Fig. 9.14, the e.m.f. must be thought of as

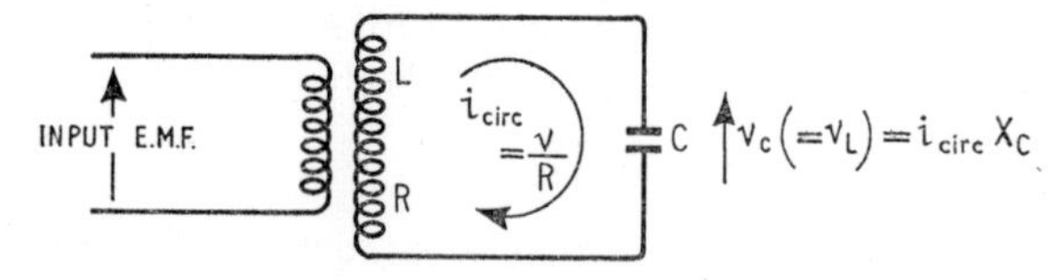

Fig. 9.14. Coupled circuits: induced voltage at resonance

being injected not across L but in series with the turns so that the e.m.f. acts round a series circuit as distinct from across a parallel one. As far as present results are concerned they will be found to be very similar to those outlined in the previous section but the different principle involved, however, is very important and will be met again in later work.

The current resulting from the injection of the induced voltage is large, being limited only by the series impedance of the circuit which is small at frequencies close to resonance. The voltage across L or C is relatively large and is given by the product of i_{circ} and X_L or X_C, or, it is Q times the injected voltage.

9.10. Choice of Type of Circuit to Obtain Optimum Selectivity

Whether a series or a parallel circuit is used for frequency separation depends on the other circuit conditions. If the signal source has a low value of internal impedance the use of a parallel circuit would not enable good separation to be obtained. This is because the input circuit and the tuned circuit form a potentiometer and the voltages across them are in the ratio of the impedances. Thus, in Fig. 9.15, the impedance of the input circuit is Z_1 and that of the tuned circuit is Z_2. The fraction of the input which appears across the tuned circuit is $\frac{Z_2}{Z_1 + Z_2}$. Since Z_1 has been postulated as small it is clear that practically the whole of the input will appear across the tuned circuit for a wide range of values of Z_2 thus nullifying the selective properties of the tuned circuit. For example, if the source provides two e.m.f.s of equal value but different frequency the impedance of the tuned circuit to one of these frequencies (the wanted one) may be 100,000 ohms and to the other (unwanted) 25,000 ohms. If the source impedance is low, say 1,000 ohms, then the fraction of the wanted signal which appears across Z_2 is 100/101 while for the unwanted signal the ratio is about 25/26. The ratio

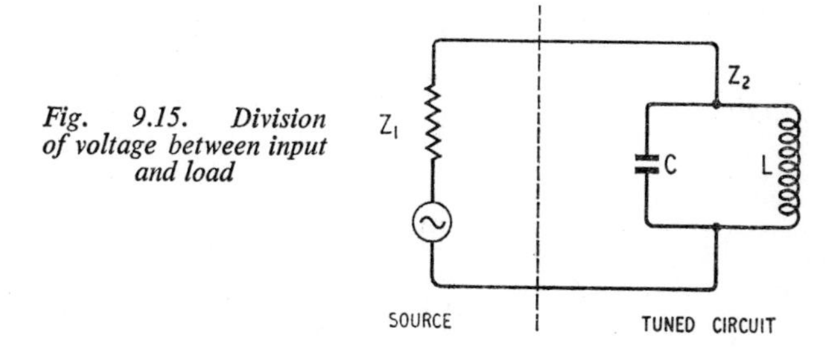

Fig. 9.15. Division of voltage between input and load

is thus about unity in each case and the inherent selectivity of the tuned circuit is not obtained.

Clearly, as similar calculations show, what is needed is a low-impedance tuned circuit (i.e. a series circuit) for a low-impedance source and a high-impedance tuned circuit (i.e. a parallel circuit) for a high-impedance source. Thus parallel tuned circuits are found in valve anode circuits and at the base of high-impedance aerials and series circuits at the base of low-impedance aerials (the whole being tuned to resonance).

9.10.1. TAPPED PARALLEL CIRCUIT

Considerations of transfer of power sometimes render it advantageous to supply a parallel tuned circuit by connection from the source across part of the inductor or capacitor instead of the whole of it. The source is normally the output of a valve or transistor

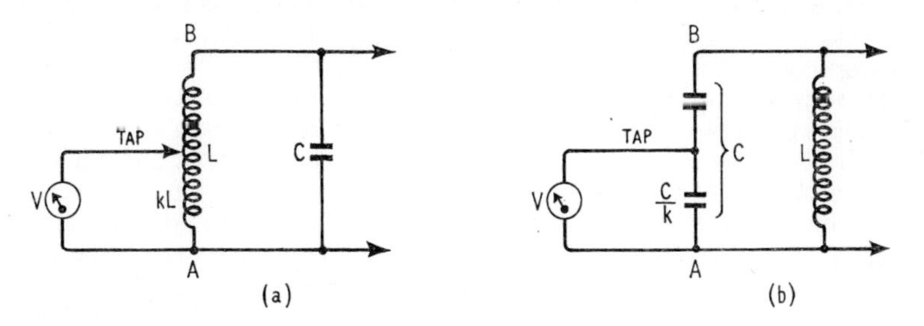

Fig. 9.16. Tapped parallel circuits

stage or an aerial circuit. Methods of doing this are shown in Fig. 9.16. In each circuit the resonance curve as measured between the tapping point and A is of the same shape as would be obtained by measuring between A and B, i.e. across the whole circuit, but the voltage is reduced in amplitude throughout by a factor k^2 where k is the ratio of the reactance included between the taps to the reactance of the whole inductor. Thus, in Fig. 9.16 (b), if the two capacitors are equal, k is equal to 0·5 and the output voltage is one quarter that obtainable across the whole circuit. In the same way, in Fig. 9.16 (a), if the inductance between the taps is two-thirds of the total the response curve ordinates are four-ninths of those across AB.

In Section 10.2.4 the use of tapped tuned circuits in amplifiers is dealt with.

9.11. Reduction of Q Value

Usually the Q value of a tuned circuit is required to be as high as possible to give high selectivity and most of the discussion so far has been on the tacit assumption that no factors external to the tuned circuit cause an effective reduction in Q.

Sometimes, however, the circuit Q is higher than is required and gives too small a pass-band; it must then be artificially reduced. Again, associated circuits may cause a reduction in the effective Q value. A few examples will be mentioned.

In certain equipments, notably television and radar receivers, a very wide bandwidth is necessary if the full range of information in the signal is to be retained. To enable the tuned circuits to accept this full width their effective Q value is often reduced by shunting them by a relatively low value of resistance (a few thousand ohms). This extra bandwidth is, of course, obtained at the cost of a reduction in signal output because of the reduction in impedance of the tuned circuit involved in the sacrifice of Q.

Sometimes the inductors of tuned circuits are deliberately wound with wire of high resistance in order to reduce the Q value. The high resistance may be obtained by the use of wire of high specific resistance or by employing very fine gauge wire.

When a tuned circuit is used in the input to a valve its Q is reduced if grid current is permitted to flow. In transmitters and certain other equipments power is developed and amplified stage by stage until the required power level is attained. This means that the tuned circuits in a given stage are not only required to pass on a voltage, as for r.f. stages in receivers, but must supply appreciable current as well: such tuned circuits are loaded as effectively as if they were providing power to heat an electric fire or light electric lamps. The effect of this loading on the tuned circuit, therefore, is the same as that of a shunt resistor and this usually reduces the Q considerably. The reduction may well be from a Q value of 100 in the unloaded state down to one of 10 when the equipment is working normally.

9.12. Summary of the Properties of Series and Parallel Circuits

For convenience some of the more important properties of low-loss tuned LC circuits at resonance are now summarised.

(1) Resonance frequency $= \dfrac{1}{2\pi\sqrt{LC}}$ (exactly for series circuits: approximately for parallel circuits).

(2) The applied voltage and the line current are in phase at resonance: i.e. the circuit is resistive and the power factor unity.

(3) In a series circuit the impedance is a minimum at resonance: $Z = R$ and this is $1/Q$ of the reactance of either L or C.

(4) In a parallel circuit the impedance is a maximum at resonance, $R_e = L/CR$ and this is Q times the reactance of either branch.

(5) The voltage across either reactance in the series circuit is Q times the applied voltage.

(6) The current in either branch in a parallel circuit is Q times the line current.

(7) The selectivity increases with Q.

9.13. Coupled Tuned Circuits

Tuned circuits may be coupled to one another by the provision of a common impedance. Coupling is most usually effected by mutual inductance (M, Fig. 9.17 (a)) or by capacitance (C, Fig. 9.17 (b) and Fig. 9.17 (c)). It is quite possible, of course, for two

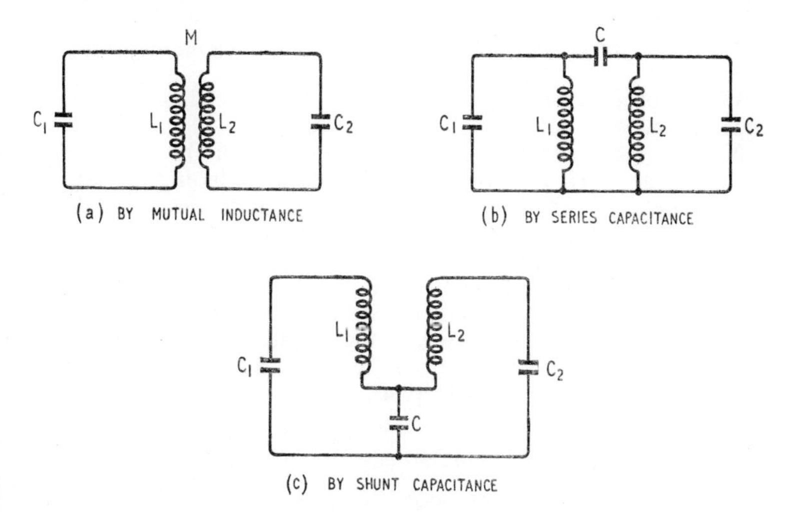

Fig. 9.17. Coupled circuits

forms of coupling to exist at the same time: it will be seen later that it may be advantageous to combine the types of Figs. 9.17 (a) and 9.17 (c).

Coupling between circuits can have far-reaching effects on the properties of the circuits because they cease to be independent. For instance, with inductive coupling, an e.m.f. in one circuit (the primary circuit) establishes a current which sets up a magnetic flux. This induces a voltage in the secondary winding as a result of which a current flows round the secondary circuit and builds up a magnetic flux. This, in turn, induces a voltage in the primary winding and so modifies the conditions in that circuit.

It is not within our province to investigate coupled circuits analytically. The effects, however, should be stated. Coupling a secondary circuit to the primary modifies the impedance of the latter by the (vector) addition of an impedance $\frac{\omega^2 M^2}{Z_s}$ where:

ω is 2π times the frequency,

M is the mutual inductance (for inductively coupled circuits),

Z_s is the series impedance round the secondary circuit.

(When it is necessary to calculate the resultant primary impedance at many frequencies in the vicinity of resonance—as, for instance to determine the overall response curve of a pair of coupled circuits—the approximation noticed in Section 9.8 is of great help in speeding the calculations.)

If the primary and secondary circuits are identical and if the coupling is by mutual inductance, $M = kL$ where L is the inductance value of the primary or secondary winding and k is the coefficient of coupling. The more tight the coupling the greater the value of k and M and the greater the influence of the coupling on the response curves.

With capacitance coupling the general effects are similar to those encountered with mutual inductive coupling.

Note that the degree of coupling in circuits of the type under discussion is much less than that between the windings of power transformers. In the latter nearly all the magnetic flux cuts nearly all the turns of both primary and secondary windings, i.e. the coupling is nearly unity. With coupled tuned circuits the coupling is usually of the order of one per cent of this value and even what is described as very tight coupling is only a few per cent.

It is usually desired to know how the output voltage—that across C_2—varies with frequency for a fixed input, i.e. to determine the response curve for the overall circuit.

9.13.1. RESPONSE CURVE: COUPLING LESS THAN CRITICAL

When the coupling is very loose the impedance and the response curve of each individual circuit is virtually unaltered by the existence of the other circuit. The overall response curve is, therefore, given by the product of the individual responses. Loosely coupled circuits are thus characterised by considerable selectivity and low output (since the voltage induced in the secondary winding is very small because of the loose coupling). This is shown in curve 1 of Fig. 9.18.

Increase of coupling increases, in approximately direct proportion, the voltage induced in the secondary winding and, therefore, the output voltage. The impedance reflected from one circuit into the other causes the overall selectivity to decrease slightly (curve 2).

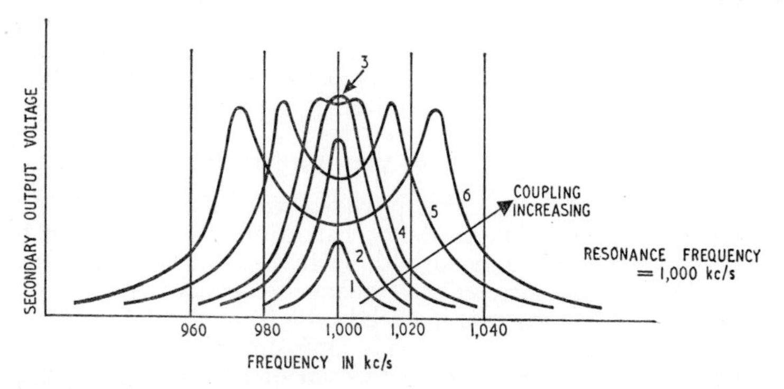

Fig. 9.18. Coupled circuits: effect of degree of coupling on response curve

Further increase in coupling is accompanied by a continuation of this trend until a particular value of coupling—critical coupling—is reached.

9.13.2. RESPONSE CURVE: CRITICAL COUPLING

With critical coupling the voltage output at resonance is greater than it is in any other circumstances; also, the peak of the response curve is rather flat (curve 3).

With identical primary and secondary circuits the critical value of coupling coefficient is $k_c = \frac{1}{Q}$, where Q is the Q-factor of either circuit. The overall bandwidth of the coupled circuit with critical coupling is $\sqrt{2}$ times the bandwidth of either.

9.13.3. RESPONSE CURVE: COUPLING GREATER THAN CRITICAL

Increase of coupling to values greater than critical results in the splitting into two of the peaks of the response curve; the greater the coupling the wider the spacing of the peaks, the greater the bandwidth and the deeper the trough between peaks (curves 4, 5 and 6). The value of response at the peaks falls slightly as the coupling is increased but for most purposes it can be assumed that the value is constant.

9.13.4. OBTAINING WIDE-BAND FLAT RESPONSE

To obtain a wide-band response over-coupling is necessary. The resulting deep trough causes very unequal response over the pass-band unless filled in. This can be done approximately by using an over-coupled circuit in conjunction with a single-tuned circuit of half the Q-value of either of the over-coupled circuits. The peak of the single circuit approximately fills the trough of the other.

Double humps may also be obtained by tuning the two circuits of a coupled pair to different frequencies. If the several tuned circuits of a multi-stage amplifier are tuned to different frequencies the result can be a substantially flat-topped response curve with steep sides. This type of tuning is known as *stagger tuning* and is applied only to fixed-tuned amplifiers such as intermediate-frequency amplifiers of superheterodyne receivers (Chapter 13).

9.13.5. VARIATION OF BANDWIDTH WITH FREQUENCY

If coupled circuits are to be provided with variable tuning any change in frequency influences the bandwidth. For coupling by mutual inductance the bandwidth increases with the resonance frequency; thus an increase in resonance frequency from 500 kc/s to 1,500 kc/s might increase the bandwidth from about 3 to 10 kc/s for a given coupling. The change is approximately linear. A similar effect is experienced using the circuit of Fig. 9.17 (b)—top or series capacitance coupling—although the increase in bandwidth is less linear.

If the coupling is arranged via a shunt capacitor (Fig. 9.17 (c)) the bandwidth decreases with increase in frequency.

A judicious combination of mutual inductance and shunt capacitance coupling can thus be arranged to provide substantially constant bandwidth over the tuning range. All that is necessary is to allow the right amount of mutual inductance coupling in the circuit of Fig. 9.17 (c)

Questions

1. A tuned circuit comprises a 1,000 μH inductor of series resistance 30 ohms in parallel with a 100 pF capacitor. An e.m.f. of 0·001 volts (r.m.s.) at the resonance frequency of the tuned circuit is induced in series with the inductor. What is the voltage developed across the tuned circuit? (*C & G*, 1951.)

2. Calculate the voltage gain between grid and anode of a single-stage intermediate-frequency amplifier operating at 465 kc/s, given that the pentode valve used has a mutual conductance of 5 mA/V and the tuned circuit consists of a 200 pF capacitor in parallel with an inductor having a Q of 100. (*C & G*, 1949.)

3. The anode tuned circuit of the r.f. stage of a medium-wave broadcast receiver has a Q factor of 100. If the circuit is damped to reduce the Q factor to 25, explain by means of sketches the effect

on the selectivity curve and on the voltage developed across the tuned circuit at resonance. (*C & G*, 1948.)

4. What is meant by the Q factor of a coil?

The Q of a coil at a frequency of 1,000 kc/s is 100 and the inductance is 100 μH. What is the series resistance at this frequency? (*C & G*, 1947.)

5. What is the meaning and importance of selectivity?

How is the selectivity of a tuned circuit affected by resistance? (*C & G*, 1946.)

6. A tuned circuit, having negligible resistance, has an inductance L and a capacitance C, and is tuned at a frequency f.

L is increased by 240 μH and the tuned frequency becomes $0{\cdot}5\,f$. Find the value of L. (*C & G*, 1946.)

7. Sketch the impedance-frequency curve for a low-loss parallel circuit in the vicinity of resonance and locate five points on the curve with reference to the circuit Q and the percentage of the resonant frequency by which the applied voltage is off-tuned.

Indicate the relative magnitudes and phases of the line and component currents at resonance.

8. What factors must be considered when deciding whether a series or parallel circuit is the most suitable for frequency selection for a given application?

What is (a) An acceptor circuit?

(b) A rejector circuit?

Give one example of the use of each.

9. What is meant by the term *dynamic resistance* of a parallel tuned circuit? Derive an expression for it in terms of the circuit constants.

The inductor of a parallel circuit which resonates at 1 Mc/s has an inductance of 159 μH and a loss resistance of 10 ohms. If the capacitor has negligible loss determine the dynamic resistance of the circuit.

10. Compare in tabular form the properties of series and parallel low-loss resonant circuits.

10

Radio Frequency Amplification

In Chapter 11 of *Radio and Line Transmission, Volume* 1, it was shown that amplification at radio frequency usually calls for:

(a) parallel tuned circuits because of their high effective resistance and selective properties, and either

(b) valves of low anode-grid capacitance, i.e. pentodes or tetrodes, or

(c) neutralisation of the feedback when devices (such as triodes) having appreciable inherent feedback are employed.

10.1. The Frequency Range and Bandwidth of Tuned Amplifiers

Tuned amplifiers may be required to operate over a very wide range of frequencies; for instance the radio-frequency stages of a communications receiver may cover from about 15 kc/s to 30 Mc/s. Such a wide range is normally catered for by the provision of a number of wave ranges, say eight or ten, each providing a frequency range of about two to one. Other amplifiers, while still requiring variable tuning, may need a much smaller number of ranges; possibly, even, only one as in v.h.f. receivers for broadcast reception.

Some amplifiers are required to operate only on two or more spot frequencies without the necessity for variable tuning. Here a number of pre-set capacitors or inductors are needed which can be switched in accordance with the tuning needs: examples are radio-frequency amplifiers in push-button broadcast radio receivers and in domestic television receivers.

Finally there are amplifiers which are required to operate on one fixed frequency band only; the most common example of this type is the i.f. amplifier of superheterodyne receivers (*see* Chapter 13).

In addition to the variety of needs as to the frequency range to be covered, amplifiers are also called on to fill many different requirements as to the bandwidth to be accepted. Usually an amplifier is designed to accept one bandwidth only; sometimes, however, a variable bandwidth must be provided. As outlined in Chapter 3 the narrowest bandwidth required is of the order of 100 c/s, the widest of several Mc/s.

With such a range of possible requirements it is not surprising that tuned amplifiers appear in many different forms.

10.2. Single Tuned Circuit

The tuned radio-frequency amplifier of Fig. 10.1 employs the simplest of tuning; a single parallel *LC* circuit in the anode. It can be represented by the equivalent circuit of Fig. 10.2 (constant voltage) or that of Fig. 10.3 (constant current). In these figures Z_a represents the effective anode load, i.e. the external load together

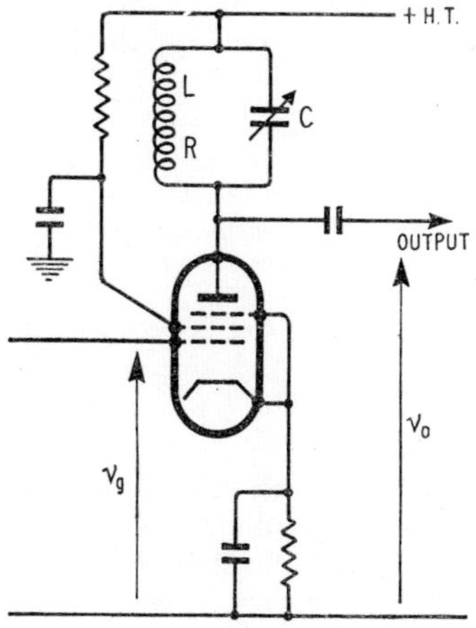

Fig. 10.1. Tuned amplifier

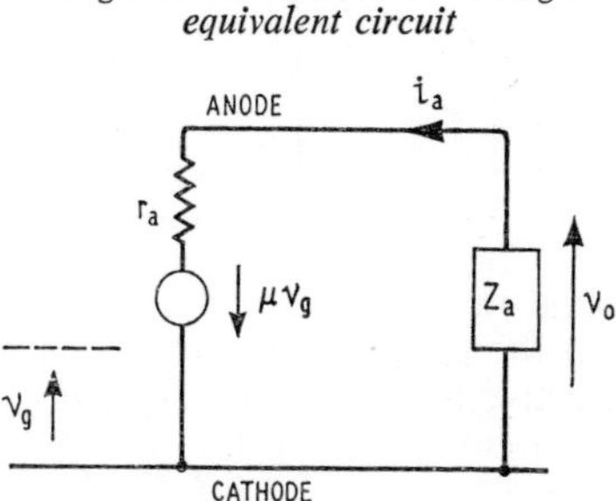

Fig. 10.2. Constant voltage equivalent circuit

with such shunting impedances as there may be, for instance the grid resistor of the next stage.

From the constant voltage equivalent circuit:

$$i_a = \frac{\mu v_g}{r_a + Z_a}$$

$$v_0 = -i_a Z_a = \frac{-\mu v_g Z_a}{r_a + Z_a}$$

in which the negative sign merely indicates the anti-phase relationship between input and output signals.

$$\text{Gain} = \frac{|v_0|}{|v_g|} = \frac{\mu Z_a}{r_a + Z_a}$$

where $|v_0|$ and $|v_g|$ represent, respectively, the magnitudes of v_0 and v_g.

From the constant current equivalent circuit: $v_0 = -g_m v_g Z_a'$ where Z_a' represents Z_a in parallel with r_a.

Therefore

$$\text{Gain} = \frac{|v_0|}{|v_g|} = g_m Z_a'$$

When pentodes are used $r_a \gg Z_a$ and the expression for the gain reduces to

$$\text{Gain} = g_m Z_a$$

with an accuracy sufficient for many purposes.

Z_a, as has been said, represents the impedance of the tuned circuit in parallel with the input impedance to the next stage. If the latter can be ignored as being much greater than that of the tuned circuit then Z_a represents the tuned circuit impedance only and, when tuned to a given frequency, varies with the frequency of the input

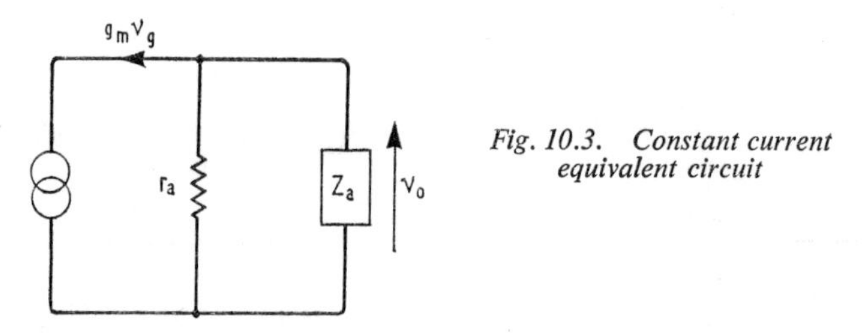

Fig. 10.3. Constant current equivalent circuit

signal in the manner indicated in Sections 9.1 and 9.5. It follows that the gain varies in the same way, that the amplifier is frequency selective and that the selectivity varies with the effective Q value of the circuit (Fig. 9.10).

At resonance $Z_a = L/CR$ (Section 9.4.2) and the gain is $g_m L/CR$. Alternatively, since the impedance at resonance is Q times the reactance of either arm of the tuned circuit (Section 9.4.2), $Z_a = \omega LQ$ and the gain $= g_m \omega LQ$.

10.2.1. VARIATION OF GAIN WITH FREQUENCY

R.F. amplifiers in which the tuning is variable tend to show variable gain over the tuning range. The gain is proportional to ωLQ; thus if the frequency increases and all else remains constant the gain increases.

If L is fixed (and the tuning therefore accomplished by varying C) the gain tends to rise with frequency because Q ($= \omega L/R$) probably decreases at a lower rate than that at which f increases. For constant gain Q would need to fall in direct proportion to the rise in frequency.

If C is fixed (and the tuning is, therefore, by variable inductance) then the changes in f are inversely proportional to the square root of the inductance; thus if f is to be increased n times L must be decreased in the ratio n^2. If Q were to remain unchanged the rise in frequency would be accompanied by a proportional fall in output. In fact Q usually rises because of the decrease in coil resistance as L is reduced and the gain tends to remain roughly constant over the tuning range, the changes in Q being approximately in direct proportion to the changes in frequency.

Example 10.1

In a capacitance-tuned parallel LC-circuit the halving of the resonance frequency was found to result in an increase of 50 per cent in the Q value. Calculate the ratio of the gain at the higher frequency to that at the lower.

$$\text{Gain} = \omega LQ$$

Let the higher frequency be $\omega_1/2\pi$ and the corresponding Q value Q_1.

The lower frequency is, therefore,

$$\frac{1}{2}\frac{\omega_1}{2\pi} = \frac{\omega_1}{4\pi}$$

and the Q value, $1{\cdot}5Q_1$.
Thus the gain at the higher frequency $= \omega_1 LQ_1$.

$$\text{Gain at lower frequency} = \frac{\omega_1}{2}L(1{\cdot}5Q_1)$$

$$= 0{\cdot}75\omega_1 LQ_1$$

$$\text{and the ratio } \frac{\text{gain at higher frequency}}{\text{gain at lower frequency}}$$

$$= \frac{\omega_1 LQ_1}{0{\cdot}75\omega_1 LQ_1}$$

$$= 1{\cdot}33$$

10.2.2. VARIATION OF SELECTIVITY WITH FREQUENCY

For a given Q the fall in response of a tuned circuit to off-tune frequencies is dependent on the change in frequency expressed as a percentage of the resonance frequency. Thus, other things being comparable, a 10 kc/s change from a resonance frequency of 1 Mc/s causes the same fall in response as a 20 kc/s change from a resonance frequency of 2 Mc/s. This is not what is usually wanted. The normal requirement is that the absolute selectivity should be the same at all points of the tuning range; that is a given frequency deviation from resonance should result in the same reduction in response at all parts of the tuning range.

For constant selectivity over the range, Q is required to change in direct proportion to the frequency.

When tuning is effected by variable capacitance Q decreases with rise of frequency so that the selectivity degenerates considerably. Steps can be taken to apply correction for this if necessary.

When tuning is by variable inductance Q varies as the frequency so that the selectivity, like the gain, remains substantially constant over the range.

In most equipments the majority of the gain takes place at a fixed frequency (*see* Chapter 13) so that these considerations may not be important.

Example 10.2

A single-stage amplifier has for its anode load a parallel LC-circuit resonant at 1 Mc/s, of effective $Q = 100$ and $L = 200\ \mu$H. Calculate

(a) the bandwidth between half-power points,

(b) the approximate gain to a signal of (i) 1,000 kc/s and (ii) 990 kc/s if the mutual conductance of the valve at the point of operation is 1 mA/V.

(c) the bandwidth between half-power points if the circuit were to be made resonant at 1·5 Mc/s and Q remained unchanged in the process,

(d) the necessary value of Q at a resonant frequency of 1·5 Mc/s if the bandwidth at the half-power points is to be the same as at 1 Mc/s.

(a) The half-power points are those at which the response is reduced to $\frac{1}{\sqrt{2}}$ of the value at resonance. To effect this reduction the frequency needs to be shifted $f_r/2Q$ c/s from resonance (*see* Section 9.5).

The bandwidth between half-power points is therefore

$$2 \times f_r/2Q = f_r/Q$$
$$= 1{,}000/100 \text{ kc/s}$$
$$= 10 \text{ kc/s}$$

(b) (i) Gain at resonance (1,000 kc/s)

$$= g_m Z_a$$
$$= g_m \omega L Q$$
$$= \frac{1}{1{,}000} \times 2\pi \times 10^6 \times \frac{200}{10^6} \times 100$$
$$= 40\pi$$
$$= 126$$

(ii) At 990 kc/s the signal is 10 kc/s off resonance, i.e. it is f_r/Q off resonance. At such a frequency the response, and hence the gain, is reduced to about 45 per cent of that at resonance. Therefore the gain is

$$0{\cdot}45 \times 126 = 57$$

(c) Bandwidth between half-power points,

$$f_r/Q = 1{,}500/100 \text{ kc/s} = 15 \text{ kc/s}$$

(d) For the same bandwidth at 1,500 kc/s as at 1,000 kc/s Q must be such that

$$f_r/Q \text{ kc/s} = 10 \text{ kc/s}$$

i.e.

$$1{,}500/Q = 10.$$

Therefore $Q = 150$.

10.2.3. THE EFFECT OF INCREASING THE NUMBER OF STAGES OF AMPLIFICATION

The properties of an amplifier having a single tuned circuit as its frequency-determining factor have just been considered. When there is more than one stage of amplification the resultant response curve is equal to the product of the individual ones. Thus if one stage reduces the relative strength of a signal off-tuned by a certain amount in the ratio of three to one then a second similar stage would effect a like reduction so that as a result the two stages would give a total relative reduction of nine (three times three). After n stages the relative reduction would be 3^n.

Example 10.3

A given amplifier stage has a gain of 50 at the frequency to which it is tuned. To a signal off tune by a certain amount the gain is 10. What is the gain of two such stages to signals of the same frequencies? What is the relative discrimination in favour of the wanted signal?

In two stages the wanted signal is amplified 50×50 times $=$ 2,500 times.

The unwanted signal is increased in the ratio $10 \times 10 = 100$.

The relative discrimination in favour of the wanted signal is

$$\left(\frac{50}{10}\right)^2 : 1 \left(\text{or } \frac{2{,}500}{100} : 1\right) = 25 : 1$$

Example 10.4

The gain of a given amplifier stage is reduced to one half of its maximum value when the input signal is off-tuned by a certain amount. What is the relative reduction in (a) three and (b) n similar stages to a signal of the same frequency?

(a) In each stage the gain is one half of the maximum possible. In three stages, therefore, the gain is $(\frac{1}{2})^3$, i.e. one-eighth of the maximum possible.

(b) In n stages the gain is $(\frac{1}{2})^n$ of the maximum.

Using decibels (*see* Chapter 6) the calculations are even more easy. It is simply necessary to multiply the dB reduction, or increase, in each stage by the number of stages. Thus if one stage provides a discrimination of 4 dB against a certain off-tune frequency three stages will provide 3×4 dB $= 12$ dB discrimination.

Clearly the effect of adding additional stages is to narrow the response curve, i.e. to increase the selectivity. However, in order to obtain the necessary rejection of strong unwanted signals which are fairly remote from resonance it may well be that the fall off in response to frequencies close to resonance thus caused is so excessive as to cut out essential side frequencies of the wanted signai.

Example 10.5

Consider a single tuned circuit of $Q = 100$ resonant at 470 kc/s. Let the response at 470 kc/s be unity.

(i) Calculate the response at 435 kc/s and at 466 kc/s.

(ii) How many similar stages would be needed to reduce the response at 435 kc/s to 0·03 per cent of that at 470 kc/s?

(iii) What reduction in response to the frequency of 466 kc/s, compared with that for the frequency of 470 kc/s, would result from the use of the number of stages calculated in (ii) above?

(iv) What conclusions can be drawn from the answer to (iii) above?

(i) From Section 9.5

$$N^2 = 1 + 4S^2 \text{ where } S = \frac{Q\delta f}{f_r}$$

If

$$f = 435 \text{ kc/s} \qquad \delta f = 35 \text{ kc/s}$$

and

$$S = \frac{100 \times 35}{470} = \frac{350}{47}$$

$$N^2 = 1 + 4\left(\frac{350}{47}\right)^2 \text{ and } N = 15$$

$\therefore$

$$\text{response at 435 kc/s} = \frac{1}{15}$$

If $$f = 466 \text{ kc/s}, \delta f = 4 \text{ kc/s and } S = \frac{40}{47}$$

∴ $$N^2 = 1 + 4\left(\frac{40}{47}\right)^2 \text{ and } N = 2$$

∴ $$\text{response at 466 kc/s} = \tfrac{1}{2}$$

(ii) Reduction in response is 0·03 per cent = 0·0003. Reduction in one stage is in ratio 15

∴ the number of stages needed, n, is given by

$$\left(\frac{1}{15}\right)^n = 0{\cdot}0003$$

∴ $$n = 3$$

(iii) The reduction in one stage is by one half. In 3 stages it is by $(\frac{1}{2})^3$, i.e. by one-eighth.

(iv) It is concluded that single tuned circuits cannot be used to provide both adequate selectivity and a pass-band of reasonable width for the carrier and the side frequencies.

10.2.4. THE TAPPED PARALLEL TUNED CIRCUIT

The tapped parallel tuned circuit was introduced in Section 9.10.1. It was there mentioned that tapping a fraction k of the total inductance reduces the response curve ordinates in the ratio k^2 compared with those obtaining across the whole circuit.

There are several advantages to be derived from the use of a tap or taps:

(1) Tapping down reduces the voltage swing across the tapped part of the coil in the ratio k^2, while the net output of the stage, other things being comparable, is reduced only in the ratio k. It is the voltage across the coil in the valve or transistor output which causes instability; this is therefore reduced in the ratio k^2 at the cost of a reduction in output in the ratio of only k.

(2) The loading of the circuit by the valve or transistor is reduced, i.e. the effective Q is increased. Or, the use of an untapped circuit might require the use of an inordinately low L/C ratio with the necessity for an exceptionally and inconveniently large value of C; the use of a larger, tapped, coil enables the value of C to be reduced considerably. For example, a transistor stage using an untapped coil might need a capacitor of 3,000 pF: this could be reduced about nine times by increasing the size of the coil and introducing a tap at one-third the number of turns.

(3) Although the use of a tap does not change the selectivity of the tuned circuit itself, it improves the overall selectivity of the whole system. The greater the tapping down the greater the selectivity of the complete circuit.

(4) If the generator resistance is lower than that of the tuned circuit a tap can be found which gives maximum output.

In transmitting equipments the tapping of coils is by no means uncommon. In receivers, apart from transistor circuits, it is less commonly encountered because similar, and often better, effects can be obtained by the use of r.f. transformers.

10.3. Transformer-coupled Circuits

The advantages which may accrue from the use of a tapped inductor can also be derived from the use of a transformer.

Indeed the use of transformer coupling instead of a tapped coil has the following advantages:

(a) a tap need not be used,

(b) the following circuit is isolated from the H.T. supply and hence there is no necessity to use a blocking capacitor and grid resistor,

(c) if both windings of the transformer are tuned a much improved response curve may be obtained (*see* Section 9.13).

Transistor circuits often employ a transformer with a tapped primary (*see* Item 2 of Section 10.2.4) and with a secondary winding of considerably fewer turns in order to match into the low impedance of the following stage.

10.3.1. UNTUNED PRIMARY, TUNED SECONDARY CIRCUIT

If a transformer is employed and the primary is left untuned the circuit appears as in Fig. 10.4. Compared for instance with that of a power transformer in which it approaches unity, the coupling between the windings is relatively small so that the resonance frequency of the anode load is given by $f_r = \dfrac{1}{2\pi\sqrt{LC}}$.

At resonance the anode load is $\dfrac{1}{T^2}\cdot\dfrac{L}{CR}$ where T is the effective ratio of secondary turns to primary turns. The Q of the circuit, and hence the selectivity, increases with T and approaches that of the secondary alone when the coupling is very small. However, if T is adjusted for maximum gain, $T = \sqrt{\dfrac{r_a}{L/CR}}$, the selectivity is only that of a tuned circuit of half the Q value of LC. For this reason, to increase the selectivity T is often made greater than that value which gives maximum gain. This also improves the stability of the circuit—Item (1) of Section 10.2.4.

It is the effective value of T which has been quoted; this is the value which results in T volts appearing in the secondary for each volt in the primary. This value can be achieved with a few primary turns with tight coupling or by more primary turns with loose coupling. The former is better because a large primary inductance influences and shifts the secondary response curve. As usual in all

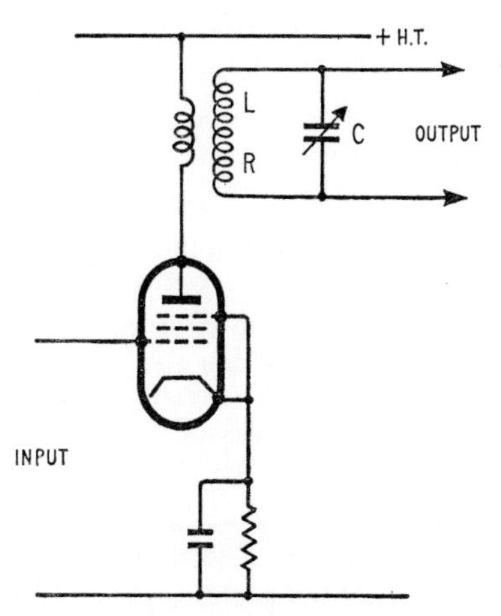

Fig. 10.4. Untuned primary transformer-coupled anode circuit

such circuits the coupling should be at the earthy end of the secondary so as to avoid adding capacitance to the secondary circuit which might reduce the waveband coverage.

The employment of an anode load of this type can bring to the circuit the benefits outlined in Section 10.2.4. As far as the response curve is concerned, however, the results are the same in character as those to be obtained from a single tuned circuit of lower Q than that which constitutes the secondary.

10.3.2. TUNED PRIMARY, TUNED SECONDARY CIRCUIT

The employment of tuned primary and tuned secondary circuits can result in a fundamental change in the character of the circuit properties as has been discussed in Section 9.13 and illustrated in Fig. 9.18.

Circuits of this kind, known as band-pass circuits, can be arranged to provide satisfactory response curves to meet most needs. The coupling and tuning need to be adjusted in the various stages of the amplifier with, perhaps, the inclusion of one or two single-tuned stages to fill in the trough between the humps, or, by stagger tuning (Section 9.13.4) when a very wide bandwidth is needed. Refinements can be added as appropriate; as, for instance, in television receivers to enable the sound signal to be separated from the vision.

The circuits may also be shunted by relatively low resistances, a few thousand ohms, to reduce the Q-value and broaden the characteristic (Section 9.11).

The variation in bandwidth which occurs when simple band-pass circuits are used with variable tuning has already been noticed in Section 9.13.5 and an approximate method of reducing the variation mentioned.

10.3.3. VARIABLE SELECTIVITY

Variable selectivity is often a necessary feature in equipments, particularly in communication receivers. It can be accomplished in various ways as, for instance:

(1) By providing variable coupling between the coils of the coupled circuits (by physical movement of one of the windings).

(2) By the introduction into the coupled circuit arrangement of a third winding through which the transfer from one circuit to the other is effected. A variation in the number of turns in use on this coil (by means of a tapping switch) or of the value of a resistor across its ends, gives a change in the coupling.

It can also be arranged that the selectivity shall vary automatically according to the strength of the signal or of the interference level.

The variable bandwidth feature, when provided, is included in the fixed-frequency circuits of the equipment, i.e. normally in the i.f. stages of a superheterodyne receiver (*see* Chapter 13).

10.4. Radio-frequency Amplifiers Employing Valves and Transistors

10.4.1. FIXED-FREQUENCY AMPLIFIERS

In Fig. 10.5 is shown a stage of fixed-frequency amplification (e.g. the i.f. stage of a superheterodyne receiver) using a transistor. The performance of transistor stages is comparable with that of valve circuits but when designed for maximum gain at these frequencies it is necessary to annul the inherent feedback. When this is done completely, in magnitude and phase, it is termed *unilateralisation* and is effected by two components (R_1 and C_1 in the figure). Less complete cancellation, not involving phase correction, is carried out by the use of one component (often a capacitor) only. This is termed *neutralisation*.

R_2 and R_3, in Fig. 10.5, provide the correct bias for the base while R_4 sets the emitter current at the required figure and acts as a stabilising resistor. C_2 and C_3 by-pass radio-frequency signals and prevent the losses which would otherwise take place in the resistors which they shunt. LC is the resonant circuit, tapped in order to avoid the need for an exceptionally large value of C (*see* Section

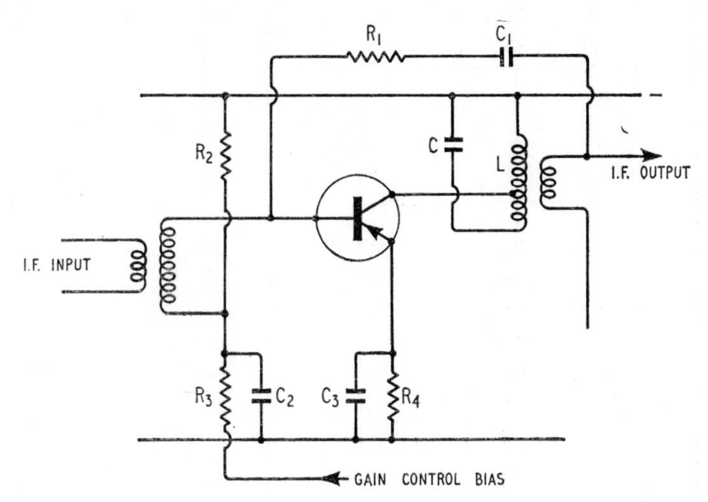

Fig. 10.5. I.F. amplifier: transistor

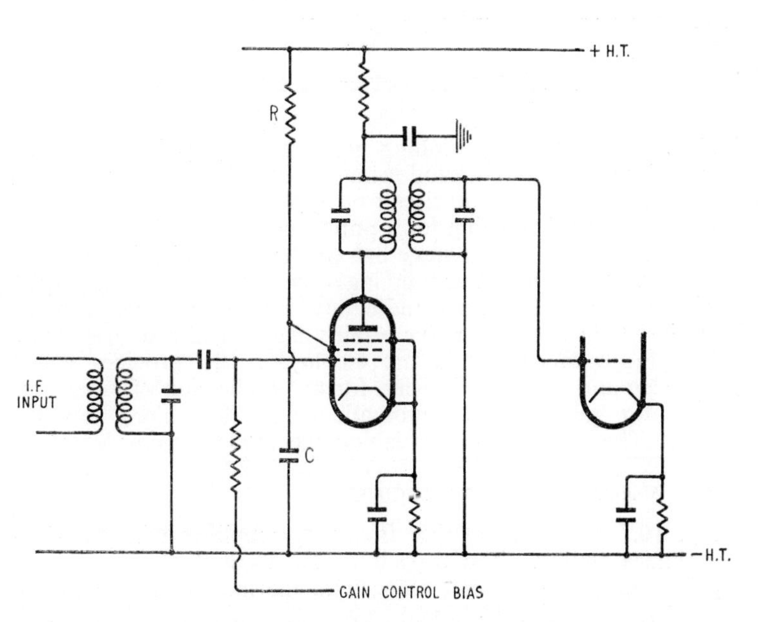

Fig. 10.6. I.F. amplifier: thermionic valve

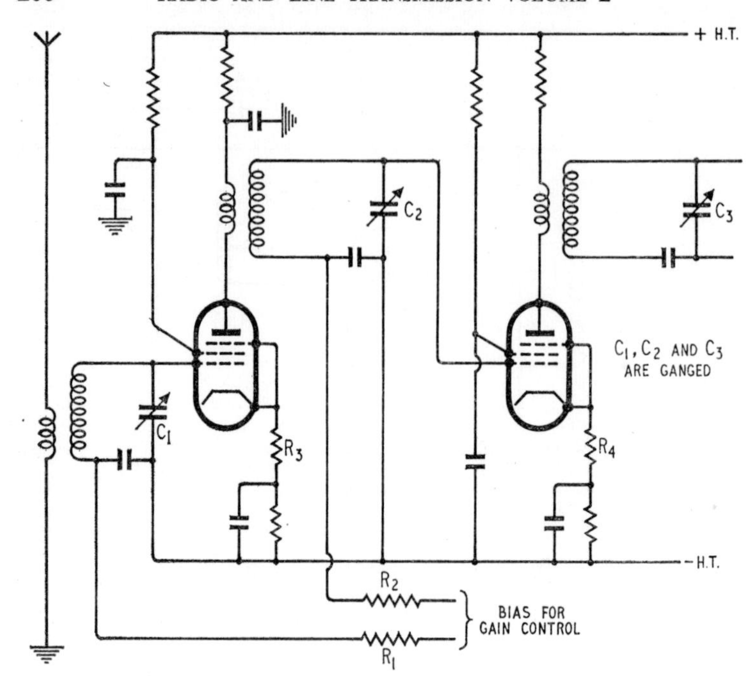

Fig. 10.7. R.F. amplifier: thermionic valve

10.2.4). The secondary winding provides the step-down required for matching the load to the following low-impedance input transistor stage. Bias for gain control can be applied along the line shown and must be positive-going for a pnp transistor.

In Fig. 10.6 is a similar stage but employing a valve instead of a transistor. The tuning and coupling of the tuned circuits are chosen to give the correct frequency and bandwidth, the coupling probably being just tight enough to give a fairly flat-topped response curve, unless stagger-tuned circuits are employed. *RC* provides screen decoupling and there is provision for the application of a negative voltage for gain control, automatic or manual, along the line shown. The valves must be variable-mu pentodes if gain-controlled.

10.4.2. VARIABLE-FREQUENCY AMPLIFIERS

Fig. 10.7 shows a variable frequency amplifier with variable-capacitance tuning. The capacitors C_1, C_2 and C_3 are ganged together on the same shaft along with the local oscillator tuning capacitor (*see* Section 13.4.3) and any other tuning capacitors which may be included in the set.

Only the more ambitiously made receivers include two stages of r.f. amplification. The section of circuit shown is such as would be found in a communications receiver, probably switchable to cover a number of frequency ranges and possessing extremely high sensitivity. It is thus shown equipped with gain control on the r.f. valves in order to enable adjustment to be made to provide the best signal-to-noise ratio (*see* Chapter 7) and to prevent overloading on strong signals. The valves are variable-mu pentodes and their gain varied manually by a potentiometer-controlled voltage which is taken to the line marked gain and thence through resistors, R_1 and R_2, to the grids. Resistances R_3 and R_4, which are un-bypassed, can be included to provide negative feedback.

Questions

1. A class-A radio-frequency amplifier operates at a frequency of 1 Mc/s and employs a pentode valve with a mutual conductance of 8 mA/V. The anode load is a parallel resonant circuit the inductor of which has a value of 50 μH. If the bandwidth of the amplifier for a reduction of 3 dB is 9 kc/s what is the Q factor of the inductor?

If a sinusoidal signal of amplitude 0·5 volts r.m.s. at a frequency of 1 Mc/s is applied to the grid calculate:

(a) the power dissipated in the inductor, and

(b) the peak voltage developed across the load. (*C & G*, 1958.)

2. Explain how the effective amplification of a valve is dependent on the external load. (*C & G*, 1946.)

3. A single-stage amplifier has a tuned anode circuit having a capacitance of 400 pF, an inductance of 300 μH and a resistance of 12 ohms. If the plate resistance is 40,000 ohms and the amplification factor is 10 what is the r.f. voltage across the tuned output circuit when 0.12 volts. is applied to the grid of the valve? (*C & G*, 1946.)

4. Explain briefly why triode valves are unsuitable for r.f. amplification in a radio receiver and describe how these difficulties were overcome by the development of the tetrode and later the pentode valve. (*C & G*, 1956.)

5. Why is it usually necessary to neutralise a triode valve when used in a r.f. amplifier?

Describe, with a simple circuit diagram, one method of neutralisation. (*C & G*, 1958.)

6. Outline the design and construction of a two-stage tuned r.f. amplifier suitable for use at about 5 Mc/s.

If the input power is —40 dB relative to 1 watt and the power output is 10 volts across 100 ohms, what is the gain of the amplifier in decibels? (*C & G*, 1949.)

7. What advantages may derive from the use of a tapped, instead of an untapped, parallel resonant circuit in a radio receiver or transmitter?

Compare the performance of a tapped resonant circuit with that of a transformer-coupled circuit with untuned primary.

8. Why is the performance of single-tuned circuits usually inadequate for the various effects required in the tuned circuits of radio receivers?

9. Sketch the circuit of a r.f. amplifier employing a tuned primary tuned secondary arrangement for frequency selection. Draw curves and explain how the effects vary as the coupling between primary and secondary windings is altered.

10. Sketch the circuit of an i.f. amplifier employing a transistor and state the function of each component included.

11

Oscillatory Circuits

11.1. General

11.1.1. INTRODUCTION

The topic of oscillatory circuits has already been introduced in Chapter 12 of *Radio and Line Transmission, Volume* 1. In the present chapter these early remarks will be extended and summarised as appropriate.

An oscillator is a generator of electrical oscillations (Fig. 11.1). To be maintained at constant amplitude the oscillation requires the supply of power in the correct manner from an external source.

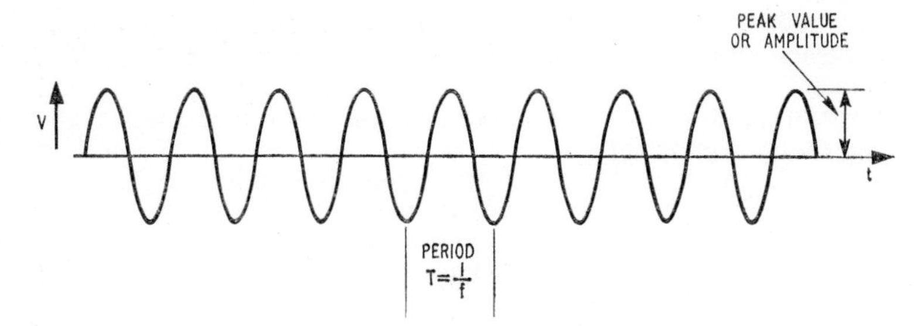

Fig. 11.1. Undamped oscillation

If external power is not available, or is not available in sufficient quantity, the oscillation will decrease in amplitude because of the power loss in the inevitable circuit resistance. An oscillation which is decreasing in this way is known as a *damped oscillation* (Fig. 11.2). The greater the resistance in an oscillatory circuit the sooner will the oscillation die out. Excessive resistance will prevent any oscillation taking place at all. In an *LC*-circuit of resistance *R*, for instance, in the absence of the supply of external energy, no oscillation can occur at all if *R* is greater than $2\sqrt{\frac{L}{C}}$, *R* being the resistance round the circuit in ohms, *L* measured in henrys and *C* in farads. In a low-loss *LC*-circuit the frequency of oscillation is about $\frac{1}{2\pi\sqrt{LC}}$

c/s and is little affected by the value of the resistance. If the resistance is very large then the frequency may differ widely from the value quoted.

11.1.2. FACTORS AFFECTING THE QUALITY OF OSCILLATION

Oscillators for use in telecommunication systems are normally designed to give a good quality of oscillation at low level, any power which may be needed being obtained by amplification following the oscillator. This is because an oscillator which is itself delivering

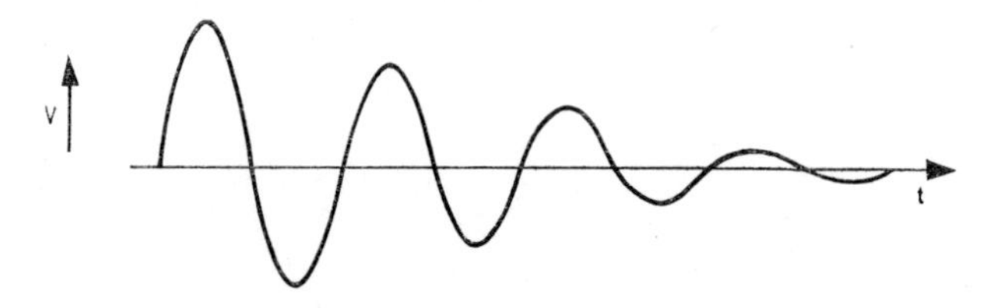

Fig. 11.2. Damped oscillation

high power cannot also fulfil the requirement of good quality oscillation.

By good quality oscillation is meant one of the correct and stable frequency, of constant amplitude and free from noise and spurious content; one substantially unaffected by changes in humidity and temperature, in mains voltage and frequency and by changes in the load on the equipment.

For r.f. purposes the waveform is usually required to be sinusoidal. This is because the closer the wave approximates to a sine wave the fewer other, undesired, frequencies will be present. For certain purposes, e.g. frequency multiplication, a non-sinusoidal waveshape may be preferable.

Additional spurious frequencies are a source of trouble. In receivers, they can cause overloading and, by interaction with other frequencies, set up interference whistles in the output. In transmitters they can lead to overloading and to the radiation of frequencies other than the design frequency and so cause unnecessary interference.

For good waveform the minimum amount of feedback necessary to maintain oscillation should be applied and *LC*-circuits, where used, should be of high effective *Q*. An absolutely pure waveform is impossible to obtain with a feedback oscillator but the smaller the losses and the less the feedback the more nearly can the oscillation approach the pure form.

For constant amplitude of output the oscillator should be provided with stable power supplies free from hum: also any possibility of hum modulation should rigorously be excluded. Modulation of this type can occur, for instance, by coupling across the cathode-heater capacitance of a thermionic valve the cathode of which is at r.f. (the heater, of course, receiving its supply from an a.c. source). This is referred to in Section 11.3.2.

For good frequency stability temperature changes should be kept to a minimum; the frequency-determining unit may be placed in an oven maintained at constant temperature, or, at any rate, isolated from sources of heat (e.g. power supplies) or cold. Humidity changes likewise should be avoided. Frequency-determining components should, as far as possible, have zero temperature coefficient. Where this is not possible, or where an unbalanced positive temperature coefficient necessarily exists in one part of the circuit, it may be expedient to balance this out by the deliberate introduction of a negative coefficient in a suitable place. The influence of parts of the circuit which inevitably affect the frequency and for which a low temperature coefficient cannot be arranged (wiring and valves, etc.) should be kept to a minimum. The operating voltages should be stabilised since changes here affect the valve parameters. H.T. circuits should be adequately decoupled to prevent hum. Valves should be as small and rigid as possible so that mechanical vibration shall not cause undue movement. Other important parts of the circuit, inductors and capacitors, should, of course, also be as rigid as possible.

When the frequency is determined by an LC-circuit this is inevitably shunted by a valve or transistor, by the equivalent impedance of the load and by stray capacitances. To reduce as far as possible the effect of these factors on frequency, the valve or transistor resistance should be as large as possible and the LC-circuit of sufficiently high C and low L as to leave the effective Q virtually unchanged by the damping. Too large a value of C is inconvenient so that damping is often reduced by tapping across only part of the tuned circuit (*see* for instance Figs. 11.10 and 11.21). Grid current should be as small as possible; feedback should only be as much as is needed to maintain oscillation: when the feedback is provided by mutual induction better results are obtained with a small coil tightly coupled than with a larger coil more loosely coupled.

For precise control of frequency, circuits including crystals are employed (*see* Section 11.6). Circuits of this type provide only as many basic spot frequencies as there are crystals but by a process known as frequency multiplication, frequencies which are an exact multiple of these can also be obtained. Often a variable frequency is required, as in the superheterodyne receiver (Section 13.4). Whether fixed or variable the general requirements for frequency stability are the same.

11.2. Conditions for Oscillation

An oscillatory circuit consists broadly of two parts:

(1) a frequency determining section,

(2) a circuit which in some way provides the replacement of the losses inevitable in the frequency determining section.

Whether or not a circuit will oscillate can be decided in a number of ways. For example, from a consideration of

(a) power replacement;
(b) voltage feedback
  (i) by calculation,
  (ii) by vector diagram;
(c) negative resistance;

and the list is not exhaustive.

In the first method power is regarded as lost in the frequency-determining circuit and calculations are used to show whether the rest of the circuit can replace this power. For example, in a parallel

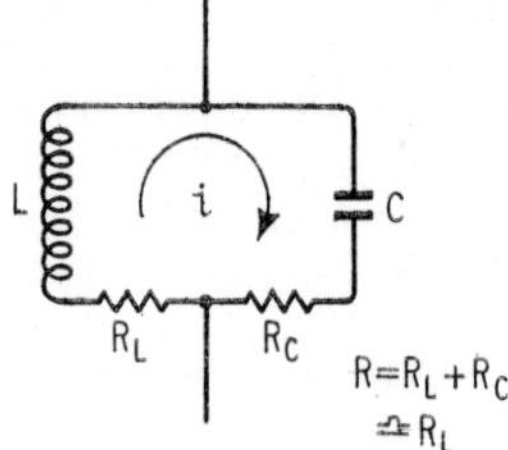

Fig. 11.3. Power loss in parallel LC circuit

Fig. 11.4. Radio frequency amplifier

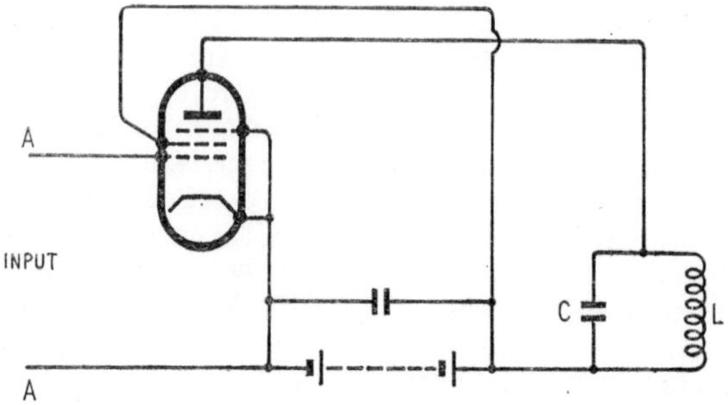

LC-circuit of total series loss resistance R the power loss is i^2R where i is the r.m.s. current circulating round the circuit (Fig. 11.3). If by some means this power can be fed back into the circuit to replace what is lost then sustained oscillations will take place in *LC*. The tuned-anode and tuned-grid oscillators can readily be considered in this way.

The second method treats the oscillator circuit as an amplifier which provides its own input. Consider Fig. 11.4 which shows an amplifier having as its input a sine wave of frequency f. If the output circuit is tuned to f an amplified version of the input appears across L. Suppose now that the feedback coil L' of Fig. 11.5 is coupled to L and that there appears across *BB* a voltage of the same phase and the same amplitude as (or greater amplitude than) that across *AA*: if *BB* are connected to *AA* the original input can be dispensed with because it has been replaced by the feedback from L'. Thus, it can be said that if the feedback to the input is made sufficiently large and in the correct phase to sustain the output then the circuit will oscillate.

Calculations based on this method give a precise indication of the conditions necessary for oscillation. The vector-diagram approach only shows whether, in a given circuit, it is possible for the phase of the feedback to be conducive to the maintenance of oscillations.

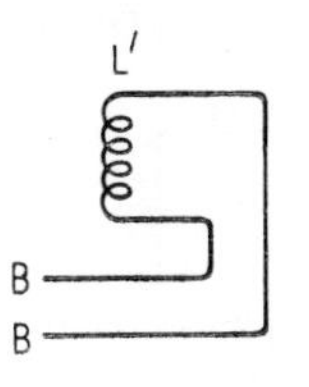

Fig. 11.5. Feedback coil

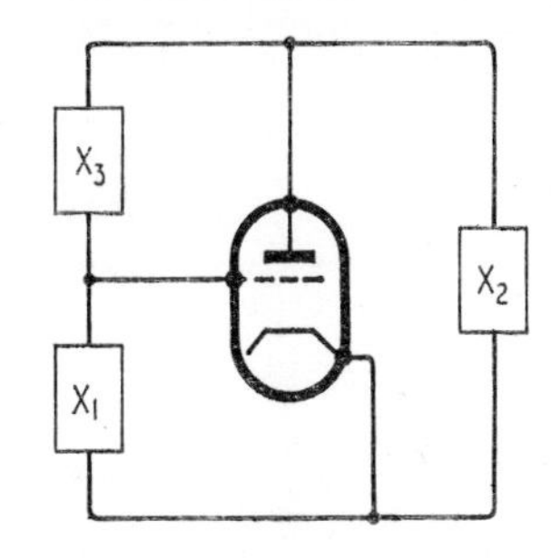

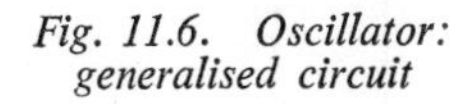

Fig. 11.6. Oscillator: generalised circuit

Whether the magnitude of the feedback is sufficient is another matter.

In the third method resistance is regarded as the cause of oscillations dying down in a circuit. Thus a spring, fixed at one end and supporting a weight at the other, will oscillate when set in motion. The oscillation dies down because of the frictional resistance in the material of which the spring is made and in the air. Similarly in the parallel LC-circuit of Fig. 11.3 the oscillation dies down because of the electrical resistance. In each example the arrangement would oscillate indefinitely if the resistance could be eliminated. Mathematically, one way to dispose of an unwanted quantity is to add to it a similar quantity which is numerically equal but of opposite sign. Thus to get rid of 2 it is necessary to add -2 to it. Similarly, to dispose of an unwanted resistance R it is necessary to add $-R$.

In this method of approach, therefore, the maintaining section of the circuit is regarded as a source of negative resistance. If it can supply a negative resistance to cancel the positive resistance of the frequency-determining section then oscillation is possible.

For completeness it is probably worth mentioning here, without proof, that if the circuit reactances between cathode and grid, between cathode and anode and between grid and anode are X_1, X_2 and X_3 as indicated in Fig. 11.6 then oscillation is impossible unless X_1 and X_2 both have the same sign, i.e. are both inductive or both capacitive, and X_3 is of opposite sign. If these conditions are satisfied then the circuit is potentially capable of oscillation and will, in fact, oscillate if the sum of X_1, X_2 and X_3 is zero and if the feedback is adequate, i.e. if the total effective resistance round the circuit is zero (which means that sufficient negative resistance is introduced to neutralise the positive resistance inevitably present).

11.3. Hartley Oscillator

11.3.1. BASIC CONSIDERATIONS

The fundamental circuits are in Fig. 11.7 (for a triode) and Fig. 11.8 (for a transistor). Consider these by the feedback method, that is by method (2) of Section 11.2. For each circuit, considered as an amplifier, the output is approximately in antiphase with the input (a characteristic of common-cathode valve circuits and of common-emitter transistor circuits when equipped with a resistive

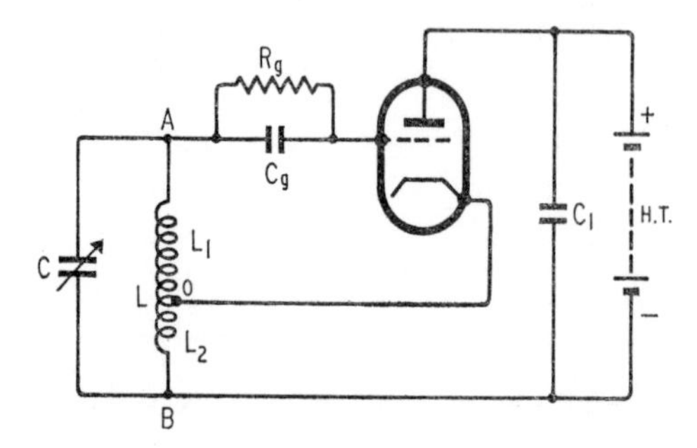

Fig. 11.7. Hartley oscillator: thermionic valve

load, as is a resonant LC-circuit). Regarded as an oscillator the input for each circuit is the e.m.f. across L_1 (at A with respect to 0), while the output is that across L_2 (at B with respect to 0). The two voltages are themselves in anti-phase since they are the voltages at the opposite ends of an inductor. Thus the phase of the feedback is approximately correct for the maintenance of oscillations and it would appear probable that, given sufficient feedback, oscillations would be maintained in this circuit. This treatment is not rigorous; it is an elementary example of the vector diagram method of approach already mentioned. It can, however, be stated as a fact that the Hartley circuit is one readily capable of oscillation and is in very wide use because it is simple to operate, will oscillate over a very wide range of frequencies and has conditions for oscillation which are easy to satisfy. The frequency of oscillation is a little over $f = \frac{1}{2\pi\sqrt{LC'}}$ where C' is the total capacitance in parallel with the inductor L.

11.3.2. PRACTICAL CIRCUIT

The circuit of Fig. 11.7 is termed the series-fed Hartley and it is sometimes used in practice. It has the advantage of requiring a minimum of components and of having one end of L and C at earth potential. On the other hand the cathode is at r.f. potential and, if the heater is earthed, the cathode-heater capacitance shunts L_2;

since the heater is normally supplied by a.c. the appearance of hum in the output of the oscillator is likely. If this difficulty is overcome by earthing the cathode at r.f. (instead of earthing the anode) the whole H.T. supply must have an r.f. potential to earth. This drawback can be overcome by the use of choke and capacitors but this is more complicated than the method of parallel feed to be described below.

C_1 provides a low-reactance path for the r.f. signals across the H.T. supply while $C_g R_g$ gives automatic bias and provides for the self-starting of oscillations (giving zero bias until after the circuit is switched on and has started to oscillate) and for amplitude stability as discussed in *Radio and Line Transmission*, *Volume* 1, Section

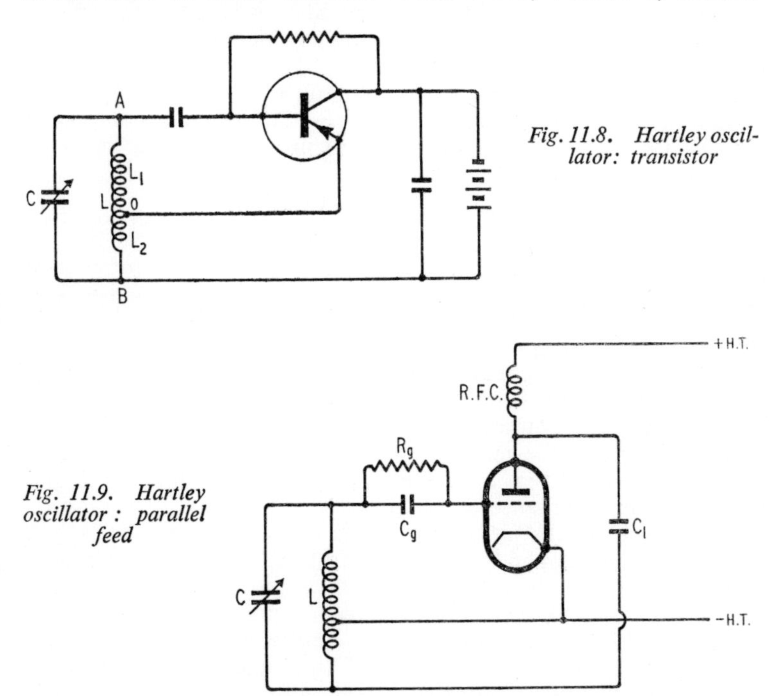

Fig. 11.8. Hartley oscillator: transistor

Fig. 11.9. Hartley oscillator : parallel feed

12.4.3. The magnitude of the feedback is determined by the position of the tapping point: a small amount is all that is normally needed, that is, the cathode tap is well towards the anode end of the coil.

In Fig. 11.9 is shown the parallel-feed connection (parallel because the H.T. circuit is completed independently, through the r.f. choke, and not by series connection through part of the oscillatory circuit as in Fig. 11.7). The r.f. choke allows the H.T. to be supplied to the anode without passing through L. If the choke were not included the r.f. signal across the lower end of L would be short-circuited by the H.T. supply. C_1 permits ready passage of r.f. current from the lower end of the tuned circuit to the valve anode and yet keeps H.T. out of the *LC*-circuit. As drawn the circuit has the advantage that the cathode is at earth potential but it suffers from the drawback that both fixed and moving plates of the tuning capacitor are at r.f. potential to earth. Other details are as for the series circuit. For frequencies of the order of 1 Mc/s C_g and R_g are commonly about 100 pF and 100,000 ohms respectively. C_1 can be 0·01 μF and the r.f. choke about 2 mH.

Transistor circuits can also be series or parallel fed. Points of difference are that the H.T. supply is of much lower voltage and there is no heater to introduce hum. Also the transistor input resistance is much lower than that of the valve and so is shunted across only part of the *LC*-circuit to prevent undue damping. This has no effect on the resonance frequency of the circuit (Section 9.10.1). Alternatively it can be used with a circuit of lower L/C-ratio than would normally be used with a valve. Appropriate component values are included in Fig. 11.10, although these are liable to considerable variation depending on the circumstances.

11.4. Colpitts Oscillator

The Colpitts oscillator depends for its action on the same general principles as does the Hartley, viz. that if a continuously-wound coil is joined with one end to the grid (or base), one end to the anode (or collector) and a suitable tap in between taken to the cathode (or emitter) then the circuit will oscillate. In the Hartley circuit the tapped connection is made to a point on the coil; in the Colpitts an electrically equivalent point is found by the use of a capacitor potential divider across the coil (Fig. 11.11).

In the circuit of Fig. 11.11 the frequency is determined by L and the resultant capacitance of the network C, C_1 and C_2. C_1 and C_2 are commonly equal but the circuit will oscillate for ratios of C_2/C_1 up to about the amplification factor of the valve. At high frequencies C_1 and C_2 can be dispensed with, the grid-cathode capacitance serving as C_1 and the anode-cathode capacitance as C_2.

The other components function as described for similar circuits.

For good frequency stability C_1 and C_2 should be large enough to swamp the effects of other capacitances and should themselves have a zero-temperature coefficient.

Fig. 11.12 shows a Colpitts circuit using a transistor. R_1, R_2 and R_3 provide stabilising voltages; they will have values of the order of 400 Ω, 20 kΩ and 5 kΩ respectively; C_3 acts as a by-pass at the frequency of oscillation.

11.4.1. EQUIVALENT CIRCUIT

The equivalent circuit of the Colpitts oscillator can readily be drawn as is shown in Fig. 11.13 for the valve circuit, the tuning capacitor being omitted for simplicity.

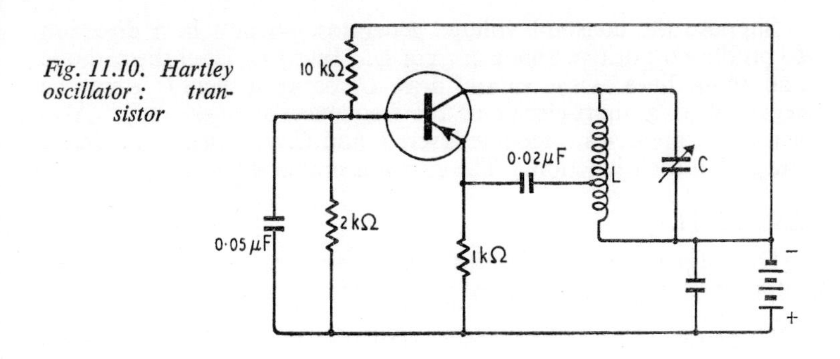

Fig. 11.10. Hartley oscillator: transistor

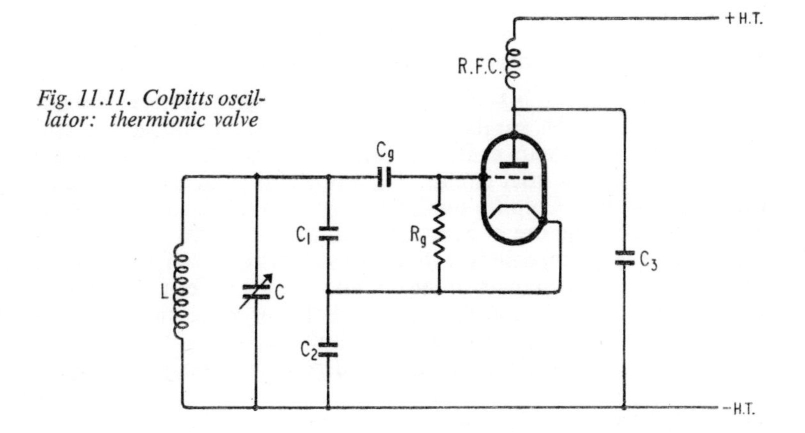

Fig. 11.11. Colpitts oscillator: thermionic valve

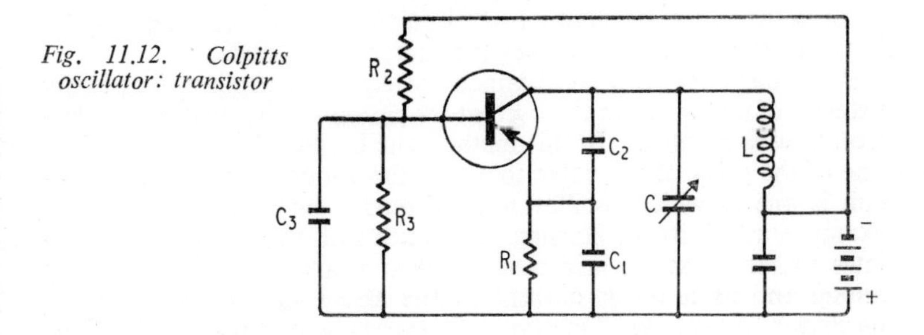

Fig. 11.12. Colpitts oscillator: transistor

Suppose the constant voltage generator μv_g acts in a direction to produce a positive anode current i_a. Insert r_a, label the cathode and anode lines as shown and draw C_2 between them (C_3 may be regarded as a short-circuit at the frequency of operation). Also across the anode-cathode line insert L and C_1 in series and indicate the grid at their junction. The circuit is now complete.

Example 11.1

Draw the equivalent circuit for a valve-maintained Hartley oscillator indicating the position of grid, anode and cathode.

Using reasoning and procedure similar to that applied to the Colpitts oscillator the equivalent circuit may be drawn as in Fig. 11.14 in which L' is that part of the coil between anode and cathode (the lower part of the coil in Fig. 11.9) and L'' is the remainder. There is some inductive coupling between L' and L''. C is the tuning capacitor.

11.4.2. SERIES-TUNED COLPITTS, OR CLAPP, CIRCUIT

The Colpitts circuit is sometimes designed to have a series resonant circuit instead of the parallel one of Fig. 11.11. In this form, and using a valve, it appears as in Fig. 11.15, and is sometimes called the Clapp or Clapp–Gouriet circuit. It was observed in the account of the parallel-tuned Colpitts oscillator that frequency stability could be improved by the use of zero-temperature-coefficient fixed capacitors of such large capacitance that they swamped the effect of other, notably valve, capacitances in the circuit. In the series-tuned circuit C is small while C_1 and C_2 are much larger (and can be zero-coefficient capacitors). Hence this circuit is characterised by a relatively small tank circuit current and good frequency stability.

11.5. Electron-coupled Oscillator

It was stressed that one of the essentials in the search for frequency stability is the isolation of the oscillator from any appreciable load and in particular from any change in load. In transmitters it is common to carry out this separation by the interposition of a buffer stage between the oscillator and circuits in which the load may be heavy or variable. Sometimes the buffer is untuned.

A similar effect can be obtained by electron coupling, that is, by transfer of energy from the oscillatory circuit to the next by the electron flow in a valve. In this method the oscillatory circuit employs one of the grids of the valve to act as the anode. Provided the circuit is one in which oscillations can readily be maintained, as, for example, a Colpitts or Hartley, the circuit still oscillates. The oscillatory voltages at the grid modulate the electron stream to the valve anode and as a result current in the anode circuit varies at the oscillator frequency. One type of arrangement is indicated in Fig. 11.16. The circuit is a Hartley, the screen grid acting as the

Fig. 11.13. Colpitts oscillator: equivalent circuit

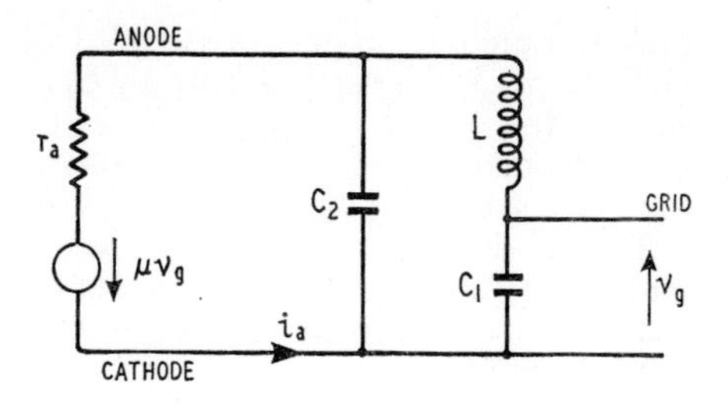

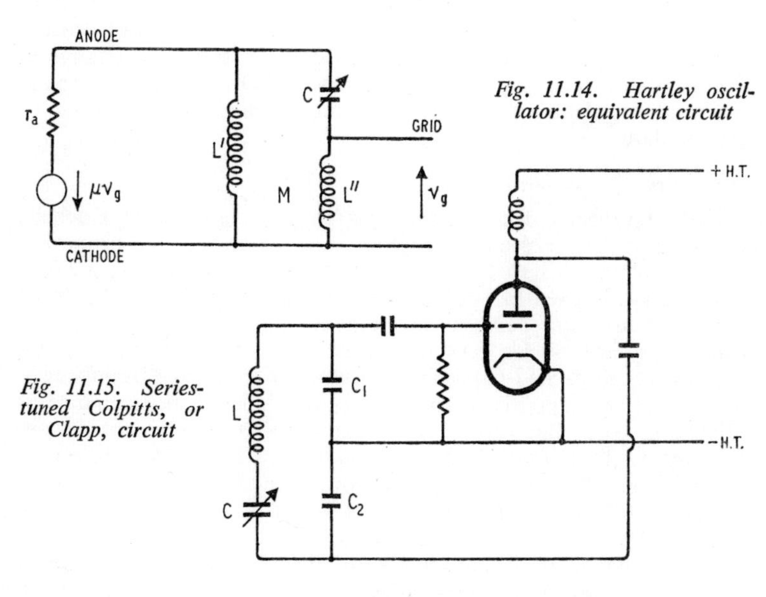

Fig. 11.14. Hartley oscillator: equivalent circuit

Fig. 11.15. Series-tuned Colpitts, or Clapp, circuit

Fig. 11.16. Electron-coupled oscillator

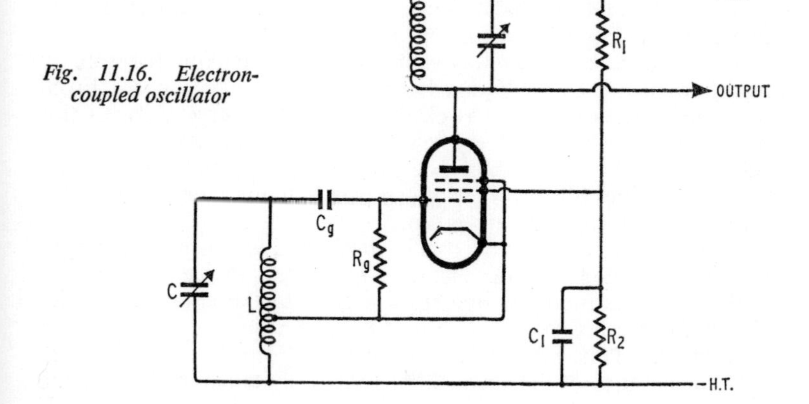

oscillator anode. It derives its H.T. from a potentiometer across the H.T. supply and is earthed at r.f. by C_1 (of the order of 1,000 pF). The smaller the values of R_1 and R_2 in the potentiometer the greater the independence of the screen voltage on the screen current and the more stable the operation.

The anode circuit can be tuned to the same frequency as the oscillator or to a harmonic. The anode is well screened from the oscillator by the screen and suppressor grids so that changes in the output circuit have little effect on the oscillator (changes over a wide range in the anode voltage of a pentode have little effect on either the anode current or grid circuit). Such effects as do result from any interaction between anode and grid circuit are seen as a change in the oscillator frequency and amplitude. These consequences are less (and negligible) if the anode circuit is tuned to a harmonic of the oscillator.

11.6. Crystal-controlled Oscillator

Certain crystals, for instance quartz, tourmaline and Rochelle salt, exhibit the piezo-electric effect. A discussion of the properties of materials of this sort has already been given in Section 3.2.6.

The crystals used in communications equipments are usually of quartz; tourmaline being unduly expensive and Rochelle salt somewhat unstable. Such a crystal is equivalent to a resonant *LC*-circuit as shown in Fig. 3.12 (b). The mechanical resonance frequency of the crystal corresponds to the frequency of electrical resonance as calculated from the values of L and C in the equivalent circuit. Frequencies in common use range from a few tens of kc/s to ten or more Mc/s. An ordinary *LC*-circuit, resonating at, say, 1 Mc/s would have values of L and C of the order of 200 μH and 200 pF respectively. In the crystal equivalent circuit the L and C values for this order of frequency are about 2H and a few pF respectively. As a result, and although the effective resistance is much higher in the crystal than in the *LC*-circuit, the Q value is exceptionally high, of the order of tens of thousands, i.e. about one hundred times that of a low-loss *LC*-circuit.

The importance of a high Q value as a factor affecting the stability of the frequency of oscillation has already been mentioned. From what has just been said it is clear that a quartz crystal meets this need very well. In fact crystal-controlled oscillators can readily achieve a long term stability of several parts in one million and higher stabilities (of the order of one part in ten million) are possible when great precautions are taken.

A wide variety of crystal oscillator circuits are in use; a simple one is illustrated in Fig. 11.17. This is the Pierce circuit. To understand the mode of operation substitute the equivalent series circuit for the crystal, *LC* in Fig. 11.18, and then re-arrange as in Fig. 11.19. This will be recognised as a Colpitts type of circuit with the cathode of the valve characteristically taken to the junction of a series arrangement of capacitors across the inductor L. It

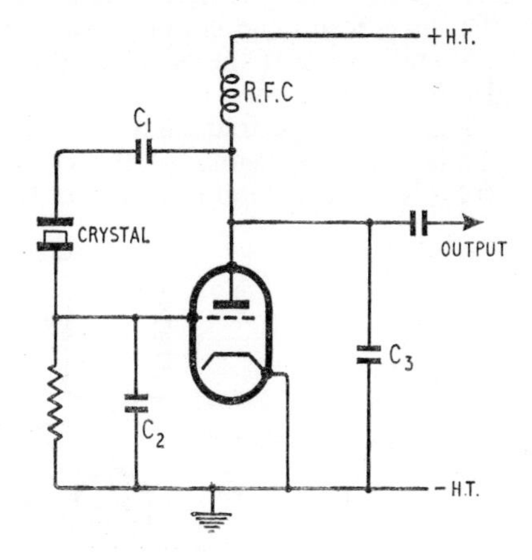

Fig. 11.17. Pierce crystal oscillator

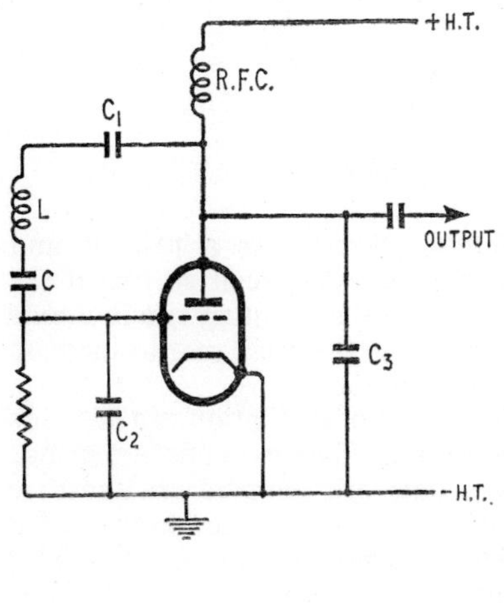

Fig. 11.18. The circuit of Fig. 11.17 with crystal equivalent circuit substituted for the crystal

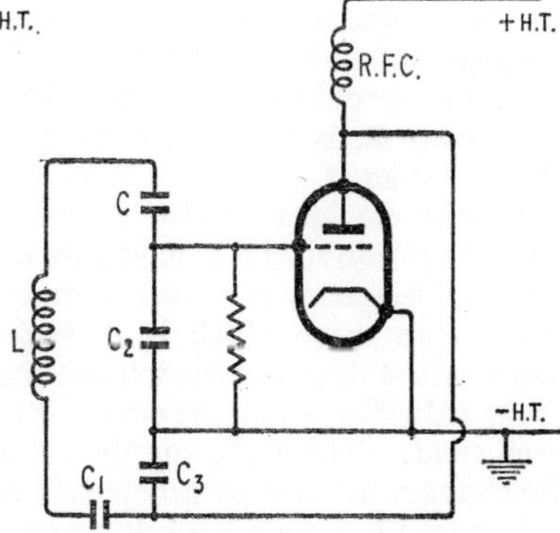

Fig. 11.19. Fig. 11.18 rearranged

has already been mentioned that C is of very small value, only a few pF; as a result the coupling between the valve and the crystal is very small so that the effect of the valve on the crystal Q value is also very small.

The higher the frequency at which the circuit is to oscillate the smaller the dimensions of the crystal. At frequencies above, say 10 Mc/s, crystals tend to be rather fragile and susceptible to fracture

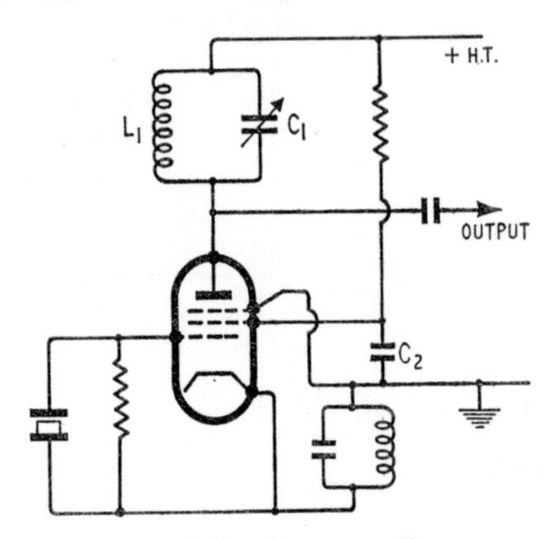

Fig. 11.20. Tri-tet oscillator

if allowed to oscillate too violently. Sometimes a resistor and parallel capacitor are included in the cathode circuit so that if the amplitude of oscillation tends to increase the resulting rise in anode current will cause the voltage drop across the resistance to increase also and hence to exercise a stabilising influence.

Slices of crystals suitable for the control of oscillation can be made in various ways from a complete crystal. The cut chosen depends on the frequency required. Most cuts are temperature sensitive but one cut shows almost zero variation in resonance frequency for temperature changes even so wide as from zero to 100° C. This cut, however, is unsuitable for crystals required to operate above about 500 kc/s.

To give an idea of size it can be said that a crystal for use at 500 kc/s is about one inch square and one quarter inch thick.

If the frequency is to be controlled very closely the crystal may be kept in an oven which is thermostatically controlled so that the temperature does not vary more than one or two degrees either side of the value for which the crystal has a zero or very small temperature coefficient. The H.T. supply voltage may be stabilised to prevent the changes in valve parameters caused by changes in applied voltage.

Changes in anode load do not have much effect on the frequency of crystal oscillator circuits. Such effects as do exist can be minimised by using an electron-coupled type of circuit. A typical example is shown in Fig. 11.20; this is the tri-tet circuit.

As in the electron-coupled circuit described in Section 11.5 the tri-tet circuit uses the screen grid of a pentode as an oscillator anode; the suppressor acts as a screen between the oscillator and the anode circuit L_1C_1. This latter is often tuned to a multiple of the crystal frequency so that the stage acts as a frequency multiplier. When tuned as a frequency multiplier the anode circuit has little effect on the oscillatory circuit.

The crystal has a high d.c. resistance and so is shunted by a grid resistor R—about 100 kΩ. C_2 is the blocking capacitor and is of about 1,000 pF.

The circuit LC requires to be set up initially to provide the correct conditions for oscillation, after that of course it need not be altered.

Transistor circuits can also be crystal controlled. A circuit is shown in Fig. 11.21. Oscillation is maintained by feedback between L' in the emitter and L in the collector circuit. In series with L' is the crystal so that at all frequencies, except that of series resonance of the crystal, the impedance in the feedback circuit is considerable;

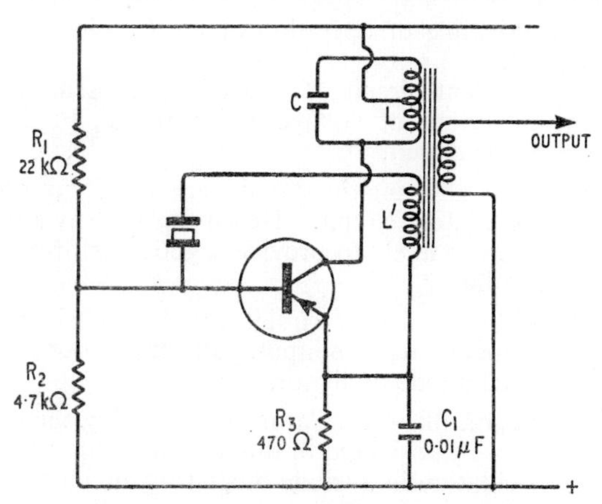

Fig. 11.21. Crystal oscillator: transistor

at resonance the impedance is small enough to permit sufficient feedback to initiate and maintain oscillations. The collector circuit, LC, is tuned to the frequency of oscillation and, as in the other examples (*see* Section 11.3.2) the collector is in effect connected to a tapping point on the inductor. It will be recalled that if a tap is not employed a much smaller L/C-ratio is necessary leading to an inconveniently large value of C.

R_1 and R_2 provide the correct bias for the base. The emitter stabilising resistor is by-passed to the oscillatory frequency by C_1. Typical values are shown on the diagram but, of course, considerable variations from these figures will be met in practice depending on the frequency of operation.

Questions

1. What properties of quartz enable it to be used for controlling accurately the frequency of a valve oscillator?

Give a circuit diagram of a crystal-controlled oscillator and explain its action. (*C & G*, 1958.)

2. Give the circuit diagram of a variable-frequency oscillator covering the range 1 to 2 Mc/s and indicate typical values for the components.

State the precautions necessary to obtain good frequency stability and indicate the order of stability you would expect. (*C & G*, 1952.)

3. Explain the causes of frequency instability in a valve oscillator.

Describe how the instability effects can be minimised for (a) an oscillator for use on a single frequency and (b) an oscillator to cover a 3–1 frequency range.

Give the circuit diagrams of the two oscillators. (*C & G*, 1947).

4. Sketch the circuit of an electron-coupled oscillator. Explain its operation and state the advantages of this type of coupling.

5. Explain the principle of operation of the Colpitts and Hartley types of oscillator.

Draw the equivalent circuit of one of these oscillators clearly indicating the directions of voltage and current. Explain briefly the derivation of the circuit.

6. Detail the factors affecting the stability and quality of oscillations generated in an oscillatory circuit. Hence explain why a modulated power oscillator is not likely to provide a good performance under either of these headings.

7. Explain how an oscillatory circuit can be considered as an amplifier which provides its own input. Illustrate your answer with reference to a tuned-anode oscillator.

8. Explain the action of an oscillatory circuit using the concept of negative resistance. Considered in this way what are the necessary conditions for oscillation in a triode oscillatory circuit?

9. Give one example to illustrate the use of an oscillatory circuit in (a) a radio transmitter, (b) a telephone line terminal equipment and (c) a radio receiver.

Explain briefly the function of the oscillator in each example.

10. What is a damped oscillation? How is a damped oscillation affected by the amount of resistance in the circuit?

What is needed to convert a circuit which can provide a damped oscillation into one which can maintain continuous oscillations?

12

Modulation and Detection

The subject of amplitude modulation is introduced in Chapters 3 and 13 of *Radio and Line Transmission, Volume* 1 and is considered further in Chapter 3 of the present volume.

12.1. Production of an Amplitude-modulated Wave

12.1.1. BASIC REQUIREMENTS

To produce an amplitude-modulated wave a process which effectively results in the multiplication of the constituent wave magnitudes must be employed. This can be done by:

(a) straightforward multiplication, each wave being caused to exert an influence in direct proportion to its own instantaneous magnitude so that the ultimate wave-shape is proportional in magnitude at any instant to the product of the magnitudes of the original waves. This is known as *multiplicative mixing* and, unlike method (b) below, does not require non-linearity in the modulation circuit,

(b) adding the two waves and then applying them to a non-linear device such as a diode detector. This is known as *additive mixing*.

As will be seen in the next section, however, the mere addition of two waves without application to a non-linear device does not result in amplitude modulation. The non-linear action does, in fact, cause a multiplicative effect between the two waves and it is this effect which gives amplitude modulation.

12.1.2. MODULATION USING NON-LINEAR CIRCUIT

Fig. 12.1 shows a valve circuit biased to a steady voltage V_s and supplied with two, series-acting, e.m.f.s, $V_c \sin\omega t$ and $V_a \sin pt$. If the valve acts linearly the change in anode current is proportional to the change in grid voltage

$$\therefore \qquad \Delta I_a \propto \Delta V_g$$

or

$$\Delta I_a = k\Delta V_g$$

where k is a constant and indicates the slope of the anode current-grid voltage characteristic. Thus

$$I_a = I_s + kV_g = I_s + k(V_c \sin\omega t + V_a \sin pt)$$

where I_s is the standing current appropriate to the bias V_s. No modulation has occurred, the output waveform having the shape shown in Fig. 12.2. The waveforms have merely been superimposed on each other.

Suppose the valve is arranged to operate over a non-linear part of the characteristic. To determine the type of output now obtained a graphical construction, such as employed in Chapter 15 of *Radio and Line Transmission, Volume* 1, can be used. This gives a result similar to that shown in Fig. 12.3.

Alternatively an analytical approach may be made as follows. Assume that the non-linearity of the characteristic derives chiefly from the square-law component. That is, the anode current varies partly in proportion to the grid voltage and partly in proportion to its square. This assumption is often approximately true, particularly when operation takes place over a small sweep of the characteristic.

It can now be said that

$$I_a = I_s + k_1V_g + k_2V_g^2$$

in which k_2 is a constant which gives appropriate weight to the square-law component of the characteristic.

Expanding:

$$\begin{aligned} I_a &= I_s + k_1(V_c \sin\omega t + V_a \sin pt) + k_2(V_c \sin\omega t + V_a \sin pt)^2 \\ &= I_s + k_1(V_c \sin\omega t + V_a \sin pt) + k_2(V_c^2 \sin^2\omega t + V_a^2 \sin^2 pt + \\ &\quad + 2V_aV_c \sin\omega t \sin pt) \qquad (1) \end{aligned}$$

Recalling that

$$\sin^2 A = \frac{1 - \cos 2A}{2}$$

and that

$$\sin A \sin B = \frac{\cos(A - B) - \cos(A + B)}{2}$$

we may put

$$I_a = I_s + k_1(V_c \sin\omega t + V_a \sin pt) + k_2\left[\frac{V_c^2}{2}(1 - \cos 2\omega t) + \right.$$

$$\left. + \frac{V_a^2}{2}(1 - \cos 2pt) + V_cV_a\{\cos(\omega - p)t - \cos(\omega + p)t\}\right]$$

These terms represent:

(1) The standing current, I_s, appropriate to the bias.

(2) Outputs at the frequencies of the applied voltages,

$$k_1 V_c \sin\omega t + k_1 V_a \sin pt.$$

(3) A d.c. component which increases with the square of the applied voltage,

$$\frac{k_2}{2}(V_a^2 + V_c^2)$$

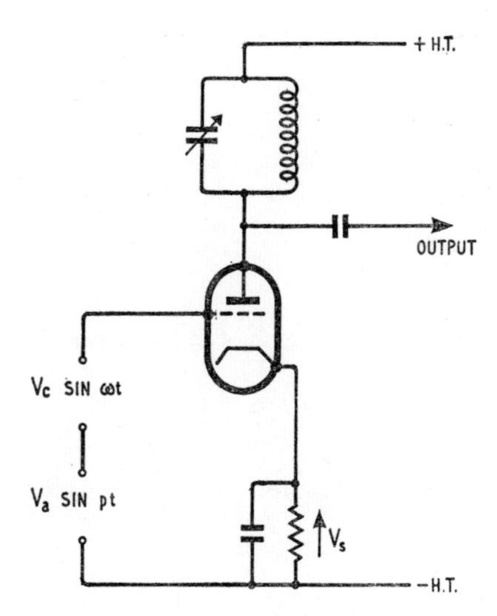

Fig. 12.1. Additive mixing

(4) Second harmonics of each of the input frequencies and proportional to the square of their magnitude,

$$-\frac{k_2}{2}V_c^2 \cos 2\omega t - \frac{k_2}{2}V_a^2 \cos 2pt$$

(5) Side-frequencies,

$$k_2 V_c V_a \cos(\omega - p)t - k_2 V_c V_a \cos(\omega + p)t$$

If ω is the angular frequency of a radio-frequency carrier and p that of an information-frequency, for instance, sound or vision, then the anode circuit of Fig. 12.1 is tuned to $\omega/2\pi$ c/s and is arranged to be of a type suitable for accepting signals with angular frequencies

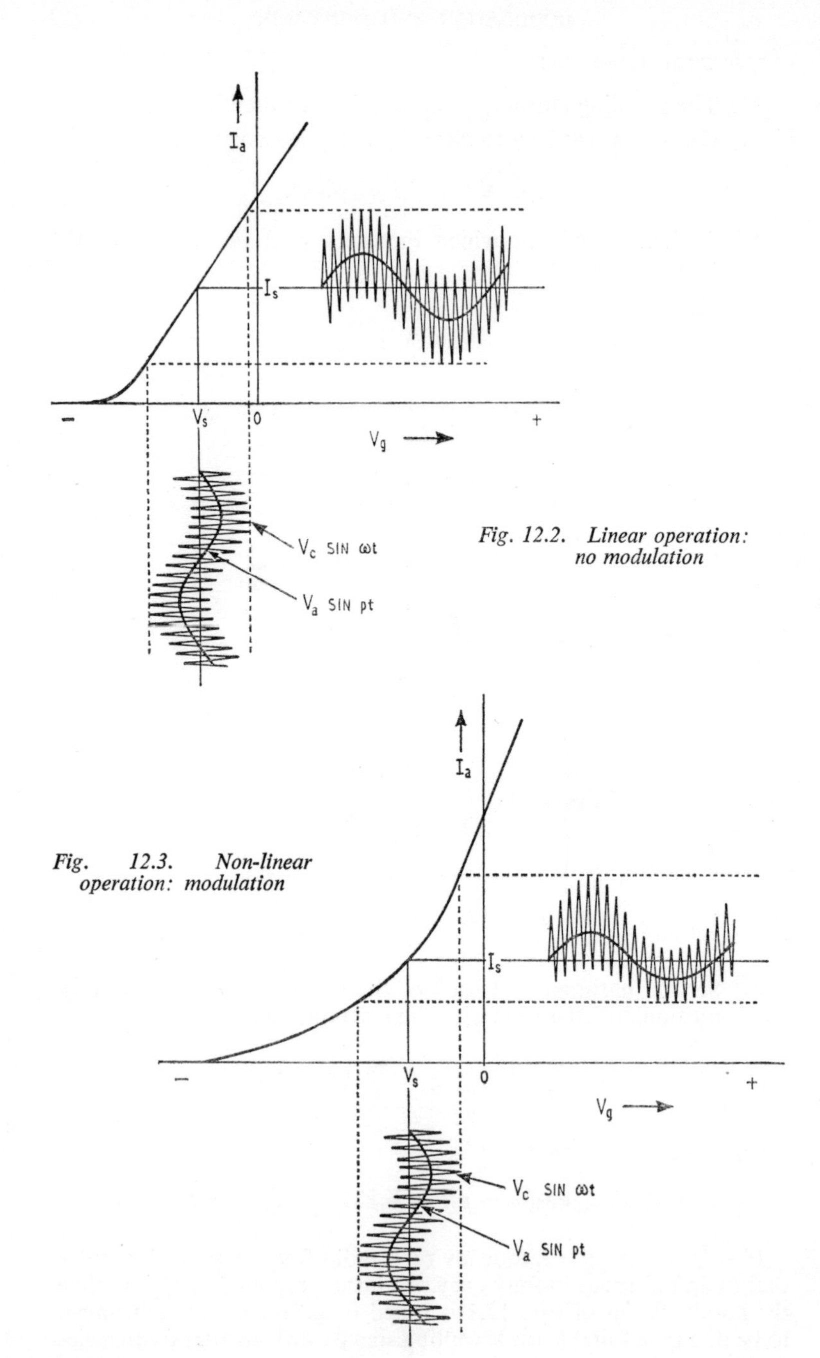

Fig. 12.2. Linear operation: no modulation

Fig. 12.3. Non-linear operation: modulation

ω, $\omega - p$ and $\omega + p$ while rejecting the remainder. The output, therefore, consists of

$$k_1 V_c \sin\omega t,\ k_2 V_c V_a \cos(\omega + p)t,\ -k_2 V_c V_a \cos(\omega + p)t$$

an expression which represents a carrier and two side-frequencies, the side-frequencies having an amplitude proportional to that (V_a) of the information-frequency signal.

Note that the side-frequency signals, which convey the information, are the result of the product term $2V_a V_c \sin\omega t \sin pt$ of Equation 1. Without the product term there is no amplitude modulation; that is, when the process of additive mixing is employed non-linear action must be introduced so that the multiplicative effect is obtained.

The same general effects take place when several frequencies are involved. Given one carrier signal $V_c \sin\omega t$ and several information-frequency signals $V_{a1} \sin p_1 t$, $V_{a2} \sin p_2 t$, etc., the current in a device possessing square-law properties contains:

(a) signals at all the original frequencies,

(b) a d.c. component,

(c) second harmonics of the original signals,

(d) signals with frequencies equal to the sum and difference of the frequencies of the original signals.

All that will survive a suitable tuned circuit in the output of the device will be components at the carrier frequency and at the sum- and difference-frequencies of the carrier and each modulation-frequency, i.e. the carrier and the upper and lower sidebands. All the other frequencies will be so far removed from the resonance frequency of the tuned circuit as to evoke no response.

Thus it can be said that a device which includes a square-law term in its characteristic can be used to effect amplitude modulation. The effectiveness of such modulators depends on the magnitude of the side-frequency terms, i.e. on the magnitude of k_2 (and, of course, on V_a and V_c). Consideration of the factors affecting k_2 is rather complicated but in general it can be said that k_2 increases with the curvature of the characteristic. Even with k_2 at its optimum value, however, square-law modulation does not give appreciable output without an unduly high proportion of distortion; also its efficiency is low.

To represent non-linear characteristics adequately we need a much more complicated expression than we have assumed in order to correlate the changes in anode current with the changes in grid voltage which initiate them. Consideration of such expressions by analytical means is beyond the scope of this volume. It can be added, however, that the outline given above is a representative indication of the operation of such modulators and that the higher the order of curvature of the characteristic the greater the number of spurious frequencies introduced.

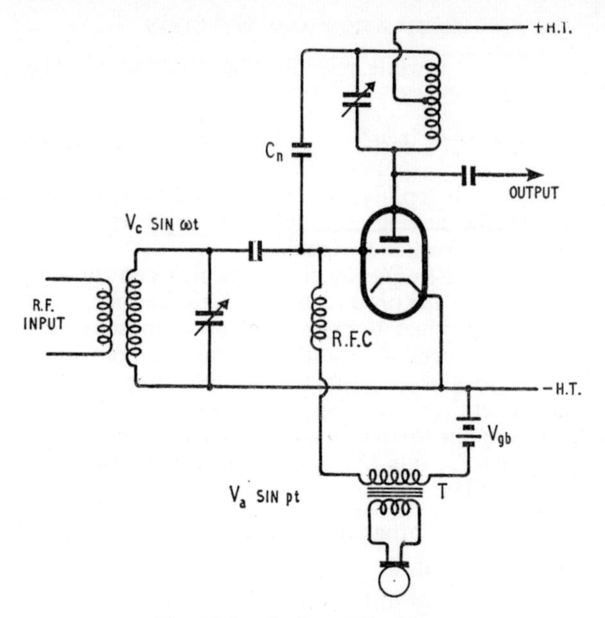

Fig. 12.4. Grid-modulated input

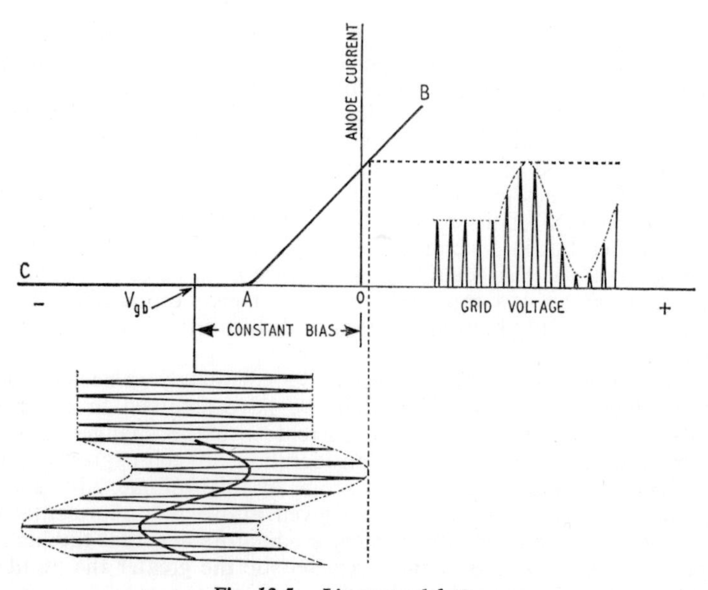

Fig. 12.5. Linear modulation

Example 12.1

A radio frequency of 1 Mc/s is applied in series with a modulating wave containing frequencies of 400 c/s and 500 c/s to a valve possessing a characteristic which is part linear, part square-law.

(a) What components will appear in the valve anode current?
(b) Which of these will cause appreciable output across a resonant circuit tuned to 1 Mc/s in the valve anode circuit?

(a) All signal frequencies:
1,000 kc/s, 500 c/s and 400 c/s.
Second harmonics:
2,000 kc/s, 1,000 c/s and 800 c/s.
Sum and difference frequencies:
1,000·4 kc/s, 1,000·5 kc/s, 999·6 kc/s, 999·5 kc/s
900 c/s
100 c/s.

(b) Of the above frequencies a circuit tuned to 1,000 kc/s will accept:
1,000 kc/s, 1,000·4 kc/s, 1,000·5 kc/s, 999·6 kc/s, 999·5 kc/s.

12.1.3. GRID-MODULATED AMPLIFIER

As an example of the type of modulation which depends for its action on the changing slope in the characteristic of the device in use consider the grid-modulated amplifier of Fig. 12.4. Both the carrier and modulating voltages are applied between the same pair of electrodes (the grid and cathode) and modulation takes place only if the characteristic is curved or if advantage is taken of a discontinuity. Let the valve be biased somewhat beyond cut-off (Fig. 12.5). If we ignore the curvature at the foot of the characteristic and assume the rest of it to be straight, the anode current variations are a true representation of the input variations which caused current to flow. Distortion is introduced to the extent to which the curvature of the characteristic modifies the output. The output contains components at the original frequencies, their sum and difference frequencies and such r.f. components as are needed to represent the distorted version of the r.f. input which appears in the output. The tuned circuit, which is shown with a centre tap and neutralising capacitor, C_n, to prevent instability, removes all components except those with frequencies close to the carrier $\frac{\omega}{2\pi}$; thus the predominant frequencies in the output are those of the carrier and the side-frequencies. Note from Fig. 12.5 that a constant bias is needed; this is shown in Figs. 12.4 and 12.5 as V_{gb}. The r.f. choke prevents the modulation-frequency transformer, T, from presenting a low impedance at radio frequency across the grid-cathode circuit of the r.f. amplifier.

The amplifier normally operates with a little grid current for in this way efficiency is appreciably higher than if the bias and input are arranged to avoid grid current. In these circumstances the loading on the r.f. source varies; if the source is an oscillator, changes in frequency may result. The oscillator, however, is normally isolated from the modulator amplifier by at least one intervening stage, often called a *buffer*.

Modulation is achieved because the input is applied to a non-linear device. (Because the characteristic is not simply *AB*, it is *CAB*, with a discontinuity at *A*.) However, the operative parts of the characteristic are substantially straight and because of this the operation is often termed *linear modulation*, in contradistinction, for instance, to square-law modulation (Fig. 12.3).

12.2. Demodulation, Detection

12.2.1. BASIC REQUIREMENT

At the sending end of a communication system the intelligence-frequency may be translated to another frequency by the process of modulation. At the receiving end the intelligence must be extracted, i.e. the wave must be *demodulated*; or, *detection* must be effected. This process can be thought of, as in the method employed in Chapter 15 of *Radio and Line Transmission, Volume* 1, as that of extracting the envelope of the modulated wave. Alternatively it can be considered as one of suitably mixing the components at the side-frequencies with the carrier components to produce, among other frequencies, the difference-frequency, which is the required intelligence. Using the latter viewpoint it can be seen that the conditions for demodulation are the same as for modulation, that is, the frequencies concerned must be caused to act in a multiplicative manner either by applying them, for instance, to separate electrodes in a valve so that each exercises a proportional influence on the anode current, or by applying them to a non-linear device such as a diode.

12.2.2. DEMODULATION USING NON-LINEAR CIRCUITS

Regions of non-linearity are found at the foot of the anode current-grid voltage characteristic of a thermionic valve, at the foot of the anode current–anode voltage curve of a thermionic diode and of the corresponding curve for a semiconductor diode. By far the most commonly used circuits are those employing diodes.

12.3. Diode Detector

In Chapter 15 of *Radio and Line Transmission, Volume* 1 the action of the diode detector circuit together with considerations affecting choice of component values were discussed in some detail. The circuit is of the general form shown in Fig. 12.6 for circuits embodying thermionic valves, or Fig. 12.7 for those using semiconductors.

The *CR* values are selected in accordance with the basic principles and for sound reception are of the order shown on the diagrams.

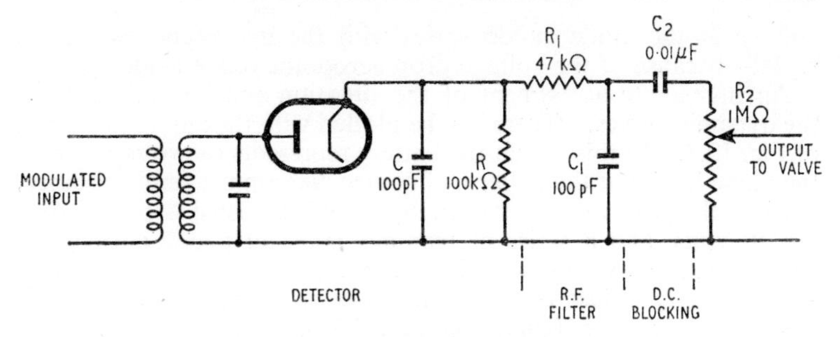

Fig. 12.6. Thermionic diode detector

For vision the values are about 20 pF and 3 kΩ respectively. The valve detector includes an r.f. filter R_1C_1, a d.c. blocking capacitor, C_2, and volume control R_2. The semiconductor, feeding as it does the low input impedance of the following transistor stage, has for its CR combination a small value of resistance and large capacitance; additional r.f. filtering is, therefore, sometimes dispensed with. The series resistance of about 5 kΩ prevents wide variations of impedance with change of volume control setting. The coupling

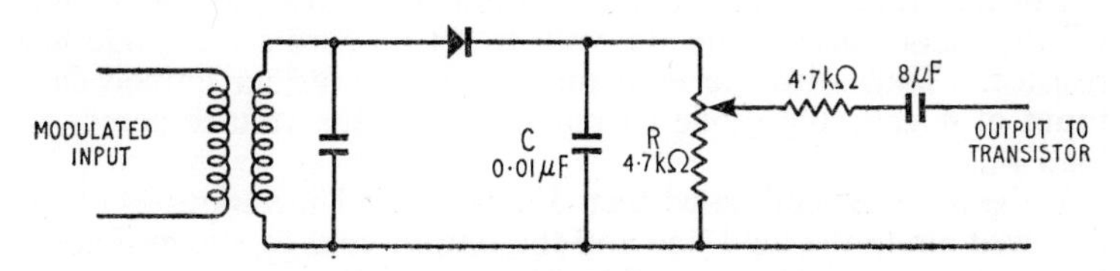

Fig. 12.7. Semiconductor diode detector

capacitor must be large because the feed is to a low-impedance stage. The diode is often given a slight forward bias to improve weak signal detection. The bias, obtained by means of a potentiometer across the supply (not shown) moves the point of operation somewhat away from the bottom bend of the characteristic.

At very high frequencies detection is effected by means of a semiconductor diode. At lower frequencies either thermionic or semiconductor diodes may be used.

Semiconductor diodes have a small capacitance, need no power supplies and introduce no hum problems. They are small and easy to mount and have a life even longer than that of thermionic diodes. On the other hand the characteristics of semiconductors depend on temperature and they pass a far greater reverse current than do thermionic diodes—the reverse current for which is practically zero.

12.3.1. DIODE DYNAMIC CHARACTERISTICS

Just as the anode voltage of a triode varies with the anode current, and hence with the grid voltage, so in the case of a detector, the voltage at the diode anode varies with the magnitude of the r.f. signal—because of the voltage drop across the diode load.

An immediate assessment of the situation can be derived from the dynamic curves. These can be plotted with the aid of the circuit of Fig. 12.8. In this circuit the battery represents the voltage across the diode load, but the battery, of course, has almost zero resistance.

V_{in} is the peak value of the r.f. input; it is adjusted to some value, say 10 volts, and a series of readings taken of anode current for various battery voltages. Clearly, in this instance, when $V_0 = -10$ volts, I_a is just zero and increases as V_0 is increased, i.e. is made less negative (curve (1) Fig. 12.9). The same process is repeated for other values of V_{in} and this, for each determination, is the peak value of an r.f. input (not a d.c. as for all the other curves which have been drawn for diodes and other thermionic valves).

In practice the diode is used in conjunction with a load: to take account of this a load line may be drawn across the curves at an angle appropriate to the load resistance, R, and used in a manner similar to that described for triode and pentode valve applications. (The effect of the coupling to the following circuit is considered in the next section.)

For any given unmodulated r.f. input it is now possible to read off the diode current and the voltage set up across the diode load resistor. Thus with the load line as shown and an unmodulated input of 4 volts the diode current is I_1 and the voltage across the load V_1.

If the input is modulated then V_{in} varies. The variations can be followed along the load line and the output voltage, i.e. that across the load, read off for every desired value of V_{in}.

Example 12.2

A certain diode in the circuit of Fig. 12.8 yielded the dynamic characteristics of Fig. 12.10. Sketch the graph of the output voltage when the same diode is used as a detector with the load for which the load line is drawn on the graph. Assume an input consisting of a 6-volt peak carrier modulated 67 per cent.

The envelope of the modulated signal increases and decreases from the mean level of 6 volts by 67 per cent, i.e. by 4 volts. Fig. 12.11 shows the waveform.

The working point for the unmodulated signal is point A on Fig. 12.10 and as the modulated wave rises and falls in amplitude the working point moves along the load line up to B and down to C. Projection downwards from the load line at as many points as are necessary shows how the output varies with time for such an input. Corresponding times of $t = t_0$, $t = t_1$, etc., are shown in Figs. 12.10 and 12.11.

The output waveform of Fig. 12.10 shows that in this example the output voltage varies between about 5 volts and 1 volt (at

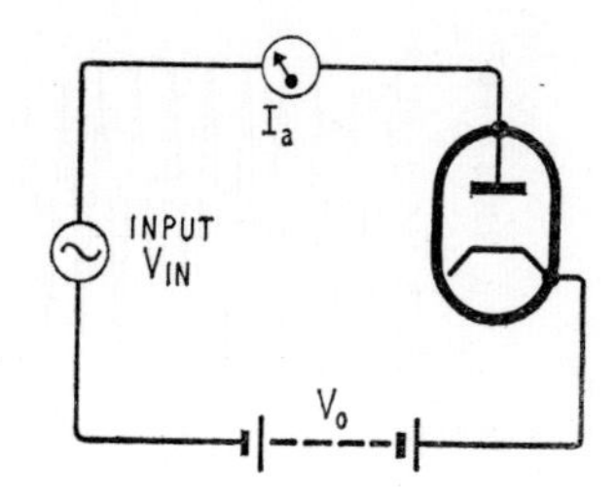

Fig. 12.8. Circuit to determine diode dynamic characteristics

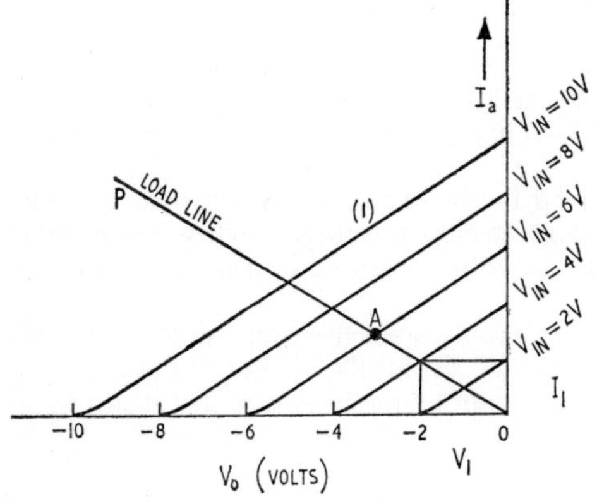

Fig. 12.9. Diode dynamic characteristics

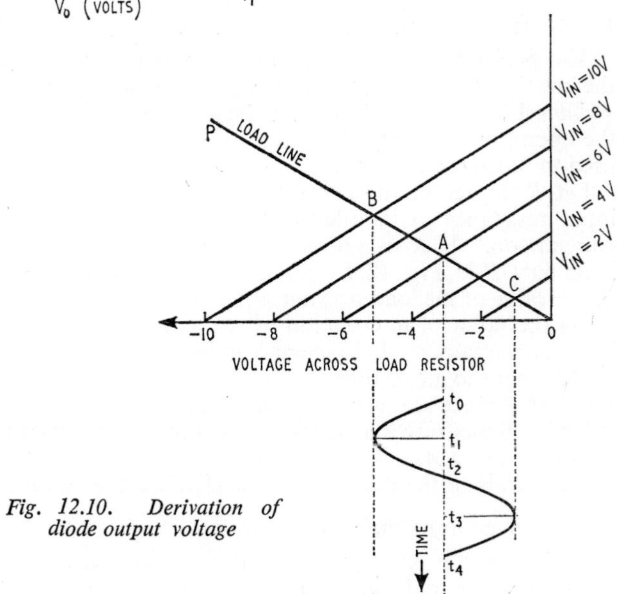

Fig. 12.10. Derivation of diode output voltage

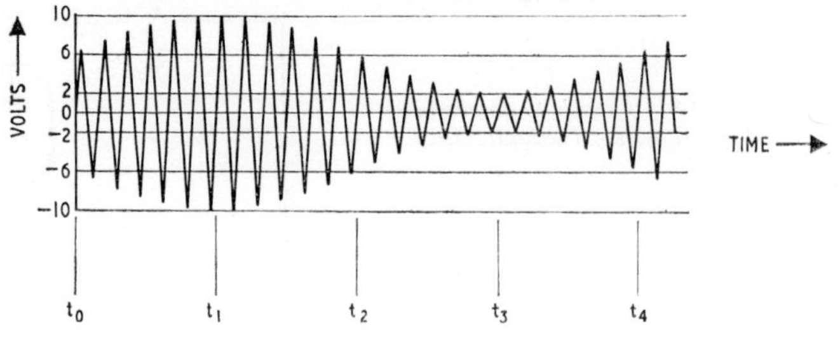

Fig. 12.11. Illustrating Example 12.2

$t = t_1$ and $t = t_3$ respectively). The figure brings out clearly how the output would be increased by the use of a larger load resistance (a more nearly horizontal load line) and would decrease with decrease of load resistance.

Use of the figures derived in the example show a detection efficiency of about 50 per cent (which is rather low), a 4-volt peak modulation envelope (10 — 6 volts) resulting in a 2-volt peak output (5 — 3 volts).

If the dynamic curves are evenly spaced along the load line the modulation envelope is followed faithfully by the output voltage. To the extent that the curves are not equally spaced distortion exists in the output.

12.3.2. EFFECT OF THE COUPLING TO THE NEXT STAGE

The performance of an amplifier is modified by the coupling to the next stage (*see* Chapter 10 of *Radio and Line Transmission, Volume* 1). Similar effects must be considered in connection with the operation of diode detectors. The circuit of Fig. 12.6 will be used to illustrate the events.

Let R' represent the effective value of R_2 and of any following input resistance in parallel; assume that R_1 can be neglected by comparison. If R' is infinitely large then operation of the diode is along OP in Fig. 12.9. OP being drawn at a slope appropriate to the value of the load resistance R. The operating point is fixed as being the point of intersection of the load line and the appropriate input carrier curve. In practice R' is finite and the effective resistance of the circuit to a.c. is that of R' in parallel with R. Operation, therefore, takes place along a load line, DE in Fig. 12.12, of slope appropriate to the effective load resistance and which passes through the operating point A.

The effects of the reduction in effective load resistance are:

(1) A given input yields less output.

(2) When the r.f. signal amplitude falls below a value corresponding to D the diode current becomes zero and remains at this level for all smaller values of r.f. signal. Hence the load voltage cannot follow the modulation envelope in this region.

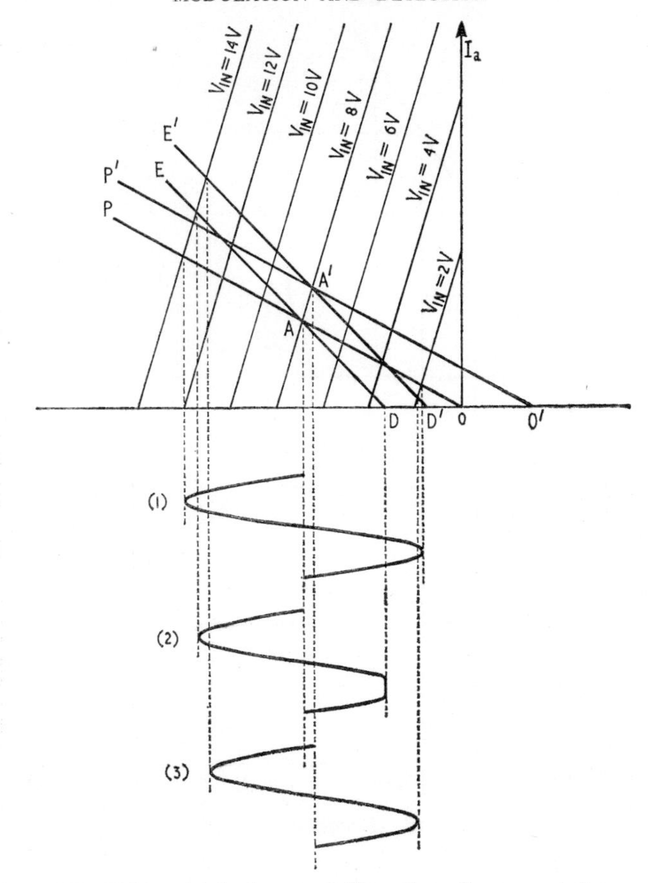

Fig. 12.12. Diode detector: effect of coupling to next stage

The outputs for the two load lines for a given modulated input (between 2 volts and 12 volts) are shown in curves (1) and (2) in Fig. 12.12.

The fact that the output is reduced is not important, for this can be put right by the introduction of a little extra gain elsewhere. The second point mentioned represents distortion and is important.

Clearly what is needed is as high a ratio R'/R as possible so that the a.c. and d.c. resistances shall be as nearly equal as possible. This cannot be achieved by reducing R, however, for this reduces the efficiency of the detector circuit; moreover, there is also an upper limit to the value of R'. If all else fails it is possible to reduce the clipping effect of R' by tapping R_2 across only part of it. This obviously also has the effect of reducing the output.

Or, the whole load line may be shifted bodily to the right by applying a positive bias to the anode. If the bias is OO' the load line OP

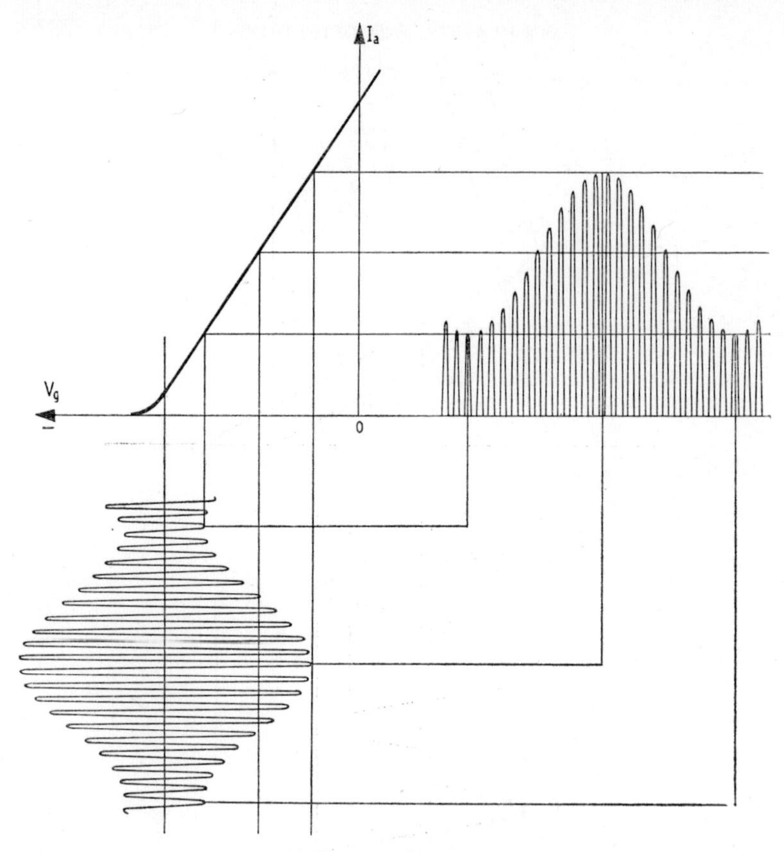

Fig. 12.13. Anode-bend detection

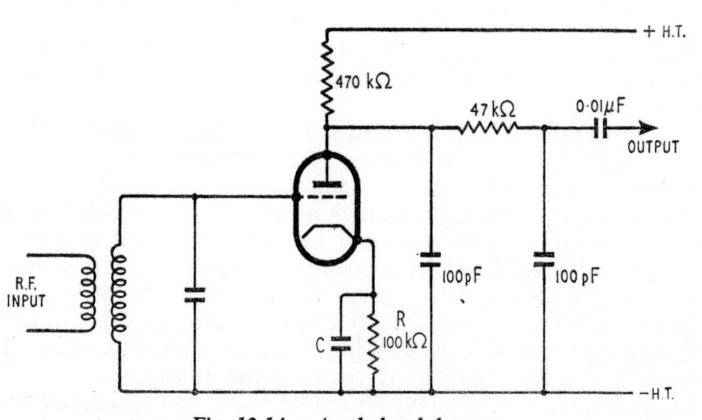

Fig. 12.14. Anode-bend detector

becomes that shown as $O'P'$ and shifts the operating point to A'. As a result the a.c. load line becomes $D'E'$. The output for the same modulated input as before becomes that shown in curve (3). This is not as great as number (1) but it is not clipped as in number (2). This method, however, introduces additional difficulties in its train because of the increased damping which results and is hardly used.

12.4. Other Forms of Detector

As already stated, detection is nearly always carried out using diode detectors. When the input is necessarily low, as in certain line circuits, for instance, it may be advantageous to employ the grid detector described in Section 12.4.3. This circuit may also be useful when it is desired to apply reaction or regeneration (Section 12.4.4).

12.4.1. ANODE-BEND DETECTION

This type of circuit is not often used. It takes advantage of the curvature at the foot of the anode current–grid voltage characteristic of a valve (or the collector current–base current characteristic of a transistor) and is biased nearly to cut-off, so that the operation takes place as shown in Fig. 12.13 for a valve. A similar type of action takes place when a transistor is used.

The circuit has the form shown in Fig. 12.14 with component values of the order indicated. The anode load value is larger than that normally appropriate to the valve in use because of the large grid bias and consequent small value of anode current. Values of about one half or one megohm are typical. The bias may be obtained from a battery or from a cathode *RC* combination as shown. *R* is about 100 kilohms as shown in the figure. The larger the capacitance value the greater the gain and the greater the distortion: *C* is usually either about 50 μF (for maximum gain) or about 100 pF (for minimum distortion).

The following points can be noticed:

1. If the crests and troughs of the modulation envelope are restricted to the straight part of the characteristic, operation is linear.
2. If the signal is excessively large the valve draws grid current presenting a varying impedance to the previous stage and distorting the peaks. For small inputs there is no grid damping with this type of circuit.
3. Feedback across the grid-anode capacitance causes some damping of the input although there is little r.f. signal in the anode circuit because of the 100 pF capacitor across the valve output.
4. If the modulation depth is so great as to take the operation into the curvature of the characteristic then distortion results.

5. If the signal is very small, operation is over the curved part of the characteristic. This has two effects—
   (a) inefficiency because the slope of the curve is small,
   (b) distortion and the introduction of spurious components. If the operation can be regarded as square-law there are components at second harmonic and sum- and difference-frequencies as shown in Section 12.1.2.

12.4.2. INFINITE-IMPEDANCE DETECTOR

Provided that grid current is avoided the input impedance of the circuit just discussed is very high apart from any damping caused by the feedback of any remnants of r.f. via the anode-grid capacitance. If the circuit is modified to that shown in Fig. 12.15 the anode-grid feedback is entirely removed. This is the circuit diagram of the so-called " infinite-impedance " detector and it has the following features:

1. There is no grid current unless the input is exceedingly high; the input impedance, therefore, is very large. (As stated later, it can be negative.)
2. There is no voltage gain, the circuit being of the cathode-follower type.
3. The output is taken from the cathode which is not at earth potential. There is thus the possibility of the introduction of hum from the heater.
4. The output impedance is low (cathode follower) and distortion of the form dealt with in Section 12.3.2 is less likely than for a diode detector.
5. Such feedback as there is is from the cathode (not the anode) and the phase is such as to tend to make the input impedance negative, i.e. it is such as to tend to sharpen the tuning of the previous circuit.

12.4.3. GRID DETECTOR

An extension of the diode type of detector is the grid detector. This is shown in Fig. 12.16 from which it can be seen that the control grid-cathode circuit functions in the same manner as the anode-cathode of a diode detector. At the grid, therefore, there is developed a voltage similar to that resulting from diode detection. The grid variations, which include r.f., are amplified in the valve and appear at the anode where the r.f. variations are filtered out in the choke-capacitance filter. If the r.f. signal is not removed not only are the usual disadvantages of its presence evident (possible overloading of the next stage and tendency to instability) but the detector output is less than in the absence of the r.f. signal. The d.c. component is not filtered out as it is following a diode detector circuit and acts as a negative bias.

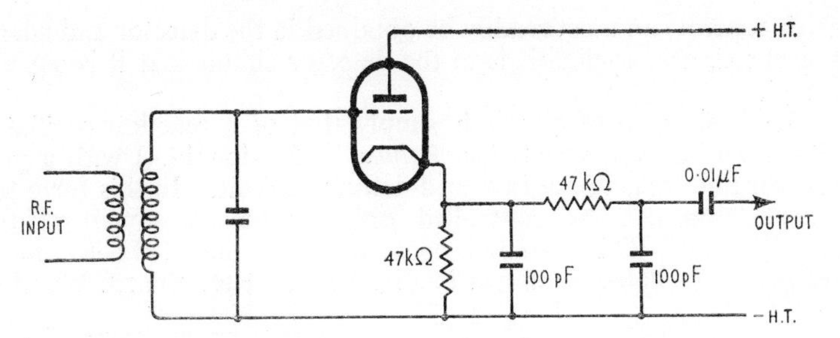

Fig. 12.15. Infinite impedance detector

Values of C_g and R_g are determined as for the diode detector (Chapter 15 of *Radio and Line Transmisson, Volume* 1). C_g is about 100 pF and R_g between about 100 kΩ (for minimum distortion) and 500 kΩ (for high efficiency). R_L is chosen to suit the valve used.

The following points can be noticed:

1. The r.f. signal cannot be removed before it reaches the anode circuit. There is, therefore, feedback via the anode-grid capacitance with resultant damping of the tuned circuit.
2. When the input signal is small the grid bias is virtually zero and the valve is at its most sensitive. That is, the gain is greatest when it is most needed.
3. Other general properties are the same as those of a thermionic diode followed by an amplifier.

12.4.4. APPLICATION OF REACTION OR REGENERATION TO A DETECTOR

The application of positive feedback to a circuit results in an output larger than otherwise would be obtained. This is termed *regeneration.* If too much feedback is applied the result is instability or oscillation. The use of regeneration can enable useful increases in output to be obtained provided the circuit is kept under control.

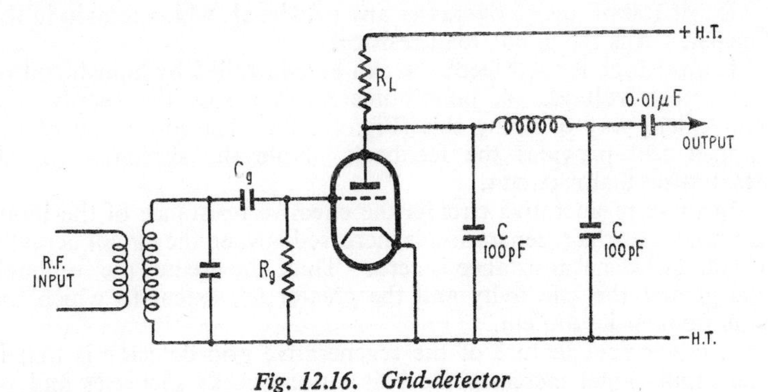

Fig. 12.16. Grid-detector

Such control can most readily be obtained in the detector and where regeneration is applied it is in the detector circuit that it is usually found.

The basic type of circuit is simply that of a reaction oscillator (*Radio and Line Transmission, Volume* 1, Section 12.4) with a grid resistor and capacitor as in a grid-detector circuit. In this form the regeneration can be controlled only by the movement of the coupling coil. This is seldom convenient. A modified form, more readily controllable, is shown in Fig. 12.17. Here the r.f. signal in

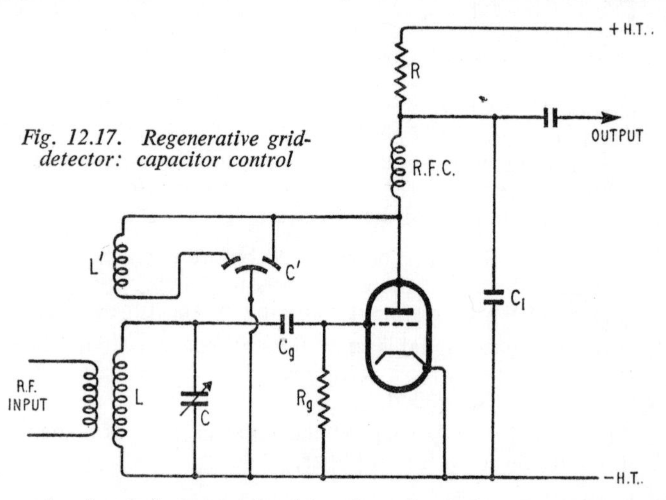

Fig. 12.17. Regenerative grid-detector: capacitor control

the anode circuit is constrained by the r.f. choke to return to the cathode through the differential type capacitor C' and the coupling coil L'. There is fixed mutual coupling between L and L', the feed-back being controlled by C'. The feedback capacitor is arranged in the manner shown so that even when little reaction is applied there is still a low reactance path to ground for the r.f. through the other part of the capacitor.

C_1, of 100 pF or so, by-passes any r.f. signals which remain in the output. R is the anode load resistor.

If a pentode is used feedback can be controlled by adjustment of the screen voltage. A potentiometer, P, across the supply is a convenient way of doing this (Fig. 12.18). The lower end of the tapped coil provides the feedback. Note the similarity to the Hartley oscillator circuit.

In these regenerative circuits the effective resistance of the input is reduced as the regeneration is increased—when the circuit actually oscillates the net resistance is zero. Thus, the greater the feedback the greater the selectivity and the greater the extent to which the side frequencies are cut.

A convenient feature of the regenerative grid detector is that if the input signal increases the bias automatically increases and so

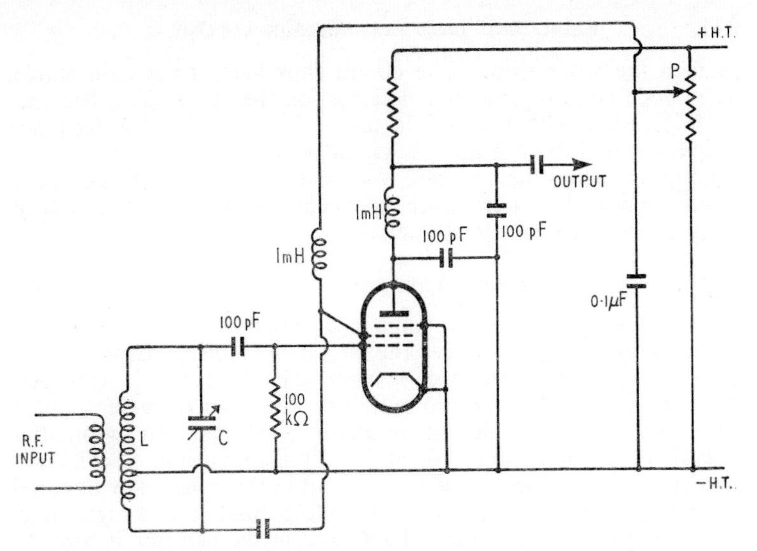

Fig. 12.18. Regenerative grid-detector: screen control

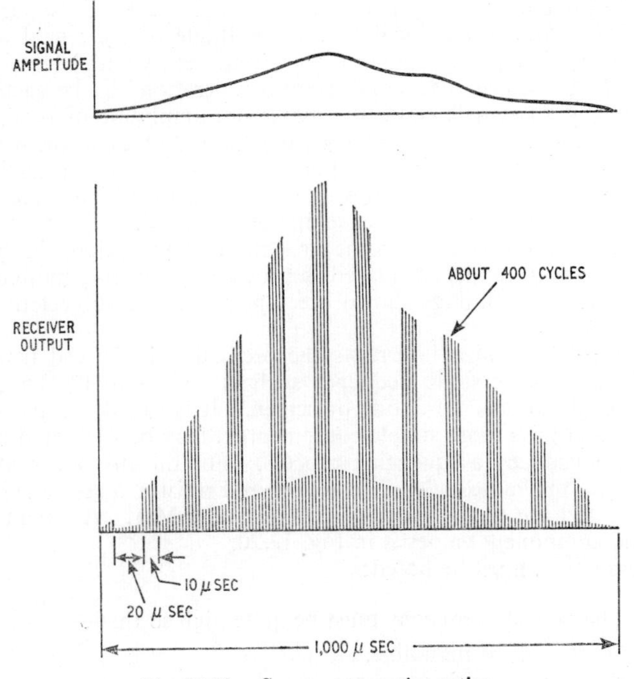

Fig. 12.19. Super-regenerative action

reduces the valve gain. The circuit thus tends to remain stable. The anode bend regenerative detector, on the other hand, becomes more sensitive with increase in input signal so that the feedback automatically increases and the circuit tends to instability.

Circuits such as these radiate when in the oscillatory state and can cause considerable interference, particularly, but by no means only, when connected direct to the aerial.

12.4.5. SUPER-REGENERATIVE DETECTOR

The regenerative detector gains in sensitivity because the positive feedback appreciably reduces the circuit resistance. In the super-regenerative detector feedback is increased still further so as to permit self-oscillation with rapid build up. In order to preserve intelligibility it is arranged that the oscillation is stopped very soon after initiation and then allowed to begin to build up again. Each burst of oscillation occupies about 10 μsec and in this time a few hundred cycles of r.f. signal take place and therefore build up to a high value. At the end of the burst the voltage level in the detector is brought down to that due to the signal alone until, about 20 μsec later, the inhibition is removed and the oscillation again allowed to build up. This process goes on continuously. Since the build up always begins from the signal level then existing and since it always goes on for the same length of time it follows that the peaks of the bursts of oscillation are proportional to the amplitude of the signal at the beginning of each burst and thus constitute a fairly faithful reproduction of it. Alternatively, the mode of operation can be such that practically all signals cause the maximum amplitude of oscillation to be reached. In this case the greater the amplitude of the original signal the longer the time each burst of oscillation is at a maximum and the greater the average output of the detector which is thus very roughly proportional to the envelope of the signal.

Fig. 12.19 shows how, in the first mode of operation, the bursts of oscillation render a pulsed reproduction of the modulation envelope. The timings shown are typical but are subject to wide variation.

The voltage which interrupts the oscillation is derived from an oscillator of supersonic frequency such as 20 kc/s to 100 kc/s, and is normally applied to anode or screen. It is called the *quenching voltage*. In the more simple equipments it may be generated in the detector itself by a squegging process. For this to occur at the right rate the values of grid capacitor and resistor must be suitably chosen and are of the order of 200 pF and 5 MΩ. A circuit with typical component values is in Fig. 12.20.

Two points must be noted:

1. The quench frequency must be quite high so that—
   (a) it shall be inaudible,
   (b) rapid changes in modulation waveform may be followed.

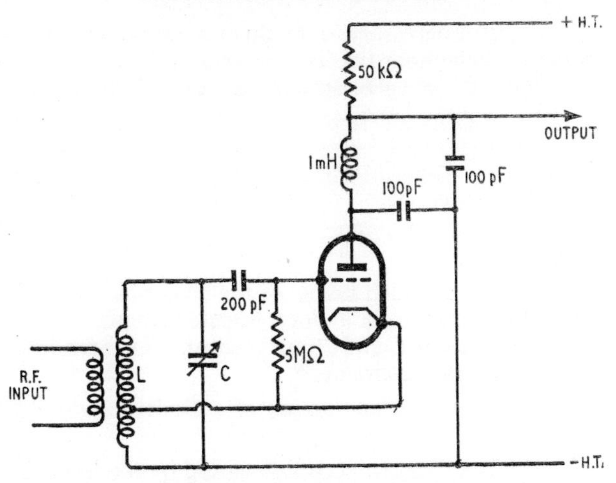

Fig. 12.20. Self-quenching super-regenerative grid-detector

2. The effectiveness of the method depends on the rapid rise of the amplitude of the oscillations: in this connection it is observed that the extent of the build up of an oscillation is a function of the number of cycles occurring; it can be said that in a typical circuit about 200 cycles are needed before the maximum is approached. Thus the higher the signal frequency the greater the amplitude of oscillation reached in a given time. Bearing in mind point (1) above, it follows that a detector of this sort is effective only at high signal frequency.

It is further observed that the detector is very sensitive; any signal, however small, will initiate oscillation. Hence the receiver is noisy for low inputs. Large noise and interfering signals, however, are limited by the fact that the oscillation cannot build up beyond a certain point however great the initial impulse.

This type of detector is useful only when some of the features of compactness, economy and utility are of supreme importance. It will deliver intelligible speech from a weak signal in the presence of quite strong interference; it will not produce a good quality output even under the most favourable of conditions.

Questions

1. Draw the circuit of a detector stage employing positive feedback (reaction) and explain its operation.

For what purposes is reaction employed in such a circuit and what are its advantages and disadvantages? (*C & G*, 1955.)

2. Explain with the aid of diagrams the action of either a leaky-grid detector or an anode-bend detector when used for the detection of amplitude-modulated waves. (*C & G*, 1957.)

3. Why is a detector necessary to obtain audio signals from an amplitude-modulated (telephony) radio carrier?

Describe one form of valve detector circuit. (*C & G*, 1947.)

4. Two e.m.f.s, at different frequencies, are applied in series between grid and cathode of a valve which can be assumed to have a square law characteristic. Determine the different components in the output circuit of the valve.

Explain why it is that a square-law device is able to act as a modulator and as a detector.

5. Discuss the factors which influence the choice of load resistor, R, and reservoir capacitor, C, in a diode detector circuit.

Why are the values of C and R not the same for thermionic diode and semiconductor diode circuits?

6. Compare the operation and action of a diode detector with that of a grid detector.

7. In what ways is a detector stage influenced by the following stage and to what extent can the performance of the preceding stage be affected by the detector?

8. Sketch the circuit of a detector which has a very high input impedance. Explain its action and discuss its merits and shortcomings.

9. Explain the action of a super-regenerative detector including in the discussion a statement of the various factors affecting component values.

What are the advantages and disadvantages of this type of circuit?

10. Sketch the circuit of a grid-modulated amplifier. Explain its action and indicate briefly the function of the components included.

13

Superheterodyne Method of Reception

A superheterodyne receiver is one in which the incoming signal frequency is changed in the receiver to a different, nearly always lower, fixed frequency before the pre-detection amplification is completed. The name arises from the fact that a heterodyne process is one in which two frequencies are suitably mixed, as in modulation or detection, to produce another frequency. This latter frequency in a superheterodyne receiver is always above the audible limit, i.e. super-sonic. Hence, the receiver is called a super-sonic heterodyne, or a superheterodyne, or, simply a superhet.

13.1. Reasons for the Use of Superheterodyne Receivers

With a few exceptions receivers are required to receive at least one band of frequencies (of maximum to minimum ratio of about two to one) and possibly half a dozen or even more. If the receiver is to be capable of (a) receiving transmissions of rather low signal strength, (b) separating such transmissions from more powerful ones on adjacent frequencies and (c) delivering an ample output to the loudspeaker, cathode ray tube or other output device it must possess the following attributes:

1. A number of resonant circuits to provide the required selectivity.
2. At least two or three stages of amplification before the detector.
3. The gain and selectivity to be approximately the same for all frequencies in the tuning range.
4. Good stability.
5. Some post-detector amplification to provide sufficient output to the indicating device employed (loudspeaker, cathode ray tube, etc.).
6. Not introduce noise or spurious signals above some acceptable minimum level.

Given a signal of reasonable amplitude from the detector requirement (5) is straightforward. If it is attempted to satisfy (1) to (3) by the use of conventional r.f. stages considerable difficulties may have to be overcome. If all the tuned circuits are given separate controls it will be virtually impossible to tune the receiver; if the various controls are mechanically linked so that one manipulation suffices for all the circuits (ganged tuning) it will be found that extreme difficulties, both mechanical and electrical, exist when more than three or four individual circuits are concerned.

Again, amplification becomes increasingly difficult to obtain as the frequency of operation rises; also, as the tuning is varied even over one waveband, both the gain and selectivity can vary considerably.

Thus, there are many advantages to be derived from carrying out much of the radio-frequency amplification in a receiver at a fixed

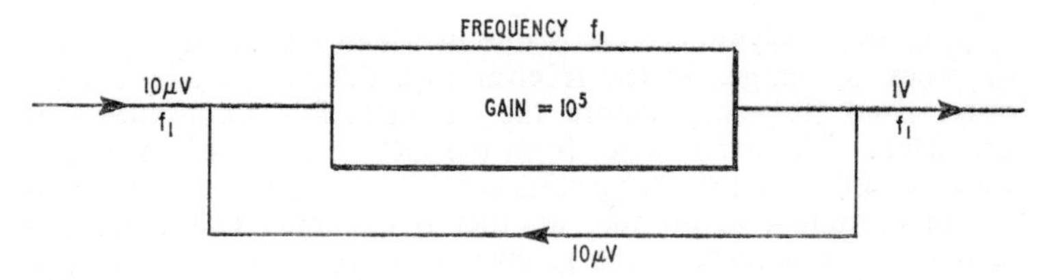

Fig. 13.1. Instability: amplifier of gain 10^5 at one frequency

frequency. The frequency can be chosen to be such that amplification is relatively easy to achieve; the tuned circuits can be selected and tuned once and for all during manufacture to give optimum gain and bandwidth; ganging and other difficulties associated with variable tuning disappear; screening becomes comparatively easy and straightforward because of the absence of variable tuning and the rather cumbersome capacitors and ganging arrangements necessitated thereby. This latter point is conducive to the satisfying of point number (4), stability, and a further aid in this respect is that if the gain is divided between two radio-frequencies less gain is needed at each frequency so that there is less chance of instability. Figs. 13.1 and 13.2 illustrate this. The former shows an amplifier accepting a signal at frequency f_1 and giving a gain of 100,000 at this frequency. If, in this amplifier, only one part in 100,000 of the output is fed back to the input in such phase as to sustain the output then the circuit will oscillate. Considerable trouble and care is necessary to avoid this possibility.

Fig. 13.2 shows the same overall gain obtained from two amplifiers, one operating at the signal frequency, f_1, and the other at some other frequency, f_2, a suitable device being assumed to exist between the two amplifiers to change the frequency from f_1 to f_2. If now one part in 100,000 of the output at frequency f_2 happens to be fed back to the input of either amplifier there will be no serious consequences. Oscillation can only occur if one-thousandth part of the output of the second amplifier, or one-hundredth part of the output of the first, is fed back to its own input. This is much easier to avoid than

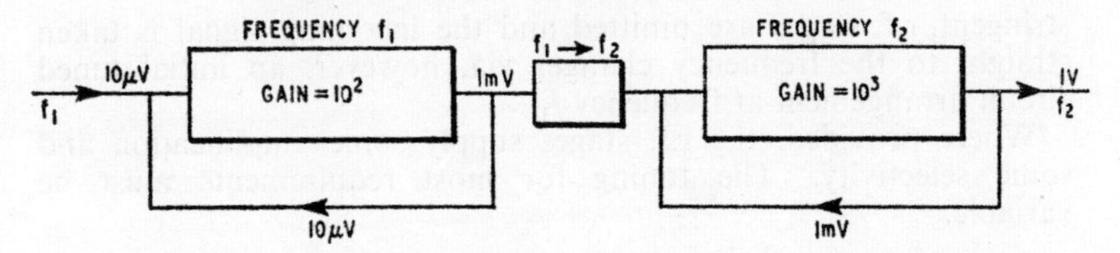

Fig. 13.2. Instability: amplifier of gain 10^5 spread between two frequencies

the small amount of feedback which is all that is needed to cause oscillation in the arrangement of Fig. 13.1.

These factors show the desirability of a superheterodyne type of receiver in which some amplification and frequency selection may take place at the signal frequency while the major part follows at some (usually) lower fixed frequency called the intermediate frequency (because it is intermediate between the signal and final frequencies). The intermediate frequency is fixed: thus the many difficulties associated with variable tuning are not encountered. The complication of arranging for this is the (relatively small) price to be paid for the many advantages.

The necessity to introduce a device to effect the change in frequency from that of the signal to that of the intermediate stages makes it

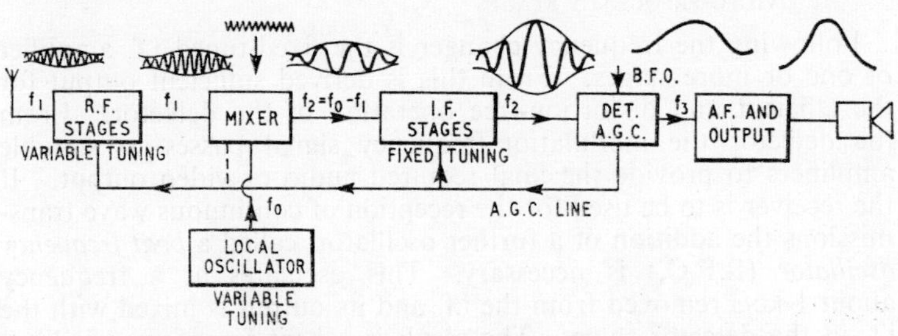

Fig. 13.3. Superheterodyne receiver: block diagram

more difficult to meet requirement number (6) regarding the introduction of noise and spurious frequencies.

13.2. General Description and Block Diagram

A block diagram illustrating the principles of operation of the superheterodyne receiver is given in Fig. 13.3. A brief description now follows; greater detail will be found in Sections 13.3 to 13.6.

13.2.1. RADIO-FREQUENCY STAGES

The wanted signal, f_1, together with unwanted ones, is delivered from the aerial to the input of the first r.f. valve. In the cheaper types of domestic radio receiver, where the specification is not very stringent, r.f. stages are omitted and the incoming signal is taken straight to the frequency changer via, however, an initial tuned circuit arrangement at frequency f_1.

Where provided, the r.f. stages supply some amplification and some selectivity. The tuning for most requirements must be variable.

13.2.2. FREQUENCY CHANGER

Following the r.f. circuits is the frequency changer. Here the wanted signal, together with such interference as has been passed by the r.f. circuits, is mixed with a locally-generated signal of frequency, say, f_0, to provide the intermediate frequency, f_2 ($\equiv f_0 - f_1$). Since f_2 is to be fixed it follows that the tuning of the local oscillator must closely follow that of the r.f. circuits in order that the required difference shall always be maintained.

Note that just as the wanted signal at a frequency, f_1, mixes with the oscillator frequency, f_0, to provide an output at the intermediate frequency so also does a signal at an unwanted frequency, f_u, such that $f_u - f_0 = f_2$. This particular frequency is as much above the local oscillator frequency as the wanted signal is below it and it is therefore called the *image signal* or *second channel signal* (Fig. 13.4).

13.2.3. INTERMEDIATE FREQUENCY, DETECTOR AND AUDIO-FREQUENCY OR VIDEO-FREQUENCY STAGES

Following the frequency changer is the fixed-tuned i.f. amplifier of one or more stages. From this is derived sufficient output for the efficient and distortion-free operation of the detector. From the detector the modulation-frequency signal passes to suitable amplifiers to provide the final required audio or video output. If the receiver is to be used for the reception of continuous wave transmissions the addition of a further oscillator, called a *beat frequency oscillator* (B.F.O.) is necessary. This oscillates at a frequency about 1 kc/s removed from the i.f. and its output is mixed with the i.f. in the detector stage. The result is a beat frequency of about 1 kc/s which thus renders the C.W. audible in the output. Domestic broadcast receivers, never called on to receive C.W. transmissions, are not fitted with a beat frequency oscillator.

13.2.4. MANUAL GAIN CONTROL

Depending on the complexity of the receiver the gain may be controlled manually,

(a) by one control which determines the fraction of the detector output to be passed on to the next stage, or

(b) by two controls, one as at (a) and the other controlling the gain of either the i.f. or the r.f. stages, or

(c) in very elaborate arrangements by three controls determining, respectively, the a.f., i.f. and r.f. gain.

The advantage of the more complex arrangements is that they enable the operator to adjust each section of the receiver more precisely for optimum performance in accordance with existing conditions of signal strength and interference and, sometimes, to prevent overloading in the presence of a strong signal.

13.2.5. AUTOMATIC GAIN CONTROL

A voltage proportional to the amplitude of the carrier (or some other reference level) of the received signal is developed in the detector stage or in a special stage immediately following the i.f. amplifier. To give automatic control of gain this voltage is fed back to the r.f. and i.f. amplifiers in such a way as to reduce the gain as the amplitude of the received signal increases and vice versa. In this way fluctuations in output level are minimised as is the possibility of an unduly large output when tuning through a powerful station. The absolute level of output, of course, is set by the manual control and the automatic control is included simply to prevent changes being superimposed on this level. It is usually arranged that the automatic reduction of gain does not take effect on signals of less than some predetermined value.

13.3. More Detailed Description: R.F. Stages

The main functions of the r.f. stages are:

1. To provide enough selectivity to rid the wanted signal of any interference likely to be troublesome in the i.f. stages.
2. To provide, with the introduction of as little noise as possible, sufficient amplification to ensure that at the input to the frequency changer (which itself introduces rather a lot of noise) there shall be a signal large enough to swamp the noise introduced there and form the basis for a good signal-to-noise ratio in the output.
3. Prevent as far as possible the radiation from the aerial of oscillations from the local oscillator.

Reverting to (1); it is not the function of the r.f. stages to eliminate all possible interference. Most of the elimination can be achieved in the i.f. stages. These are usually more numerous than the r.f. stages and, being tuned to a lower frequency, are, other things being equal, more selective. For instance, if the response of a circuit is reduced by 50 per cent when detuned 10 kc/s from a resonance frequency of 1 Mc/s a circuit of the same quality gives a reduction of about 75 per cent when detuned by the same amount from a resonance frequency of 0·5 Mc/s.

Obviously any contribution which the r.f. stages can make towards the reduction of adjacent-channel interference is an advantage; but what the r.f. stages must do, without fail, is to remove any interference with which the i.f. stages cannot deal. This is second-channel or image interference (Section 13.2.2). Any interference on this frequency which succeeds in reaching the i.f. amplifier receives, there and thereafter, just as much amplification as the wanted signal so that it is essential that it is removed in the r.f. amplifier. Fig. 13.4, drawn with reference to a sound broadcast receiver, illustrates what has been said. The wanted signal, f_1, of, say 1,000 kc/s, is heterodyned by the local oscillator, f_0, which at this setting must be of 1,470 kc/s to produce an i.f. of 470 kc/s. The response curve of

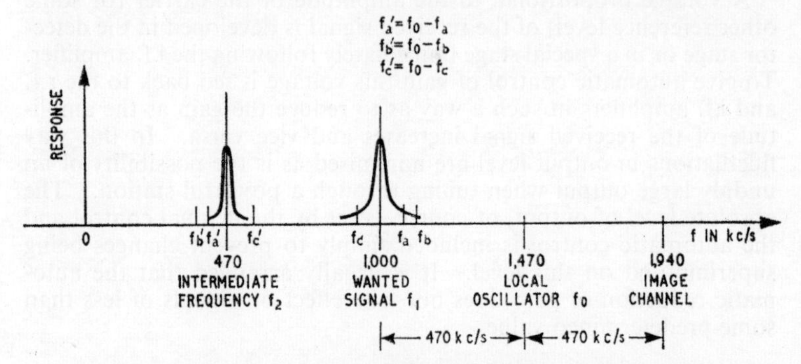

Fig. 13.4. Adjacent and image channel interference (broadcast band)

the r.f. tuning is shown approximately and it can be seen that any signals at image frequency ($1{,}470 + 470 = 1{,}940$ kc/s) are likely to be considerably attenuated.

The frequency translation of unwanted adjacent frequencies f_a, f_b and f_c (to frequencies $f_0 - f_a$, $f_0 - f_b$ and $f_0 - f_c$ respectively) is also shown on the diagram. Their absolute spacings from the wanted frequency are unchanged; it is for the i.f. circuits to complete their removal.

In considering whether a given unwanted channel is likely to be eliminated it must be remembered that it may be many times more powerful than the wanted one and that a very small interfering signal, which may well cause an interference whistle, can seriously diminish the value of an entertainment programme or telephone link. On a commercial morse circuit a surprising amount of interference can be tolerated by human operators although teleprinters are much more prone to error when confronted with interference. The skill of operators in combating interference does not mean that equipments for their use can be designed any less rigorously. On the contrary, the extreme ranges at which these equipments are often called upon to work and the severity of the interference which often exists, render the need for a very high performance indeed.

Fig. 13.4 illustrates the problem involved in the elimination of second-channel interference in broadcast receivers. The second channel signal has a frequency which is approximately twice that of the wanted station. For this reason, and because the wanted station is often near and therefore received at good strength, the cheaper broadcast receivers often do not include an r.f. stage. In such receivers image channel removal must be carried out in a single r.f. tuned circuit arrangement at the input to the frequency changer.

At higher frequencies second-channel rejection becomes more difficult. Fig. 13.5 is drawn for an i.f. of 650 kc/s and a signal frequency of 6 Mc/s. The frequency of the image channel (on $6 + 2 \times 0{\cdot}65 = 7{\cdot}3$ Mc/s) is now relatively much closer to that of the wanted signal and the difficulty of eliminating it is much greater. To illustrate this, a response curve of roughly the same relative proportions as that of Fig. 13.4 is drawn at the frequency of 6 Mc/s. Clearly the tuned circuit which this represents would not sufficiently attenuate strong second-channel interference. At this order of frequency and, of course, at still higher frequencies, where, for a given i.f., the situation becomes progressively worse, the provision of more r.f. tuned circuits is essential for good results. This requires the provision of at least one r.f. amplifier.

Fig. 13.5 also shows adjacent channel signals on frequencies f_a, f_b and f_c. These are not likely to be eliminated in the r.f. tuning but suffer considerable attenuation in the i.f. stages. The response

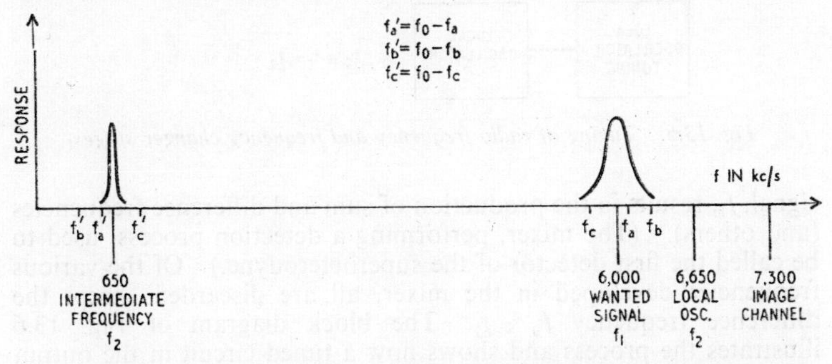

Fig. 13.5. Adjacent and image channel interference (6 Mc/s)

curve drawn against the intermediate frequency is for tuned circuits of roughly the same quality as that shown at 6 Mc/s.

The form taken by the r.f. amplifiers is as described in Chapter 10.

13.4. More Detailed Description: Frequency-changer

The frequency changer performs operations similar to those effected in the processes of modulation and detection. In amplitude modulation (*see* Chapter 13 of *Radio and Line Transmission, Volume* 1 and Chapter 12 of this volume) a radio frequency, f_1, and an information frequency, f_3, are suitably mixed to yield the sum and differences $f_1 + f_3$ and $f_1 - f_3$. In the detector the incoming frequencies $f_1 \pm f_3$ are suitably mixed with the carrier f_1 (which is usually also present in the incoming signal) to produce sum and difference frequencies. All of these are filtered out and discarded except the information frequency f_3, $(f_1 + f_3 - f_1 = f_3)$.

The frequency changer produces a similar result except that the end-product here is not the information frequency, f_3, but the intermediate frequency. A measure of the effectiveness of the frequency conversion is given by the conversion conductance which is defined as:

$$g_c = \frac{\text{anode current change at difference frequency}}{\text{input voltage at signal frequency}}$$

An oscillator in the receiver, the local oscillator, generates a frequency, f_0, which when suitably mixed with the incoming r.f.

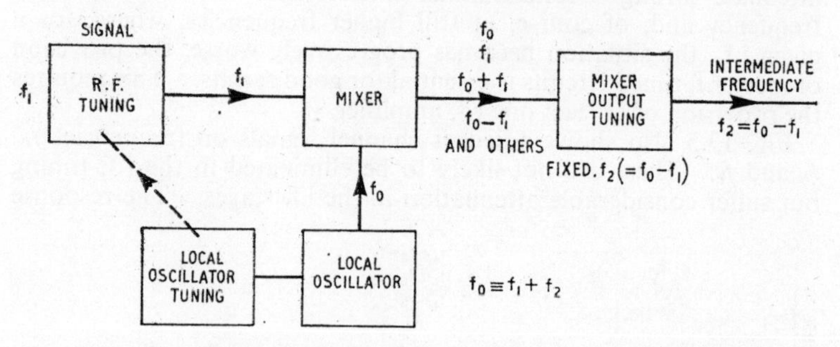

Fig. 13.6. Tuning of radio frequency and frequency changer stages

signal, f_1, results in the production of sum and difference frequencies (and others). (The mixer, performing a detection process, used to be called the first detector of the superheterodyne.) Of the various frequencies developed in the mixer, all are discarded except the difference frequency $f_0 - f_1$. The block diagram of Fig. 13.6 illustrates the process and shows how a tuned circuit in the output of the mixer enables the difference frequency to be retained while the others are rejected.

Although shown separately, the mixer and the local oscillator may be in a single valve or transistor (Fig. 13.12). The tuning of the local oscillator must always be spaced exactly f_2 c/s from the signal frequency and is usually above it (Section 13.4.2). The variable capacitor which fixes the frequency of the local oscillator is, therefore, ganged to the r.f. tuning capacitors so that they move together. The r.f. tuned circuits are made as nearly identical as possible; the local oscillator circuit, clearly, cannot be the same and for ganging to be possible with the retention of constant frequency difference between the oscillator and the r.f. tuning it is necessary to include fixed pre-set capacitors in the oscillator tuned circuit. If two capacitors C_P (padder) and C_T (trimmer) are included in the local oscillator circuit (Fig. 13.7) it is possible to arrange that the local oscillator and r.f. tunings are exactly right at three points in the range to be covered and are not more than one or two tenths of one per cent out at other points. The percentage error quoted is based on the signal frequency.

The oscillator capacitor is usually identical with those used in the r.f. stages and the oscillator inductor is, therefore, smaller than the r.f. inductors. The series capacitor, the padder, in the oscillator tuned circuit is included to reduce the maximum capacitance in that circuit while the trimmer increases the minimum capacitance. Together, therefore, these two additional capacitors compress the

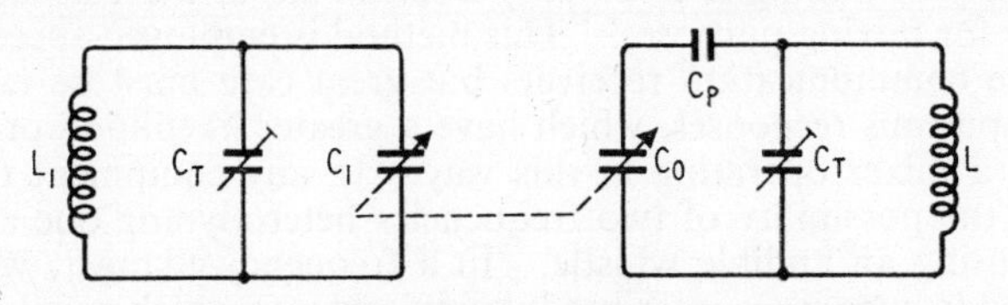

Fig. 13.7. Ganging local oscillator and radio frequency circuits

tuning range of the oscillator circuit compared with that of the r.f. stages as is required. (The absolute difference between the maximum and minimum of the range is the same in each case but the percentage difference is smaller for the oscillator because the oscillator frequency is higher than that of the r.f. circuits.)

Adjacent channel signals which have not been completely removed in the r.f. stages appear in the output of the frequency changer and are passed on to the i.f. stages where they are further attenuated.

Valves used as frequency changers are often multi-electrode types. Thus, as explained in Chapter 7, they generate more noise than simple valves. If the frequency changer were located near the output of the receiver where the signal is large, this would not be serious: in fact, however, the frequency changer is near or at the input where the signal is small. From this is seen the importance of:

(1) keeping the frequency changer as free from noise as possible and

(2) including r.f. amplification before the frequency changer when the wanted signals are anything but large, so as to build up the signal before adding to it the frequency noise (r.f. amplifiers can perform in a much more noise-free manner than do frequency-changers).

The functions of local oscillator and mixer can be combined in one valve or transistor stage or can be shared between two. The functions are often combined, even in television and frequency-modulation receivers operating at frequencies of the order of 100 Mc/s, but at high frequencies above, say, 15 kc/s the circuit difficulties in combined oscillator-mixers become extensive and the tendency for the oscillator to pull in to the signal frequency is considerable.

The less the coupling between oscillator and mixer stages the smaller is the possibility of frequency pulling occurring. Hence, in communication receivers, the use of separate stages is often found. (The term communication receiver is used to denote a receiver of considerable sensitivity, suitable for the reception of signals over a wide range of frequencies, equipped with a variety of refinements such as variable band-width, noise filters, a.f. and r.f. gain control and variable beat-frequency oscillators.) One way of removing the possibility of pulling is to employ a harmonic of the oscillator frequency for mixing purposes. This method is employed successfully in some communication receivers but great care must be taken to avoid spurious responses, which have a greater likelihood of occurring in a mixer operating in this way. In any equipment there is always the possibility of two frequencies heterodyning one another and causing an audible whistle. In a frequency-changer, which is designed to encourage inter-modulation and into which an additional frequency, the local oscillator frequency, is deliberately introduced, the chances of an audible whistle being developed are considerable. Using a lower oscillator frequency than is necessary, as when a harmonic is being used, increases the number of frequencies which may intermodulate in this way and so cause the possibility of whistles.

Further, the use of an oscillator harmonic results in the introduction of more noise than would otherwise occur and also the stage gain of a frequency-changer employing an oscillator harmonic is less than that of one operating on the fundamental.

The local oscillator itself should be as free from noise and spurious content as possible since these will attach themselves to the signal and remain with it throughout the receiver.

13.4.1. CHOICE OF INTERMEDIATE FREQUENCY

When all the factors are taken into account and particularly when the receiver specification is rigorous or the conditions are particularly difficult, as in television receivers, the choice of an intermediate frequency is one of great difficulty.

A low intermediate frequency gives the following advantages:

(1) better selectivity: this is because adjacent channel interference is spaced by an absolute and not a relative frequency difference (*see* Section 13.3),

(2) greater amplification for a given complexity of design,

(3) better stability.

Against the above not inconsiderable advantages must be set the following even weightier disadvantages of an intermediate frequency which is unduly low:

(1) the greater likelihood of second-channel interference breaking through the r.f. tuning and reaching the i.f. amplifiers (the

second channel is separated from the wanted channel by twice the i.f.),

(2) the lower the i.f. the closer the oscillator frequency is to the signal frequency and the greater the possibility of frequency pulling of the oscillator by the signal,

(3) the greater possibility of interference whistles.

Clearly the intermediate frequency should not be one used by stations whose signals might be picked up directly in the receiver wiring, or in the aerial, and thus passed direct to the i.f. amplifier.

The final choice is inevitably a compromise. Broadcast receivers use a frequency of about 470 kc/s and such a frequency is suitable also for signal frequencies up to about 7 Mc/s. Up to about twice this figure a similar i.f. may be used if the receiver is equipped with good r.f. selection but an i.f. of about 1·6 Mc/s is more likely and will be satisfactory for the reception of signal frequencies up to 30 Mc/s.

Communication receivers, those which are designed to a high standard of performance and cater for the reception of a wide range of frequencies (sometimes as great as from 15 kc/s to 30 Mc/s in about ten or even more wavebands) sometimes employ two intermediate frequencies, the first of approaching 1 Mc/s, or even of well over 1 Mc/s, in order to provide adequate second-channel rejection and the second one low, of the order of 85 to 100 kc/s, to give good rejection of adjacent-channel interference.

Television receivers, in which the required bandwidth is measured in megacycles per second, clearly require a much higher i.f. This is usually about 35 Mc/s. Radar receivers likewise need a high i.f.

13.4.2. CHOICE OF OSCILLATOR FREQUENCY

The frequency of the local oscillator may be above or below that of the signal by the correct amount. Unless the signal frequency is high and the i.f. fairly low the use of an oscillator below that of the signal results in the need for a very wide range of tuning of the local oscillator. For instance, if the tuning range covers 600 kc/s to 1,200 kc/s and the i.f. is 470 kc/s then if the local oscillator frequency is above the signal frequency its range will need to be 1,070 kc/s to 1,670 kc/s, a ratio of about one and a half to one, which is quite a feasible proposition. If it is set below the signal frequency the range will need to be 130 kc/s to 730 kc/s, a ratio of nearly six to one, virtually impossible of attainment.

The oscillator frequency, therefore, is usually above that of the signal.

13.4.3. INTRODUCTION TO FREQUENCY-CHANGER CIRCUITS

Frequency changing can be effected by applying the signal and the local oscillator output either to the same electrodes or to different ones in the mixer. A necessary requirement, of course, is that the operating conditions of the mixer are adjusted to suit the method of injection employed.

Fig. 13.8 shows a pentode mixer in which the signal and the oscillator voltages are applied to the grid. The oscillator circuit is not shown and any standard type—Hartley, Colpitts, tuned anode reaction, etc.—can be used. In all circuits the local oscillator tuning capacitor, C_0, where included in the circuit diagram is ganged to that of the r.f. tuners (labelled C_1 where included).

In Fig. 13.9 is shown a method of injecting the local oscillator signal into the cathode via a small coil L_2, the signal frequency being

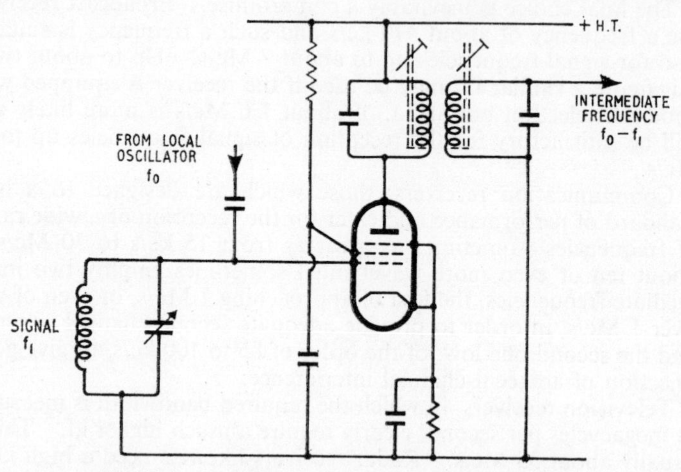

Fig. 13.8. Frequency changer: separate oscillator

taken to the grid. A separate coupling coil is shown. When a single valve with one cathode is used, for both the oscillator and for mixing, the coil L_2 can combine the functions of oscillator reaction coil and of cathode injection. Equally the feed to the cathode from the oscillator could be made by direct connection through a small capacitor in the same manner as that made to the grid in Fig. 13.8.

Fig. 13.10 shows a similar circuit employing two transistors.

A popular form of single valve circuit employs a five-grid valve, a pentagrid or heptode. Two of the grids, g_1 and g_2 (Fig. 13.11) are used as electrodes for the oscillatory circuit and a third, g_4, for the introduction of the signal frequency. Having such a large number of electrodes such a frequency changer is naturally noisy and is unsuitable for use at high frequencies. It has the advantage, however, of pentode characteristics, because g_5 screens g_4 from the anode and gives high values of μ and r_a. The oscillator section is screened by g_3 from the signal frequency electrode so that interaction is reduced; but it is not inconsiderable and there is, therefore, some tendency to frequency pulling. A disadvantage is that if the oscillator grids were made of sufficient dimensions to permit easy oscillation

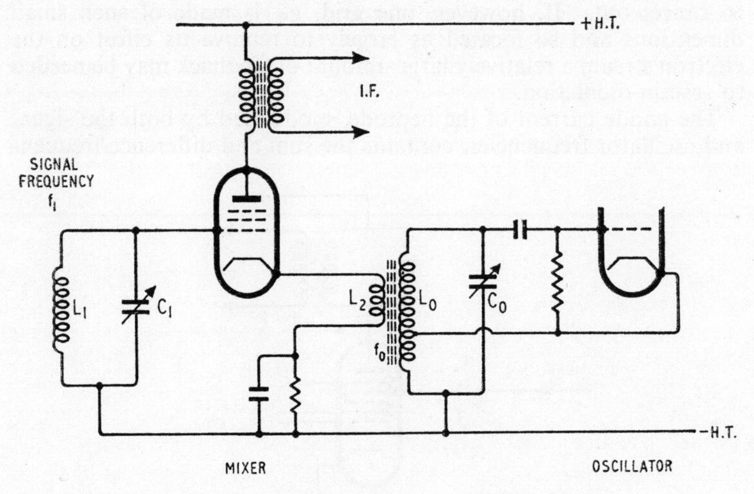

Fig. 13.9. Frequency changer: cathode injection

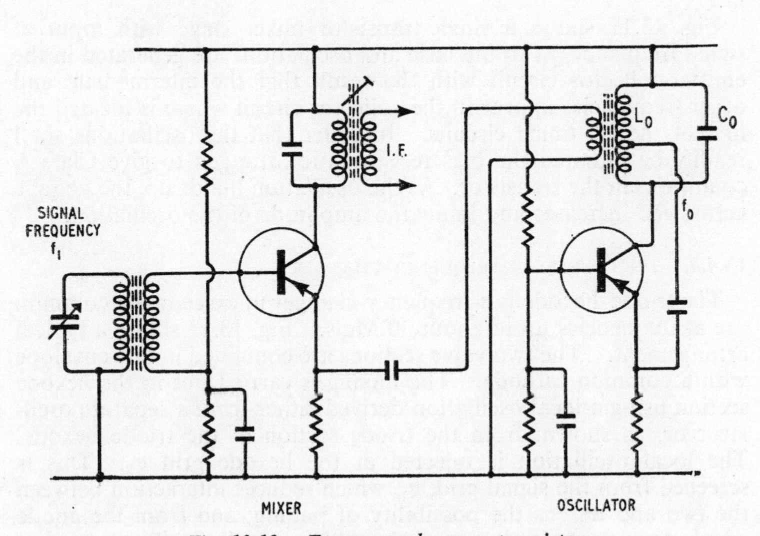

Fig. 13.10. Frequency changer: transistor

then their respective effects on the electron stream would tend to cancel out. If, however, one grid, g_2, is made of such small dimensions and so located as largely to remove its effect on the electron stream a relatively large amount of feedback may be needed to sustain oscillation.

The anode current of the heptode, modulated by both the signal and oscillator frequencies, contains the sum and difference frequen-

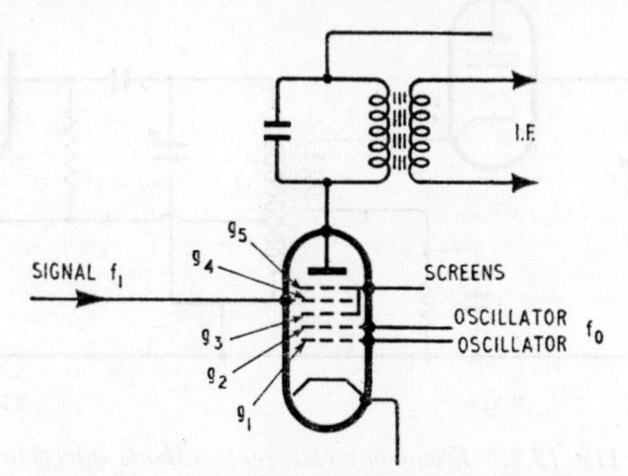

Fig. 13.11. Heptode, or pentagrid, frequency-changer valve

cies (and others) in its anode circuit where the inclusion of a circuit tuned to the difference frequency will eliminate all but the required i.f.

Fig. 13.12 shows a single-transistor mixer stage with input at signal frequency, f_1, to the base and oscillations are generated in the emitter-collector circuit with the result that the intermediate and other frequencies appear in the collector circuit where is located the first of the i.f. tuned circuits. In order that the oscillations shall readily be initiated the bias resistors are arranged to give Class-A conditions in the transistor. As the oscillation builds up, the voltage across *RC* increases and limits the amplitude of the oscillation.

13.4.4. TRIODE–HEXODE FREQUENCY-CHANGER

The triode–hexode is a frequency-changer in exceedingly common use at frequencies up to about 30 Mc/s. Fig. 13.13 shows a typical arrangement. The two valve sections are contained in one envelope with a common cathode. The mixing is carried out in the hexode section using a local oscillation derived either from a separate oscillator or, as shown, from the triode section of the triode–hexode. The local oscillation is injected at the hexode grid g_3. This is screened from the signal grid, g_1, which reduces interaction between the two and lessens the possibility of pulling, and from the anode which gives pentode characteristics—high μ and r_a. The output at the i.f. for a given signal input is large compared with that which would be obtained with triode characteristics, i.e. without g_4.

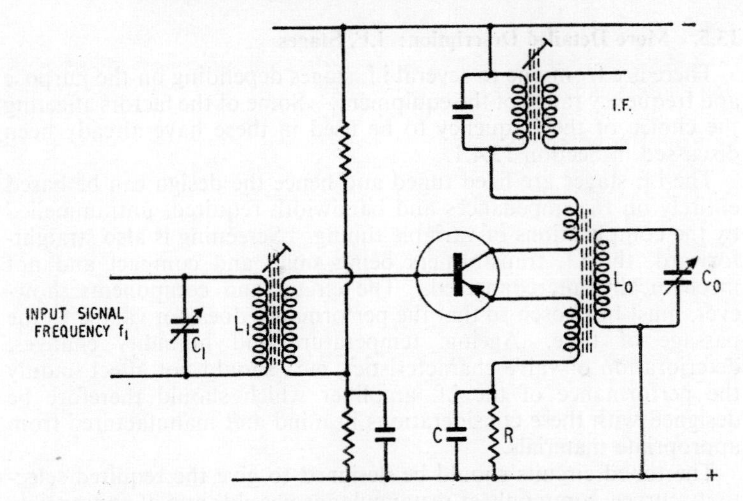

Fig. 13.12. Frequency changer: single transistor

The input at signal frequency is across L_1C_1 and applied between grid and cathode of the hexode thus modulating the electron stream at the signal frequency. The stream is further modulated at g_3 by the oscillator output so that the various modulation products appear in the anode circuit. Of these, the intermediate frequency, $f_2 = f_0 - f_1$, is selected by the tuned circuit.

The r.f. tuning capacitor is ganged with the other r.f. circuit capacitors and with the oscillator capacitor, C_0. The r.f. circuits, although made as near identical as possible, are not, in fact, identical if only because of the differences in the circuit stray capacitances, valve capacitances and so on. The inductors are thus made variable and small trimming capacitors C_T are included across each coil to enable these small discrepancies to be equalised.

The local oscillator is tuned to a different frequency and in order that the correct difference between oscillator and signal frequency tuning shall be maintained at all frequencies of signal one or both of the capacitors C_T, C_P must be included (*see* Section 13.4, Fig. 13.7) or some other means, such as shaping the plates of the oscillator tuning capacitor, adopted.

The maintenance of the required frequency difference at all settings of the tuning capacitors is termed *tracking*.

The *LC* values of the tuning circuits must, of course, be chosen to suit the frequency in use. The cathode resistor is selected to provide the correct bias for the valve and the capacitor is about 0·01 μF. The grid resistor and capacitor are about 100 kΩ and 100 pF respectively.

Provision is often made for a variable bias voltage to the grid for the purpose of gain control, manual or automatic.

13.5. More Detailed Description: I.F. Stages

There are from one to several i.f. stages depending on the purpose and frequency range of the equipment. Some of the factors affecting the choice of the frequency to be used in these have already been discussed in Section 13.4.1.

The i.f. stages are fixed tuned and hence the design can be based entirely on the impedances and bandwidth required, untrammelled by the complications of variable tuning. Screening is also straightforward, the i.f. transformers being small and compact and not mechanically interconnected. The circuit and components, however, must be chosen so that the performance does not vary with the passage of time. Ageing, temperature and humidity changes, deterioration of valve characteristics, etc., should not affect unduly the performance of the i.f. amplifier which should therefore be designed with these considerations in mind and manufactured from appropriate materials.

The tuned circuits should be designed to give the required selectivity curves, commonly as rectangular as possible and of appropriate bandwidth. In communication receivers provision may be made for varying the bandwidth in steps so that under favourable conditions a wide bandwidth may be employed with resultant good quality, while under adverse conditions of interference the use of a narrow bandwidth may still enable a readable signal to be received. Wide bandwidth might be of the order of 15 kc/s overall while very narrow could be as little as 100 c/s. In television and radar receivers

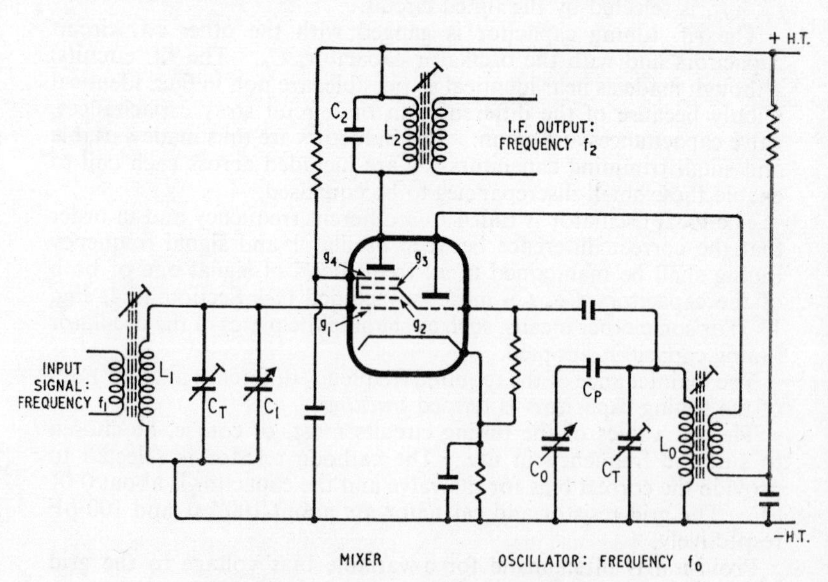

Fig. 13.13. Frequency changer: triode-hexode

the i.f. stages must be arranged to given an overall pass-band of several megacycles per second.

If valves are used they are generally pentodes with high values of μ and r_a; they are also of the variable-mu type so that the gain can be varied by manual or automatic control of the grid bias.

When transistors are used they may be tapped down the tuned circuit to avoid excessive damping which would otherwise be the result of low input and output impedances of transistors. In the common-emitter arrangement the input and output impedances are low, and despite the lower cut-off frequency of the common-emitter circuit this connection is used at high frequencies in preference to the common-base circuit. Because of the inherent feedback in transistors neutralisation or unilateralisation is often necessary. A suitable arrangement was shown in Fig. 10.5.

I.F. stages employing respectively a transistor and a variable-mu pentode valve were shown in Figs. 10.5 and 10.6 and described in Section 10.4.1.

13.6. More Detailed Description: Detector, A.F. and Output Stages

The detector of a superheterodyne receiver follows standard practice (*see* Chapter 12 of this volume and Chapter 15 of *Radio and Line Transmission, Volume* 1) but in a superheterodyne receiver it is likely to be provided with an ample signal at its input and is usually selected, therefore, for its ability to handle large signals rather than for its sensitivity. Diodes, of thermionic and germanium types, are in common use. Thermionic diodes are often in the same envelope as the valve element(s) of another stage (or stages) of the receiver, as for instance the following a.f. amplifier. In more specialised receivers a double diode is sometimes used in a full-wave type of detection circuit (similar to the full-wave rectifier circuit of Chapter 16 of *Radio and Line Transmission, Volume* 1).

The manual gain control which follows the detector precedes the a.f. amplifier (for sound reception) or the video amplifier (in a television receiver).

Audio stages are dealt with in Chapter 8 of the present volume and in Chapter 10 of *Radio and Line Transmission, Volume* 1.

Questions

1. Describe, with a sketch, the construction and working of a triode–hexode valve and explain its action as a frequency-changer. (*C & G*, 1954.)

2. Give a block schematic diagram of a superheterodyne receiver. Explain briefly the function of each stage of the receiver. (*C & G*, 1959.)

3. State the precautions to be observed in ganging the radio-frequency amplifier and oscillator sections of a superheterodyne receiver.

What do you understand by—

(a) Padding?

(b) Trimming?

(c) Tracking? (*C & G*, 1952.)

4. Why are superheterodyne receivers (a) more sensitive and (b) more selective than straight receivers?

What is an image signal? Give an example, stating the frequencies of the wanted and image signals for an i.f. of 465 kc/s. (*C & G*, 1955.)

5. State the advantages and disadvantages of superheterodyne as compared with straight receivers for use in the short-wave bands. (*C & G*, 1949.)

6. Why is it desirable to provide separate amplification at r.f., at i.f. and a.f. in a superheterodyne receiver rather than provide all the amplification in one frequency range? (*C & G*, 1954.)

7. A superheterodyne broadcast receiver, consisting of frequency-changer, single i.f. stage, detector and a.f. output stage is to be improved by the addition of a further stage of amplification. Compare the effects on performance of introducing the additional stage in the r.f., i.f. or a.f. sections of the receiver. (*C & G*, 1956.)

8. Sketch the circuit diagram of a typical stage in an i.f. amplifier for a broadcast receiver. Explain, with the aid of response curves, the effect of varying the mutual inductance between the primary and secondary windings of an i.f. transformer. (*C & G*, 1956.)

9. What do you understand by the term conversion conductance as applied to a frequency changer?

Give a circuit diagram, with suggested component values, of a triode–hexode frequency changer for use in a medium-wave superheterodyne receiver. (*C & G*, 1957.)

10. Explain the following terms as applied to the superheterodyne receiver:

(a) adjacent channel interference,

(b) image channel interference,

(c) frequency-changer oscillator radiation,

(d) automatic gain control. (*C & G*, 1959.)

14

Radio Frequency Measurements

14.1. The Cathode Ray Oscilloscope

14.1.1. PRINCIPLE OF ACTION

Since television has in many homes displaced the fireplace as the focal point of the family group, the cathode ray tube is not an unfamiliar sight to most readers. A cathode ray tube of rather smaller design has application in radio as a measuring device by which waveforms may be inspected. Before explaining the way in which it is used it is necessary to describe briefly its main parts and the principle of its action.

Fig. 14.1 shows the order of its essential parts. The tube when associated with appropriate equipment becomes an oscilloscope

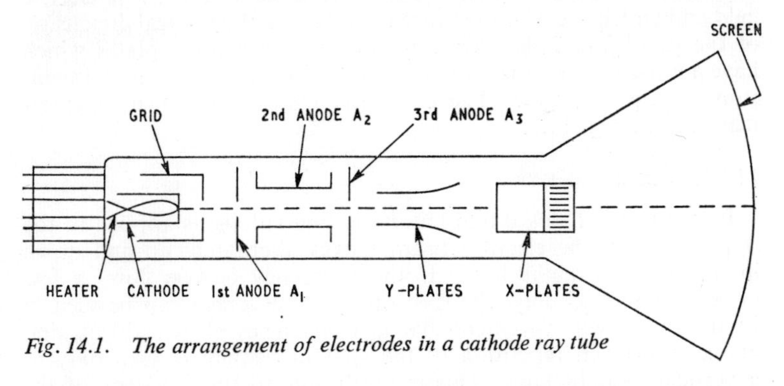

Fig. 14.1. The arrangement of electrodes in a cathode ray tube

which displays graphs produced electronically on a fluorescent screen. The " pencil " used is a sharply-focused beam of electrons which " draws " on a suitable screen which fluoresces at points where the electron beam strikes it. The screen is deposited on the inside of the glass tube but the green or blue trace is visible from the outside. The precise colour of the glow of the trace depends on the screen chemical used by the manufacturer. It may for instance be zinc orthosilicate which gives a green trace.

The source of the electron beam is a small cathode cylinder coated at its end with metal oxides which emit electrons when heated by a tungsten heater within. The electrons are accelerated along the tube axis towards the screen by three positive anodes. Of these A_1 and A_3 are usually at the same potential which may be some two or three thousand volts positive with respect to the cathode, while the middle anode A_2 is at a lower and adjustable potential. The control of the potential on A_2 is the means by which the shape of the electric fields between the three anodes is adjusted to bring the electron beam to a focus on the screen. In order that graphs may be drawn the electron beam must be capable of being moved or

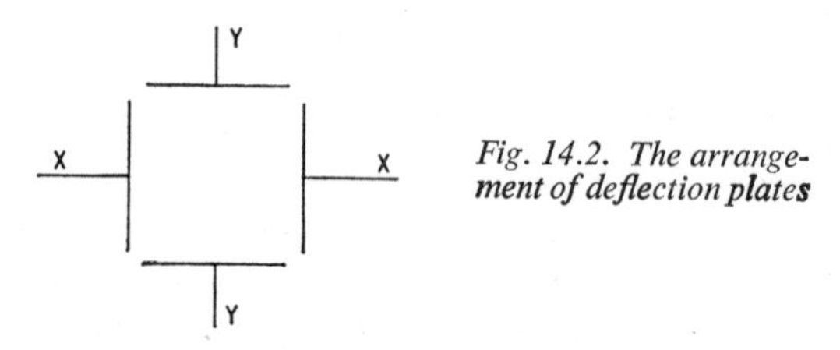

Fig. 14.2. The arrangement of deflection plates

deflected in two dimensions, viz. up and down and from side to side. From a combination of movements in these two dimensions graphs of any shape can be traced out on the screen. In cathode ray tubes in oscilloscopes the electron beam is deflected by electric fields set up by deflection plates between which the beam must pass to reach the screen. The horizontal plates which have a vertical electric field and which therefore produce deflection in the vertical dimension of the graph are called the *Y-plates* and the vertical plates which have a horizontal electric field for deflection in the horizontal dimension are the *X-plates*. The deflection plates are shown diagrammatically in Fig. 14.2.

14.1.2. OBTAINING A TRACE

If the tube is focused and an alternating voltage is applied to the *Y*-plates only, the electron beam moves alternately up and down in a vertical straight line. If the frequency is very low, a few cycles per second only, the spot on the tube screen can be seen in its up and down movement, but at frequencies of, say, 30 c/s and above, the rapid repetition of the spot movement is seen only as a vertical straight line. This is partly due to the tendency of the eye to retain an image for a brief period after the actual light stimulus has been removed (persistence of vision), and is partly the result of the screen fluorescence continuing after the electron beam has gone (after-glow). Similarly, an alternating voltage applied only to the *X*-plates draws a horizontal line on the screen.

To draw a graph of an alternating voltage of, for example, 1,000 c/s, the voltage itself is applied to the *Y*-plates and another voltage of waveform similar to that shown in Fig. 14.3 is applied to the *X*-plates. The period of the *X*-plate voltage is equal to the period of the *Y*-plate voltage or is a whole number of times greater than this period. As the *X*-plate voltage grows in amplitude at a steady rate from time t_1 to time t_2, the spot moves at constant velocity from left to right across the screen. Its distance at any instant from the left hand side of the screen is a direct measurement of the time which has elapsed since it began its left-to-right journey. During the very short time from t_2 to t_3 the electron beam flies back to the left hand side of the screen. A voltage of the waveform shown in Fig. 14.3 is called a *saw-tooth* waveform voltage and used in the manner outlined above it is a *time-base* voltage. An oscillator used to supply a time-base voltage is called a *time-base generator*. The time from t_1 to t_3 is the *time-base* of which the time t_2 to t_3 should be as small a fraction as possible. The time from t_2 to t_3 is called the *fly-back time*. Let it be assumed that the alternating voltage applied between the X-plates is of sine wave form and that a saw-tooth

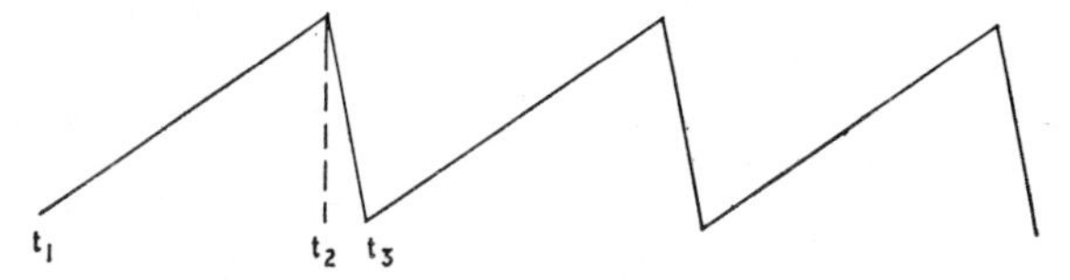

Fig. 14.3. A saw-tooth time base waveform

time-base voltage of period 1 msec is applied between the X-plates. A portion of the Y-plate voltage is also applied to the time-base generator to synchronise its frequency with that of the Y-plate voltage and ensure that the spot commences its left to right journey across the screen at the beginning of each cycle of Y-plate voltage. Throughout each movement across the screen, the electron beam is at each instant displaced vertically in proportion to the instantaneous value of the Y-plate voltage while its displacement from the left hand side of the screen represents the point in time to which the cycle of Y-plate voltage has progressed. The total movement of the spot thus executes a sine wave shape before flying back to the left hand side of the screen during the small fraction of the cycle represented by the fly-back time. If this movement occurred only once the effect on the screen would hardly be visible. The repeated movement of the spot over the same track on the screen 1,000 times a second produces the effect of a solid trace or graph of the Y-plate voltage on the screen. The fly-back of the spot may be seen as in the dotted line of Fig. 14.4 or it may be suppressed. Even if not suppressed the fly-back trace is weak because the beam velocity is high. Should the time-base period be twice the period of the Y plate voltage to be viewed, then two complete cycles are traced out during the left-to-right movement of the beam (*see* Fig. 14.5). If the time-base period is n times that of the Y-plate voltage period, n complete cycles are displayed on the screen, where n is any whole number. If n is very large then the cycles of Y-plate voltage displayed are so numerous and so close together that the screen is apparently filled with a fluorescent area having vertical lines drawn very close together.

In order to obtain a trace depicting the waveform of a current, using an electrostatically deflected cathode ray oscilloscope, it is necessary to pass the current through a resistor and apply the voltage generated across the resistor, amplified if necessary, to the Y-plates of the oscilloscope. The deflection at every instant is then

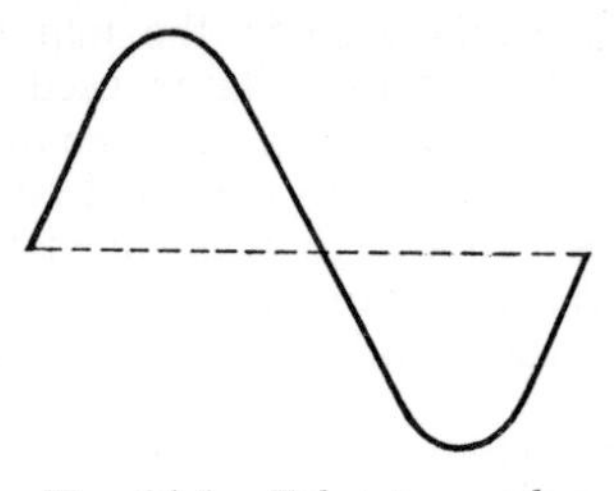

Fig. 14.4. Tube trace when time base equals Y-plate voltage period

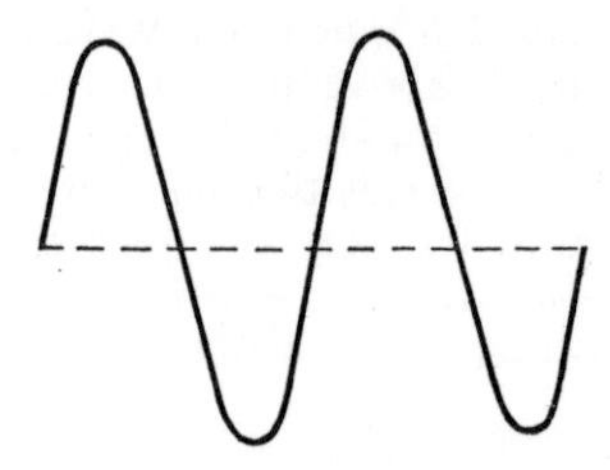

Fig. 14.5. Tube trace when time base is double the Y-plate voltage period

proportional to the voltage across the resistor and to the instantaneous value of current in it.

14.1.3. VIEWING A MODULATED WAVE

The pattern of the modulation envelope of an amplitude modulated wave can be displayed on the screen of a cathode ray oscilloscope if the time-base period is made a whole number multiple of the modulation period. The time-base generator should be synchronised at the modulation frequency. The modulated-wave voltage is applied to the Y-plates so that the cathode ray is deflected vertically

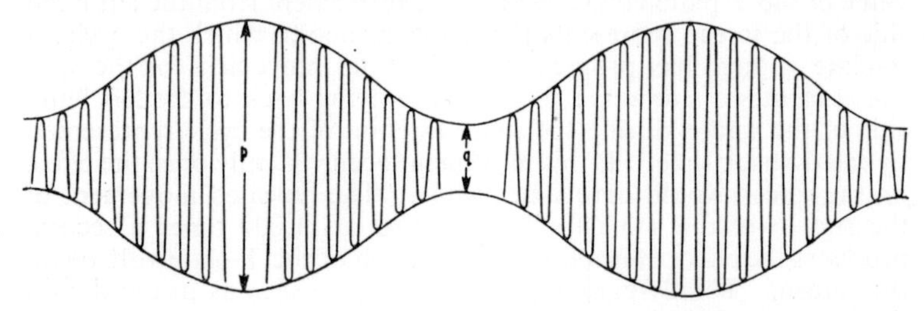

Fig. 14.6. A trace due to an r.f. modulated by an l.f. sine wave voltage

at the carrier frequency of the wave. Yet the amplitude of deflection varies at the lower modulation frequency and one cycle of variation is completed for each traverse of the screen, if the time-base frequency is equal to the modulation frequency. If the carrier frequency is f_c c/s and the modulation frequency is f_a c/s, then during one cycle of modulation the spot moves up and down the screen in its vertical deflections f_c/f_a times. Since this ratio is usually large, it is generally impossible to see individual cycles of the carrier frequency. The resultant trace is a shaded area as shown in Fig. 14.6 which depicts a 50 per cent modulation by a sine wave voltage of a radio frequency carrier.

A further example is shown in Fig. 14.7 where a carrier wave is modulated by a low frequency rectangular wave having a mark/space ratio of 2 : 1 and a carrier leak during the space period. The leak amplitude is seen to be approximately 18 dB below the mark amplitude.

Depth of modulation can be measured from the trace of the modulation envelope by measuring the distances p and q indicated in Fig. 14.6. If A is the amplitude of the carrier and B is the amplitude of the modulating voltage, then:

$$p = 2A + 2B, \text{ while } q = 2A - 2B.$$

Therefore, $p - q = 4B$ and $p + q = 4A$, so that

$$\frac{p-q}{p+q} = \frac{4B}{4A} = \frac{B}{A},$$

which is the degree of modulation.

Another way of estimating the depth of modulation is to obtain a trace of the shape shown in Fig. 14.8. For this, the time-base of

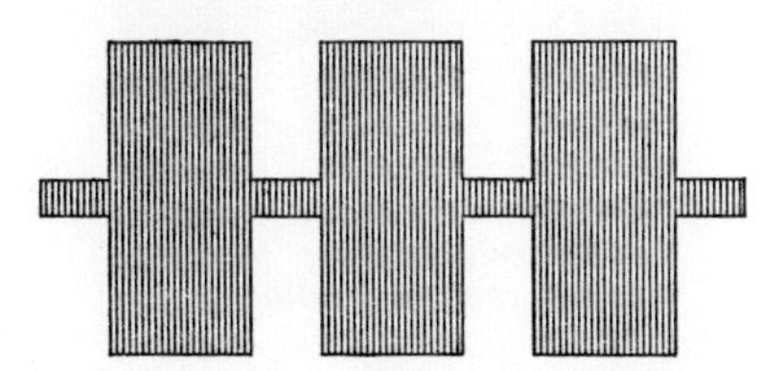

Fig. 14.7. A trace due to an r.f. modulated by an l.f. rectangular wave-form

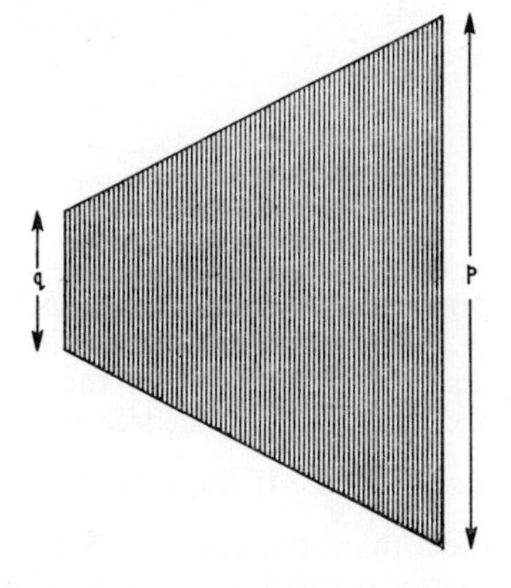

Fig. 14.8. A trapezoidal pattern displaying less than 100 per cent modulation

the oscilloscope is switched off and the modulating voltage is applied to the X-plates instead, while the modulated r.f. wave is applied to the Y-plates. The cathode ray will now travel across the tube face and back again taking the same time for both directions of travel, but not travelling at constant velocity.

The displacement of the spot from the left hand side of the screen might be written as, $x = K_1 B \sin\omega t$, where B is the amplitude of the modulating voltage.

K_1 is a constant depending on the sensitivity of the oscilloscope and $\omega = 2\pi \times$ the modulation frequency. The vertical deflection might be written as: $y = K_2(A + B \sin\omega t)$, where A is the carrier amplitude and K_2 is another constant depending upon the oscilloscope sensitivity. But,

$$B \sin\omega t = \frac{x}{K_1}$$

so that,

$$y = K_2\left(A + \frac{x}{K_1}\right)$$

or

$$y = K_2 A + \frac{K_2}{K_1}x$$

This is a linear equation relating x and y so long as the carrier amplitude A is constant. The amplitude of y deflection is directly

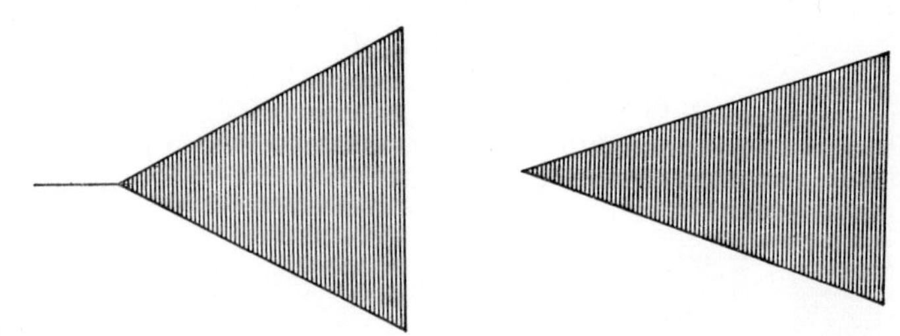

Fig. 14.9. A trapezoidal pattern displaying over-modulation

Fig. 14.10. A trapezoidal pattern displaying 100 per cent modulation

proportional to the amplitude of the x deflection and the outline of the diagram is a straight line so long as no distortion exists. The distances p and q show the maximum and minimum amplitudes of the trace and the degree of modulation can again be found from: $m = \frac{p - q}{p + q}$. If the modulation voltage should be greater in amplitude than the carrier so that over modulation occurs, then the carrier is reduced to zero during a part of the modulation cycle and a section of the trace becomes horizontal as in Fig. 14.9. 100 per cent modulation gives a trace as in Fig. 14.10.

14.1.4. LOOKING FOR DISTORTION

The oscilloscope is a most useful tool in searching for distortion in amplifiers. An amplifier is free from distortion if it amplifies all signals within its specified passband by the same factor, introduces no new frequency component and causes a phase shift of the frequencies amplified which is proportional to their frequency. If all these conditions are fulfilled the waveform of the output voltage is the same shape as the waveform of the input. A change can be detected by comparing the traces of input and output voltage waveforms on an oscilloscope. To facilitate this kind of operation, double beam oscilloscopes are available. Both beams are controlled by the same

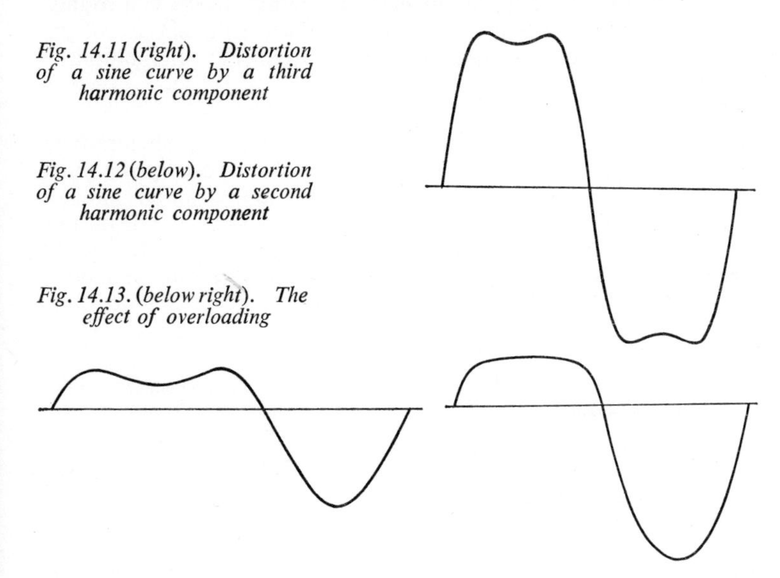

Fig. 14.11 (right). Distortion of a sine curve by a third harmonic component

Fig. 14.12 (below). Distortion of a sine curve by a second harmonic component

Fig. 14.13. (below right). The effect of overloading

time-base but the emission from the cathode is split into two and controlled in the vertical direction by two separate pairs of *Y* deflection plates. If the output voltage waveform of an amplifier is displayed by one beam and the input voltage (suitably amplified by an oscilloscope amplifier) is displayed by the other beam the two waveforms can be compared by arranging that the traces are superimposed. A deviation of the output trace from the input should then be detectable.

If the output appears as in Fig. 14.11 then a third harmonic is present and is being produced by the amplifier. If the output trace is as shown by Fig. 14.12 a strong second harmonic is present. A flattening-off of the peaks of one half cycle reveals distortion caused by overloading, insufficient bias or low emission in the valve. Any of these causes will prevent the amplifier giving adequate amplification to one half-cycle of input. This effect is depicted in Fig. 14.13.

Any parasitic oscillations can easily be detected in the trace since they are of a frequency so much higher than the frequency being examined. They may appear as "grass-like" patterns superimposed on the regular waveform or they may be so large in amplitude as to take the top and bottom edges of the trace off the screen.

The performance of wide-band amplifiers may be examined by introducing a square wave to them and examining the shape of the waveform emerging. The square wave contains, if perfect, an infinite range of odd harmonics of its fundamental frequency. Only a wide-band amplifier can give equal amplification to a sufficient range of these harmonics to make the output waveform appear as square as the input. High-frequency attenuation results in a rounding of the corners of the square wave shape, while the flatness of the tops of the square wave is destroyed by low-frequency attenuation.

14.1.5. CIRCUIT ALIGNMENT

The alignment of the band-pass circuits of a superheterodyne receiver i.f. stages is facilitated by the use of a specially designed frequency-modulated signal generator called a *wobbulator* and an oscilloscope. The arrangement is shown in Fig. 14.14. The output from the signal generator is of constant amplitude but its frequency swings above and below the i.f. of the receiver by equal excursions

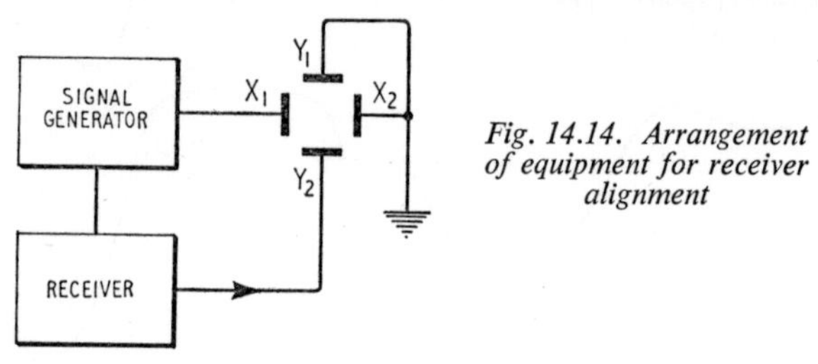

Fig. 14.14. Arrangement of equipment for receiver alignment

of possibly 15 to 25 kc/s. The voltage used for achieving this frequency modulation is also used for the horizontal deflection of the cathode ray beam. Thus the swing of the electron beam from left to right depends on the same cause as the swing of the test frequency and is proportional to it. The output from the detector of the receiver gives a d.c. voltage which is proportional to the response of the i.f. circuit under test for the frequency being applied at any particular instant. This detector output is taken to the *Y*-plates of the oscilloscope. The frequency-modulated signal centred on the i.f. of the receiver, usually 470 kc/s for a commercial broadcast set, is applied to the i.f. stage or stages to be aligned. The trace obtained is shown in Fig. 14.15 and is really a graph of detector output (d.c. voltage) plotted against frequency of receiver input. The coil slugs of the i.f. transformers can thus be adjusted for the best response while the response curve is under continuous observation.

14.2. Component Measurements

14.2.1. MEASUREMENT OF R.F. RESISTANCE

Fig. 14.16 is a circuit to illustrate a method of measuring r.f. resistance in which a known value of resistance is added to a circuit and the current measured on a thermo-milliammeter both before the resistance has been added and after, the circuit being at resonance on both occasions. O is a r.f. oscillator of good stability which is coupled as loosely as possible by mutual inductance to the circuit under test. The circuit is tuned to resonance with the frequency of the oscillator by adjusting the variable capacitor C until maximum current reading is obtained on the thermo-milliammeter. A known standard resistor of negligible self inductance and capacitance is then put into the circuit without altering the positions of the coupling coils L_1 or L_2, or the frequency of the oscillator. It is essential that the e.m.f. induced into the circuit be unchanged. The new value of current

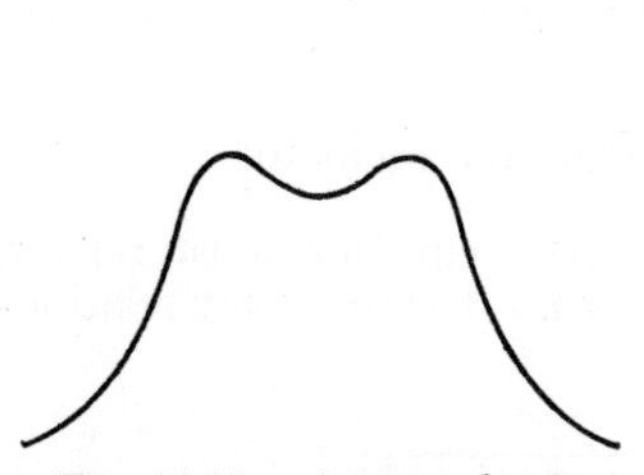

Fig. 14.15. A trace showing a "double-hump" response

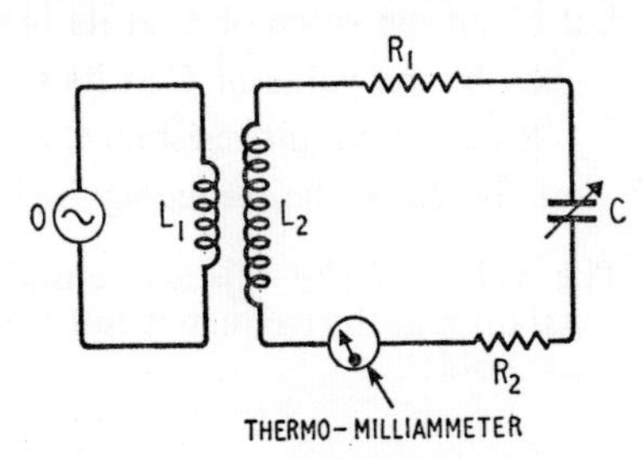

Fig. 14.16. Circuit for measurement of r.f. resistance

on the milliammeter is noted. Let the value of the circuit e.m.f. be E volts, the circuit resistance R_1 ohms and the added resistance R_2 ohms, then if I_1 and I_2 are the first and second values of current reading, we have that:

$$E = I_1R_1 = I_2(R_1 + R_2)$$

$$= I_2R_1 + I_2R_2$$

$$R_1(I_1 - I_2) = I_2R_2$$

$$R_1 = \left(\frac{I_2}{I_1 - I_2}\right) R_2$$

From the above, the circuit resistance R_1 may be found. The resistance of the thermocouple should be subtracted from R_1. This method depends for its accuracy on the accuracy of the thermo-milliammeter readings and on the accuracy with which the added resistance R_2 is known.

An alternative method of measuring resistance which makes use of an accurately-calibrated capacitor will now be described. The circuit is the same as that shown in Fig. 14.16 but the capacitor must be an accurately-calibrated standard, and no additional resistor is required. With the circuit tuned to resonance the current reading of the milliammeter is noted. C is now reduced in value until the growing capacitive reactance of the circuit reduces the current to

0·707 of its resonance value. Under these conditions the current is $\frac{1}{\sqrt{2}}$ times its resonance value and the impedance is $\sqrt{2}$ times as big as the resonant value ($\sqrt{2} \times R$) where R is the circuit resistance. The value of this capacitance is noted. The next step is to increase the capacitance until it passes through the resonance value and again increases the circuit reactance by leaving unbalanced inductive reactance sufficient to make the current once more 0·707 of the resonance current, and the circuit impedance again equal to $\sqrt{2} \times R$. This new setting of the standard capacitor is also noted.

Let C_1 be the value of C at its first setting,

C_2 be tne value of C at its second setting,

R be the circuit resistance, and

ω be $2\pi \times$ the frequency injected from the oscillator.

The value of the injected e.m.f. and its frequency must remain constant and the frequency must be known. The following relationships hold:

$$\sqrt{2}R = \sqrt{\left[R^2 + \left(\frac{1}{\omega C_1} - \omega L\right)^2\right]}$$

or squaring:

$$2R^2 = \left[R^2 + \left(\frac{1}{\omega C_1} - \omega L\right)^2\right]$$

$$R^2 = \left(\frac{1}{\omega C_1} - \omega L\right)^2$$

Taking the square root of each side:

$$R = \left(\frac{1}{\omega C_1} - \omega L\right) \quad (1)$$

Also by similar reasoning,

$$R = \left(\omega L - \frac{1}{\omega C_2}\right) \quad (2)$$

giving

$$\omega L = \left(R + \frac{1}{\omega C_2}\right)$$

Substituting for ωL in equation (1), we have,

$$R = \frac{1}{\omega C_1} - \left(R + \frac{1}{\omega C_2}\right)$$

Rearranging:

$$2R = \frac{1}{\omega} \cdot \left(\frac{1}{C_1} - \frac{1}{C_2}\right)$$

and

$$R = \frac{1}{2\omega} \cdot \left(\frac{1}{C_1} - \frac{1}{C_2}\right) \text{ ohms.}$$

In the above expression every variable excepting R has been measured and this can now be calculated.

14.2.2. INDUCTANCE MEASUREMENT

A circuit which may be used for the measurement of inductance is that of Fig. 14.17. Here the unknown inductance is joined in parallel with a variable capacitor, again a calibrated standard, and these are part of the oscillatory circuit for a simple valve oscillator. A heterodyne wavemeter is coupled to the oscillatory circuit by the

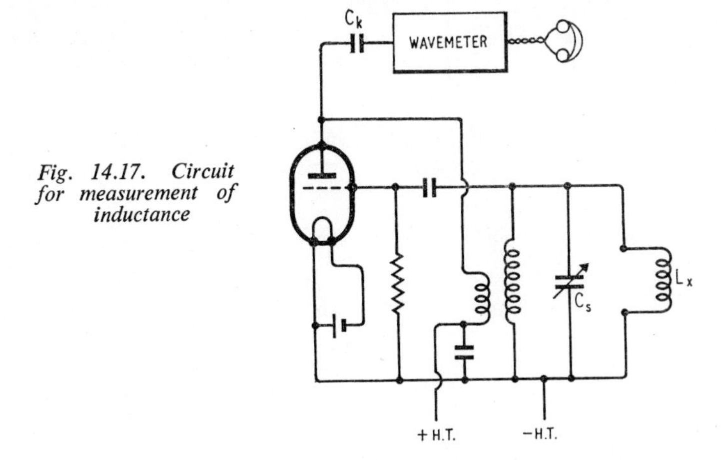

Fig. 14.17. Circuit for measurement of inductance

small coupling capacitor C_k. When the supplies to the oscillator and the wavemeter have been switched on long enough to enable all components to reach a steady temperature the standard capacitor is turned towards the maximum value of its range and its value carefully noted. The wavemeter frequency control is now adjusted until a zero beat note is heard in the phones. The frequency indicated by the wavemeter is noted. The unknown inductance L_x is now disconnected and the standard capacitor decreased in value until the frequency of oscillation is restored to its original value. Zero beat note is again detected by the heterodyne wavemeter. The alteration in the value of the standard capacitor needed to restore the frequency of oscillation to its original value is noted. The susceptance removed by the disconnection of the inductor has been made good by the change in value of the standard capacitor. Let C_1 be the first setting of the standard capacitor, ΔC be the change

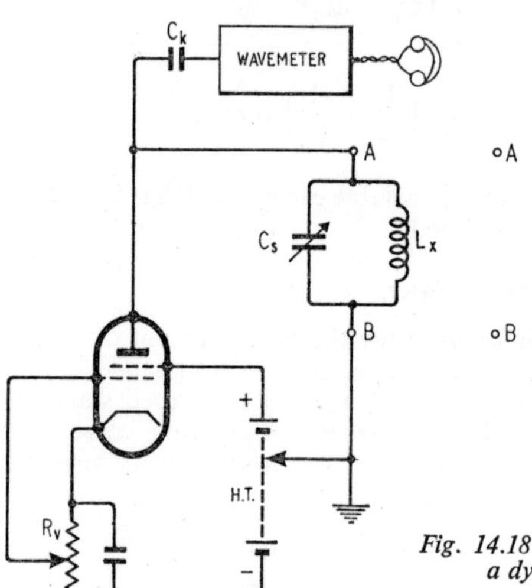

Fig. 14.18. A test circuit using a dynatron oscillator

in value of C and L_x be the unknown inductance. Let the frequency measured by the heterodyne wavemeter be such that, $f = \omega/2\pi$, then:

$$\omega(C_1 - \Delta C) = \omega C_1 - \frac{1}{\omega L_x}$$

$$\omega C_1 - \omega \Delta C = \omega C_1 - \frac{1}{\omega L_x}$$

$$\omega \Delta C = \frac{1}{\omega L_x}$$

and

$$L_x = \frac{1}{\omega^2 \Delta C} \text{ henrys.}$$

The above method depends for the accuracy of its result on the accuracy of calibration of the standard capacitor and the wavemeter.

An alternative method is possible using a dynatron oscillator and a set of standard inductors. The circuit is of the type illustrated in Fig. 14.18. The dynatron oscillator makes use of the negative resistance effect indicated by the negative slope section of the anode characteristic of a tetrode valve in order to maintain an oscillation in a two-terminal oscillatory circuit. Although a battery H.T. supply is shown for convenience, this is not essential though it is desirable that it shall be possible to apply a direct earth at point B without shorting a part of the H.T. voltage. This earth keeps the circuit between the test points A–B at a stable r.f. potential. R_v enables the grid bias to be adjusted for correct operating conditions. The coupling to the heterodyne wavemeter should be as small as possible (1 or 2 pF for example).

The unknown inductance is first used as the coil of the oscillatory circuit and the heterodyne wavemeter tuned to zero beat. The setting of the standard capacitor is noted. A standard coil of accurately known inductance is now substituted for the unknown inductance and the standard capacitor readjusted until zero beat on the wavemeter shows that the frequency of oscillation is the same as before. The LC value of the oscillatory circuit is the same for the two circuit arrangements and the value of the capacitance is known for both. Thus if the settings of the standard capacitance are C_1 and C_2, L_x is the unknown inductance value and L_s is the standard value, we have that:

$$C_1L_x = C_2L_s$$

and

$$L_x = \frac{C_2}{C_1} \cdot L_s \text{ henrys.}$$

14.2.3. MEASUREMENT OF CAPACITANCE BY SUBSTITUTION

The circuit of Fig. 14.18 can also be used for the measurement of capacitance by substitution. In this measurement only one coil of a suitable value is needed. The unknown capacitor is first connected in parallel with the variable standard capacitor which should be set towards its minimum value. The wavemeter is tuned so as to give zero beat note with the oscillation set up by the dynatron oscillator, and the setting of the standard capacitor is noted. The unknown capacitor is now disconnected and the standard value increased until the heterodyne wavemeter indicates by zero-beat note that the same frequency of oscillation has been obtained. The increase in value of the standard calibrated capacitor is equal to the value of the unknown capacitor previously disconnected. This method can only be applied to capacitances which are less than the maximum value of the standard capacitor.

14.3. Valve Voltmeters

14.3.1. DIODE

The advantages of valve voltmeters as compared with even the best moving-coil voltmeters are that they have a high input impedance, possibly of the order of 10 megohms, and that they can be used over a wide range of frequencies including high radio frequencies.

The simplest valve voltmeter is one using a small diode valve. A possible circuit is that of Fig. 14.19. The diode valve should be chosen to have low capacitance between anode and cathode. The input resistance is approximately one-third the value of R or 167 kΩ. The value of C must be large compared with the capacitance of the valve in order that most of the alternating input voltage

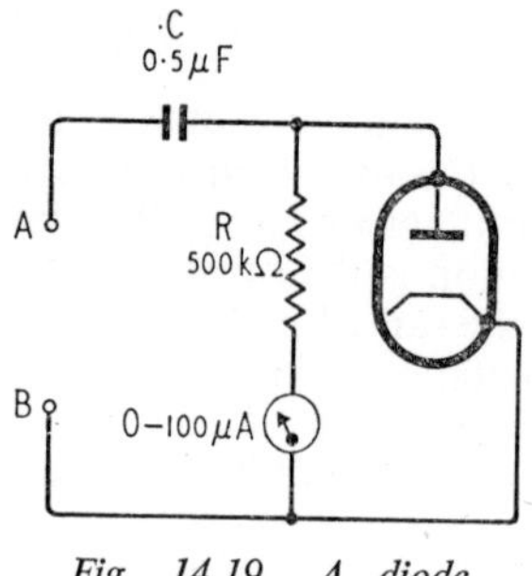

Fig. 14.19. A diode valve voltmeter

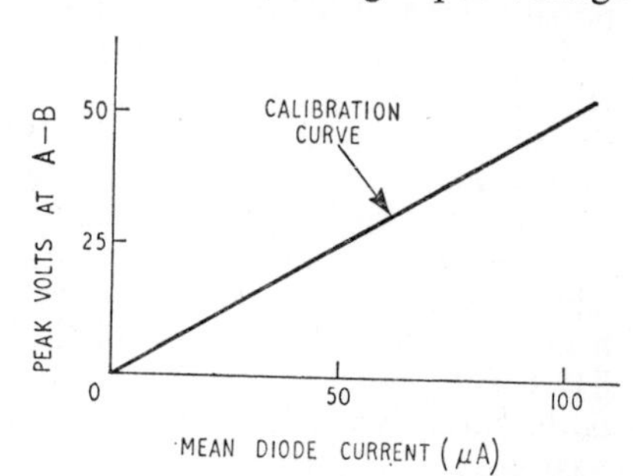

Fig. 14.20. A calibration curve for a diode valve voltmeter

appears across the valve and very little is dropped across the reactance of the coupling capacitor C. The value of C must be larger for low frequencies than for high ones because if a small capacitance were used at low frequencies, the reactance of C would be larger than the resistance of R and this would attenuate the voltage applied to R and the diode. Yet the product of RC must not be too great (implying a long time-constant) or the action of the meter is sluggish. Too short a time-constant would not permit the charge on the reservoir capacitor C to accumulate. A reasonable compromise is reached if RC seconds is approximately 100 times the period of the alternating voltage to be measured.

The current carried by the micro-ammeter is proportional to the peak alternating voltage applied and a calibration curve can be made to relate the peak voltage measured to the d.c. reading on the micro-ammeter (*see* Fig. 14.20). All the results may be divided by $\sqrt{2}$ if r.m.s. values are required and a sine waveform is assumed.

14.3.2. TRIODE CIRCUITS

The triode valve may be used as an anode bend rectifier or a grid rectifier and for each circuit the change of anode current is a function of the amplitude of the alternating voltage applied to the grid-cathode terminals. Using the valve as an anode bend rectifier, however, requires a negative bias on the grid. This bias condition allows no grid current to flow and the input impedance is therefore higher than if grid circuit rectification is employed. Fig. 14.21 is a possible circuit for a triode valve voltmeter using anode bend rectification. Battery supplies are shown. This is an advantage if portability is required but the circuit could be adapted for mains operation if necessary. B_1 provides an adjustable bias for the grid, B_2 is the H.T. supply and B_3 supplies the valve filament. B_4 in conjunction with R_2 provides for adjustable current through the meter in the reverse direction to the valve current so that in the absence of an input signal, the meter current may be reduced to zero. Any meter reading which then shows is entirely a result of the applied signal.

For large input voltages the valve should be biased to cut-off point or beyond cut-off if the signal amplitude exceeds the grid base. On one half-cycle the voltage being measured makes the grid voltage more negative and no current flows, but on the other half-cycle of

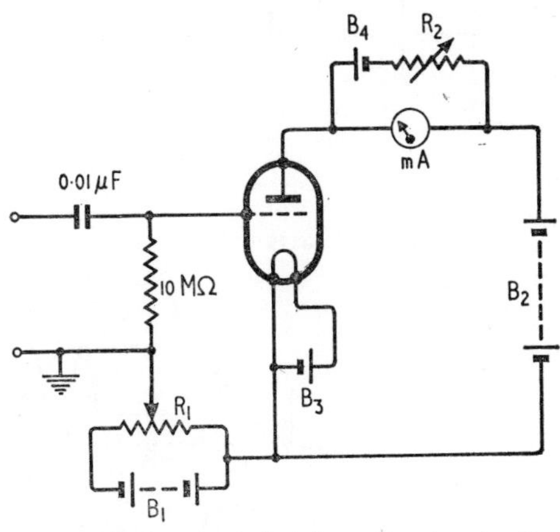

Fig. 14.21. A triode valve voltmeter circuit

input the grid is swung less negative and valve current flows for approximately half a cycle. The mean anode current resulting from the alternating voltage applied is almost in direct proportion to the amplitude of input so long as the waveform of the voltage applied remains unchanged. However, the average anode current does depend on the mean value of voltage during the half-cycle when the valve is conducting and thus depends on the waveform of the input. A square-wave input causes more current to flow than a sine-wave input of the same peak value. A calibration curve can be made to relate the valve anode current increase due to the signal at the grid, to the r.m.s. voltage, the input being assumed sinusoidal. The graphs of Fig. 14.22 illustrate the conditions of operation outlined above.

For smaller voltages of the order of 50 mV or so, a bias is used which operates the valve at a point on the lower bend of the mutual characteristic (*see* Fig. 14.23). Anode current flows on both half cycles but since the mutual conductance is less for the negative half-cycles than for positive half-cycles of input, the increase of anode current during the latter half-cycles exceeds the decrease which occurs when the grid swings more negative. An average increase of valve current res'.lts which depends approximately on the square

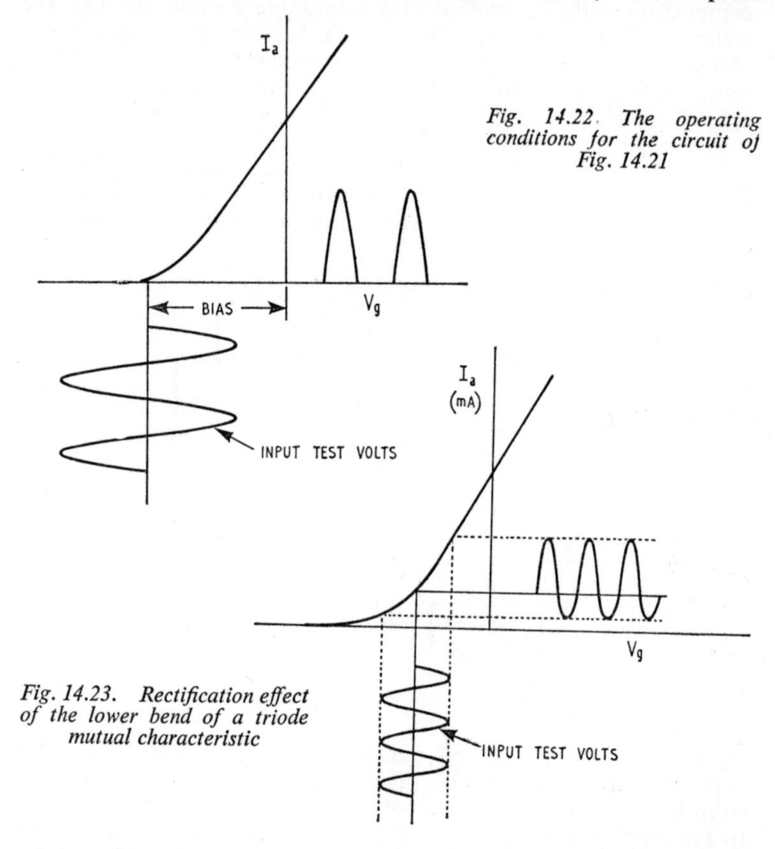

Fig. 14.22. The operating conditions for the circuit of Fig. 14.21

Fig. 14.23. Rectification effect of the lower bend of a triode mutual characteristic

of the grid voltage to be measured. A calibration curve is made which relates anode current increase to the r.m.s. voltage applied at the input. Waveform error is not involved in this square law operation.

If a triode is used as a grid rectifying valve voltmeter, the grid-cathode circuit behaves similarly to a diode rectifier with the grid playing the part of the anode. The charge stored on the grid capacitor biases the grid negatively in proportion to the amplitude of the alternating voltage applied for measurement. The anode current falls as the applied voltage is made bigger. This type of triode valve voltmeter has a much lower input impedance than one relying on anode rectification. Its chief advantage is that the valve current

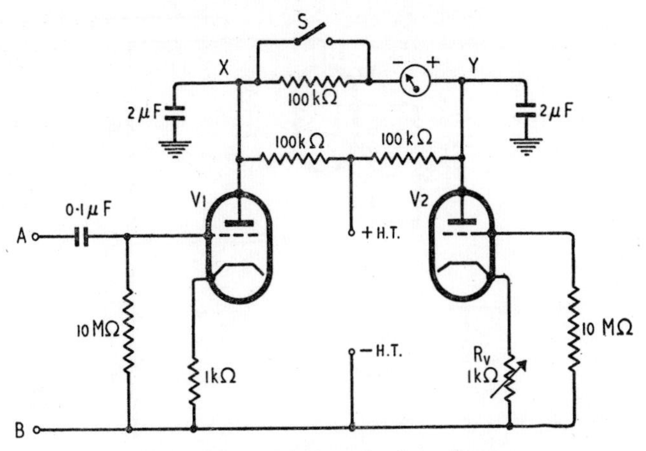

Fig. 14.24. A balanced valve voltmeter

is relatively large so that the anode current meter need not be particularly sensitive.

A further method of using a triode as a valve voltmeter is as follows: a potentiometer and bias battery are arranged to provide an adjustable bias voltage which is measured by a moving-coil d.c. reading voltmeter. The bias is increased until a milli-ammeter in the anode circuit shows the anode current to be reduced to a small datum level (almost cut off).

The alternating voltage is then applied to the grid circuit. Anode rectification causes the anode current to increase. The potentiometer which controls the grid bias is then readjusted to increase the bias and restore the anode current to its datum level. The change of bias recorded on the bias voltmeter as being necessary to off-set the effect of the alternating input voltage on the anode current, is equal to the peak value of the voltage to be measured. This arrangement constitutes a *slide-back* valve voltmeter. It is a meter with a high input impedance and is capable of giving consistent results.

14.3.3. BALANCED VALVE VOLTMETER

A balanced valve voltmeter is shown in Fig. 14.24. Triode valves are shown though pentodes could be used. With the input terminals A and B short-circuited, the variable cathode bias resistor R_v is adjusted until the points X and Y have equal potentials and the micro-ammeter carries no current. The switch S can be closed for final balance though the safeguarding resistor should be left in circuit until approximate balance has been obtained. The short-circuit can then be removed and the voltage to be measured is applied between terminals A and B. Anode rectification occurs in the valve $V1$, whose current increases. The potential of X falls below that of

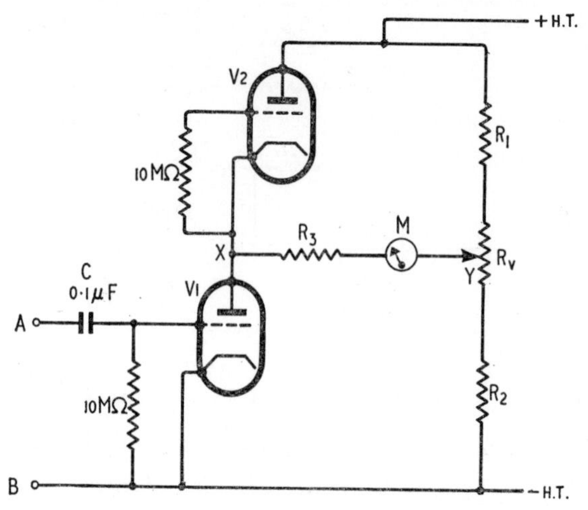

Fig. 14.25. A bridge-type valve voltmeter

Y and the micro-ammeter is deflected in proportion to the degree of the unbalance of the two valve currents. The calibration curve used relates the deflection of the meter, with the switch *S* closed, to the r.m.s. voltage applied at *AB*. The decoupling capacitors of 2 μF prevent any increase of valve input capacitance by Miller effect. There should be no alternating component of voltage on the anode.

14.3.4. BRIDGE TYPE OF VALVE VOLTMETER

A bridge type of circuit for a valve voltmeter is shown in Fig. 14.25. The two valves form two arms of a Wheatstone Bridge circuit of which the remaining arms are the resistors R_1 and R_2 and the tapped off fractions of the zero adjustment resistor R_v. Before use, the input terminals *A* and *B* are short-circuited and R_v is adjusted until points *X* and *Y* have equal potential, so that the pointer of the meter *M* shows no deflection. The input voltage for measurement is then applied across terminals *AB*, whereupon grid rectification occurs and reduces the conductance between anode and cathode of *V*1 by the negative grid bias built up on the input capacitor *C*. Since the resistance of *V*1 is now increased and is larger than that of *V*2 the potential of *X* will have become higher than the potential of *Y*. A current flows across the centre arm of the bridge through the meter. The larger the input signal, the more the bridge becomes unbalanced and the bigger the deflection of the meter *M*. The meter can therefore be calibrated to measure the input voltage applied at *AB*.

In conclusion, it should be noted that all valve voltmeters should be given time to warm up to a steady temperature before zero adjustment is made prior to taking readings.

Questions

1. Draw careful sketches to show the appearance of each of the following signals when displayed in a cathode ray oscilloscope:
   (i) a carrier wave amplitude-modulated to a depth of 50 per cent by an l.f. sinewave;
   (ii) a carrier wave amplitude-modulated by an l.f. rectangular wave having a mark/space ratio of 2 : 1. Assume the carrier leak during the space period to be —20 dB relative to the unmodulated carrier amplitude;
   (iii) a sinewave to which has been added 50 per cent second harmonic;
   (iv) a sinewave to which has been added 50 per cent third harmonic.

   Each sketch must clearly show the scales in the axes. (*C & G*, 1960.)

2. Explain how a cathode ray oscilloscope can be used in conjunction with a frequency-modulated oscillator of suitable frequency for the alignment of the i.f. circuits of a receiver.

3. In order to measure the resistance of a coil of negligible self-capacitance it is connected in series with a thermo-milliammeter of 9 ohms resistance across a variable capacitor. The circuit is then weakly coupled to a high frequency source of 1 Mc/s. When the capacitor is tuned to resonance a current of 10 mA flows in the circuit. The capacitor is then adjusted above and below the resonance frequency, until the current is 7·07 mA. The values of the capacitor to give this current are 450 pF and 650 pF. What is the resistance of the coil?

4. A coil is joined in parallel with the oscillatory circuit of a valve oscillator. The frequency of oscillation is measured to be 400 kc/s. The coil is then disconnected and the frequency restored to 400 kc/s by reducing the value of the tuning capacitor by 50 pF. What is the effective value of the inductance?

5. Describe a method of measuring capacitance by a process of substitution. State what factors will influence the accuracy of the result.

6. With the aid of a circuit diagram explain the principle of a " slide-back " valve voltmeter.

7. What advantages has a valve voltmeter over a simple moving coil voltmeter? How could the anode characteristic of a triode valve be utilised in the application of a triode to a valve voltmeter circuit?

8. With the aid of a circuit diagram explain the principle of either (a) a balanced-type valve voltmeter or (b) a bridge-type valve voltmeter.

9. An inductor of 80 μH, a standard variable capacitor and a standard variable resistor are connected in parallel and a signal voltage

of 1 Mc/s injected into the coil by a loose mutual coupling. The circuit is tuned to resonance and a valve voltmeter across the resistor reads 1 volt when the resistance value is 1 megohm.

An additional capacitor is now placed in parallel with the standard which is then readjusted to produce resonance while the standard resistor is readjusted until the valve voltmeter again reads 1 volt. The new values of the standard components are 200 pF and 3·78 MΩ. Calculate the capacitance, reactance, equivalent shunt resistance and power factor of the added capacitor. (*C & G*, 1959.)

10. Draw a diagram and explain the operation of a peak-reading diode-valve voltmeter. Such instruments are frequently calibrated to indicate the r.m.s. values of sine waves. What precautions must be taken when using instruments so calibrated? (*C & G*, 1960.)

ANSWERS

Chapter 1:

(1) Section 1.2.1;
(2) Section 1.3.2;
(3) Sections 1.4.4 and 1.5.1;
(4) Sections 1.4.1 and 1.4.4;
(5) Section 1.5.2;
(6) Section 1.5.3.

Chapter 2:

(2) Section 2.9;
(3) 60 dB;
(4) 633 ohms, 3·98 radians/mile, 1·58 miles;
(5) 548 ohms;
(6) Sections 2.2–2.3.4 and Section 2.4.2;
(7) Section 2.9;
(8) Sections 2.4.1, 2.4.2;
(9) Sections 2.5.2, 2.6.1, 2.7.1;
(11) 44·5 dB.

Chapter 3:

(4) Section 3.4.1;
(5) Section 3.2.5;
(6) Sections 3.4.2 and 3.4.3, Sections 3.3.1 and 3.3.2;
(7) Allowing for 6 dB loss in hybrid networks, total gain = 26 dB;
(8) 18 kW; (9) Section 3.2.6.

Chapter 4:

(1) 1·5 m;
(2) Parallel;
(3) 80 per cent;
(4) 1 kW;
(7) Section 4.4.2;
(8) Section 4.5;
(9) Dipole 82 cm, reflector 86 cm spaced 41·5 cm behind aerial.

Chapter 5:

(2) 1·6 kΩ, 0·96 watts, 6 mA;
(4) Section 5.1.3; (5) Sections 5.2.1 to 5.2.4;
(6) 150 μH, 25 μH;

(8) Figs. 5.20 and 5.24, frequencies f_1 and f_2;
(9) 31·8 ohms, 318 kΩ;
(10) Section 5.4.1.

Chapter 6:

(2) Section 6.4.1;
(3) 17·78 : 1, 316 : 1;
(4) —3 dB, 75;
(5) 23 dB, 26 dB, Nil;
(6) 35·4 ohms, 50 ohms, —3 dB, —6 dB;
(7) 26 dB;
(8) 8·14 A;
(9) Section 6.6;
(10) 12 dB, 1·38 nepers, 4 : 1.

Chapter 7:

(4) About 5μV.

Chapter 8:

(1) 40 kΩ, 1 W;
(4) 1,111;
(5) 6 V;
(6) 27 per cent, 2 : 1.

Chapter 9:

(1) 0·105 V;
(2) 855 or 59 dB;
(4) 2π ohms; (6) 80 μH;
(9) 100 kΩ.

Chapter 10:

(1) $Q = 111{\cdot}1$, 0·56 W, 200 V;
(3) 0·67 V;
(6) 40 dB.

Chapter 14:

(3) 45·5 ohms;
(4) 3·18 mH;
(5) Section 14.2.3;
(6) Sections 14.3.3 and 14.3.4;
(9) 118 pF, 1·35 kΩ, 1·36 MΩ, 0·001.

Index

INDEX